Wireless Interface Technologies for 3D IC and Module Integration

Synthesizing 15 years of research, this authoritative text provides a comprehensive treatment of two major technologies for wireless chip and module interface design, covering technology fundamentals; design considerations and trade-offs; practical implementation considerations; and discussion of practical applications in neural network, reconfigurable processors, and stacked static random-access memory (SRAM). It explains the design principles and applications of two near-field wireless interface technologies for 2.5–3D integrated circuit (IC) and module integration, respectively, and describes system-level performance benefits, making this an essential resource for researchers, professional engineers, and graduate students performing research in next-generation wireless chip and module interface design.

Tadahiro Kuroda is Professor and Founding Director at the Systems Design Lab at the University of Tokyo. He is a fellow of the Institute of Electronics, Information and Communication Engineers (IEICE) and the Institute of Electrical and Electronics Engineers (IEEE).

Wai-Yeung Yip is a researcher at the University of Tokyo with more than 25 years of industry experience in high-speed chip interface design.

Wireless Interface Technologies for 3D IC and Module Integration

TADAHIRO KURODA
University of Tokyo

WAI-YEUNG YIP
University of Tokyo

CAMBRIDGE
UNIVERSITY PRESS

CAMBRIDGE
UNIVERSITY PRESS

University Printing House, Cambridge CB2 8BS, United Kingdom

One Liberty Plaza, 20th Floor, New York, NY 10006, USA

477 Williamstown Road, Port Melbourne, VIC 3207, Australia

314–321, 3rd Floor, Plot 3, Splendor Forum, Jasola District Centre, New Delhi – 110025, India

103 Penang Road, #05–06/07, Visioncrest Commercial, Singapore 238467

Cambridge University Press is part of the University of Cambridge.

It furthers the University's mission by disseminating knowledge in the pursuit of
education, learning, and research at the highest international levels of excellence.

www.cambridge.org
Information on this title: www.cambridge.org/9781108841214
DOI: 10.1017/9781108893299

First published 2021

A catalogue record for this publication is available from the British Library.

Library of Congress Cataloging-in-Publication Data
Names: Kuroda, Tadahiro, author. | Yip, Wai-Yeung, author.
Title: Wireless interface technologies for 3D IC and module integration / Tadahiro Kuroda, University of Tokyo, Wai-Yeung Yip, University of Tokyo.
Description: Cambridge ; New York, NY : Cambridge University Press, 2021. | Includes bibliographical references and index.
Identifiers: LCCN 2021028684 (print) | LCCN 2021028685 (ebook) | ISBN 9781108841214 (hardback) | ISBN 9781108893299 (epub)
Subjects: LCSH: Interconnects (Integrated circuit technology) | Three-dimensional integrated circuits. | Optical interconnects. | BISAC: TECHNOLOGY & ENGINEERING / Electronics / Circuits / General
Classification: LCC TK7874.53 .K86 2021 (print) | LCC TK7874.53 (ebook) | DDC 621.3815–dc23
LC record available at https://lccn.loc.gov/2021028684
LC ebook record available at https://lccn.loc.gov/2021028685

ISBN 978-1-108-84121-4 Hardback

Contents

Preface

In the last 35 years, integrated circuit (IC) feature size was reduced by more than two orders of magnitude. However, operating voltage has been reduced by only an order of magnitude. This is because the electric field required to switch the transistor gate does not scale down linearly. Process scaling without proportional reduction in operating voltage has resulted in a three-orders-of-magnitude increase in power density. Consequently, power density has become a limiting factor in IC performance improvement.

Another challenge to the advance of IC is the integration of heterogeneous chip technologies. For example, while logic functions demand high-performance transistors, memory storage requires high-density capacitors. If both functions are integrated onto a single chip, the manufacturing cost can become prohibitively high. As a result, logic and memory are implemented on separate chips with independently optimized processes, and the interface between the two through which data flow has become a bottleneck in information processing.

Consequently, a key to further computing system performance improvement is in developing the chip interface, so as to improve power efficiency by optimizing the interconnection, and to remove the bottleneck in data transmission.

This discussion of the evolution of the IC, the chip interface, the computer, plus the connector, and the associated challenges is elaborated in Chapter 1. Its summary points to a need to shift chip interconnection from two dimensions (2D) to three dimensions (3D). The timing for a revolutionary interface is right, since the slowing of Moore's law has created an opportunity to introduce disruptive technologies to deliver more than Moore performance leaps.

Originally, the IC was invented to solve the interconnection problem of a large computer system. As an example, the Electronic Numerical Integrator and Computer (ENIAC), the first electronic, general-purpose computer, had about five million solder connections. The IC eliminated such an enormous number of hand-made connections by integrating the processing elements and interconnecting them on a piece of silicon. Modern ICs connect tens of billions of circuit elements with interconnects totaling several kilometers in length. Such connections can all be made in one fell swoop during wafer processing. However, with the slowing of Moore's law, we can no longer solely rely on integrating increasingly more functions onto a single chip for performance improvement. We need to find a revolutionary solution to the interface problem to improve multichip system performance.

Our proposal is to replace mechanical connections between chips and between modules, such as those using solder joints and connectors, with contactless, wireless connections employing near-field coupling. We have independently developed two wireless interface technologies – the ThruChip Interface (TCI) based on magnetic coupling for chip interconnection, and the Transmission Line Coupler (TLC) based on electromagnetic coupling for module interconnection – and successfully demonstrated a TCI chip interface at 1/1,000 the conventional data transmission power.

A magnetic field can penetrate through silicon. It can penetrate both metal interconnects as well as the silicon substrate. At the same time, it causes negligible interference with electronic circuits that operate using an electric field. Since our magnetic-coupling-based interface utilizes coil patterns created with metal lines, its implementation can easily be integrated into the conventional wafer process. In addition to keeping the cost overhead low, it allows the interface to benefit from the performance increase that comes with continuous process scaling. The only new requirement it introduces is the ability to scale in the third dimension, or more specifically chip thinning.

Until a few years ago, it was believed to be difficult to thin dynamic random-access memory (DRAM) chips to below 30 μm. However, recent development has reportedly achieved chip thinning down to 5 μm. As a result, development of related technologies has commenced amid high expectation from the manufacturing industry of advance in chip thinning technology.

Three challenges remain for TCI to overcome: (1) power delivery, (2) interference with DRAM operation, and (3) heat removal. We launched an ACCEL[1] project supported by the Japan Science and Technology Agency to overcome these three challenges to enable practical application of TCI. Highly doped silicon via (HDSV) for power delivery to stacked chips and a 2.5D chip interposer using TCI to facilitate heat removal were investigated, with a promising outcome. Initial test results also found no interference with DRAM operation.

In contrast to chips, modules with larger physical dimensions are more optimally interconnected using electromagnetic coupling due to the distributed nature of the interconnections. TLC is one such technology where couplers are built with transmission lines to connect modules together using nonbroadcast radio waves.

The goal of the ACCEL project was to demonstrate the feasibility of building a mobile supercomputer using TCI that achieves a power efficiency of 100 GFLOPS/W. To that end, we implemented a stacked DRAM interface that achieved a data rate of 512 GB/s at 0.5 pJ/bit, representing a 30-fold increase in data rate and 20-fold increase in power efficiency compared to state-of-the-art interface technology. Furthermore, we used TLC to connect devices such as sensors and actuators to the mobile supercomputer like Lego blocks to enable additional functionality.

[1] ACCEL is an innovation-oriented research and development (R&D) program supported by the Japan Science and Technology Agency to stimulate further company R&D, venture startup, and other publicly funded R&D based on its output.

TCI is discussed in detail in Chapter 2 and TLC in Chapter 3. Our vision for the mobile supercomputer is presented in Chapter 4.

ENIAC adopted the wired-logic architecture where programming was done by altering the physical connections between processing elements. It is easy to see that such an approach suffers from limited scalability, since the size of the largest possible program is constrained by the amount of processing hardware put in the system when it was built.

To overcome that limitation, the stored-program computer was created. In a stored-program computer, commands for the processing and movement of data as well as the data to be processed are first stored in memory. The processor then retrieves the commands and data, interprets the commands, and processes the data. Unlike a wired-logic computer, which requires physical connections of multiple processing elements, a stored-program computer requires only one processing unit which executes a different command in each clock cycle. It was hailed as a revolutionary paradigm shift that solved the problem of scalability.

Processing efficiency of modern stored-program computers is limited by how smoothly data can be moved between memory and the processor. This problem is known as the von Neumann bottleneck. TCI is a wireless IC interface technology that removes the von Neumann bottleneck to achieve a quantum leap in computer processing efficiency.

While TCI was created to solve the von Neumann bottleneck of the stored-program computer, it is our belief that it will also enable the wired-logic architecture to drive a new evolution of computer technology.

In a wired-logic computer, processing occurs as data flow through a large amount of diverse processing hardware resources. Consequently, the amount of information required to define the hardware configuration is orders-of-magnitude larger than commands used in a stored-program computer. If we can store this information in large-capacity memory, and instantly retrieve through the TCI interface the necessary hardware definition information when it is needed to configure the hardware in a specific way, we can in effect create a virtual wired-logic computer instantaneously that is optimized for the particular information processing task at hand.

The 512 GB/s stacked DRAM interface we developed in the ACCEL project enables us to remove the von Neumann bottleneck in memory access to support reconfiguration of the large-scale processing hardware in a wired-logic computer so as to implement virtualization in real time, thereby overcoming the scalability problem to allow us to take advantage of the diverse and large-scale processing resources of the wired-logic architecture.

One can think of this novel idea as an evolution from 3D to four-dimensional (4D) integration, where 3D integration in physical space solves the challenging problem of interconnecting processors and memory chips to form a hardware system, and multiple variants of such a system are implemented through reconfiguration and are virtually integrated along the time axis as the fourth dimension, resulting in 4D integration.

In a stored-program computer, the processing function is first mapped to low-level commands, then the commands are executed in sequence to complete the processing.

As a result, it is most suited for procedural information processing with deep processing sequences such as in electromagnetic simulation. On the other hand, in a virtual wired-logic computer, each processing function is first mapped to the structural space of the available processing resources, then data are processed and passed from one virtual configuration to another to complete the processing. As a result, it is most suited for structured processing with large amount of data flow such as in deep neural network (DNN) processing. If we are to map these computers for information processing to the brain, the former would be mapped to the left and the latter the right brain.

Our next goal is to explore a new paradigm of information processing as follows. We plan to build a stored-program computer as the left brain using TCI 3D IC integration to remove the von Neumann bottleneck, thereby enabling the electronic left brain to realize its full processing potential. In parallel, we will build a virtual wired-logic computer as the right brain using TCI 4D integration also made possible by the elimination of the von Neumann bottleneck. We will then use TLC to interconnect the two and other building blocks, including sensors and actuators, to create a new artificial intelligence computer, an electronic brain that exceeds the performance of the human brain to open up new possibilities in information processing.

1 Introduction

3D Integration and Near-Field Coupling

1.1 A Short History of the Computer, IC, and Connector

The advance in computer performance has been nothing short of spectacular. In just 60 years, within the lifetime of the average human being, computer performance has increased by 100,000-fold, while power consumption has decreased by a similar factor of 100,000, physical dimensions by a factor of 1,000,000, and price by a factor of 1,000. However, advance in integrated circuit (IC) technology, which has been fueling this explosive growth, is now facing severe challenges that may bring the growth to a stop. Before we ponder where we are going to go from here, it is worth looking back to see how we have come to this point.

1.1.1 History of the Computer

The 2014 British film *The Imitation Game* portrays the development of a computer by Alan Turing to break the secret code behind Nazi Germany's Enigma communications machine. Turing is considered one of the fathers of the modern-day computer due to his development of the Turing machine that provides the model for a general-purpose computer. Nevertheless, his code-breaking computer completed in 1938 had the sole purpose of breaking the Enigma machine.

Indeed, early computers were specific-purpose computers where the types of problems they could solve were determined by how their components were wired together. In other words, it was a wired-logic computer architecture. Therefore, what we think of as programming today was done by rewiring the components, if at all possible. This hardwired architecture has two severe challenges.

(1) Challenge of scale: the size of problems that can be solved is limited by the scale of the network of components.
(2) Challenge of wiring: as the system scale goes up, the wiring complexity grows exponentially and beyond what can be handled manually.

These two challenges were overcome with two groundbreaking inventions: the von Neumann computer with stored programs and the IC. In a von Neumann computer, the data to be processed and a list of commands in the form of a program to control how the hardware processes the data are stored in memory. The computer operates by retrieving input data and the program from memory. Then in each operational cycle, a different command is executed resulting in a different set of operations. Different types of problems can thus be solved by feeding different programs with different sets of commands to the same piece of hardware with the same wiring. A von Neumann computer is therefore a general-purpose computer. More complex problems can be solved with more complex programs, overcoming the challenge of scale.

Meanwhile, invention of the IC, chronicled in the next section, has enabled a large number of circuit elements to be integrated and interconnected all on a single chip. Using photolithography, all interconnections can be made in parallel as part of the chip-making process, overcoming the challenge of wiring. The IC enables integration and parallelization of simplified and progressively downsized computing resources, driving rapid advance in chip computational performance.

However, as computational performance continues to rise, a new challenge in interconnecting chips is emerging. In particular, economics requires that the processing unit and the bulk of the memory it uses be implemented on separate chips, and connecting these two chips has become challenging as the amount and required speed of data flow in the chip interface grows, resulting in what is known as the von Neumann bottleneck in the interface, limiting overall system computational performance. Consequently, continuous advance in computer performance requires increasing both chip and interface performance.

On the other hand, the rapid advance in artificial intelligence technology based on neural network and deep learning is renewing interest in the wired-logic computer architecture, where the function performed is implemented through hardwiring like the neural network in the human brain. As a result, history has come full circle and the challenges of scale and wiring are once again pushed to the forefront, which necessitates innovative solutions to drive the next phase of computing technology advancement.

1.1.2 The Four Seasons of IC Development

1.1.2.1 The "Big Bang": Invention of the IC

One of the first general-purpose computers built was ENIAC (Figure 1.1), developed in the United States in 1945 and introduced in 1946. It was a gigantic piece of machine built with 100,000 components and weighed 27 tons. It consumed so much power – 150 kW – rumor has it that streetlights in the neighborhood dimmed when it was being operated. It had about 5,000,000 solder connections made by hand. Its heat-generating vacuum tubes had such poor reliability that a few of them broke and the system went down every day.

The transistor, a more reliable replacement of the vacuum tube, was invented by William Shockley et al. in 1948, based on the groundwork laid by Bell Labs in the prior year. With much less heat generation and higher reliability, it enabled construction of much larger-scale systems. However, the increasing scale of computer systems led to the exponential growth in the number of interconnections between components and hence wiring complexity. It was such a challenging problem that it became known

Figure 1.1 Programming ENIAC. Photo by US Army, public domain

as the tyranny of numbers. The winning solution that emerged, among various proposed, was the IC.

The IC was invented by Jack Kilby and Robert Noyce in 1958–1959. At the time, Texas Instruments (TI) was working on the Micro-Module solution to the problem of the tyranny of numbers that enabled circuit components to be densely packed on a printed circuit board (PCB). Kilby, despite being a TI engineer, developed an alternate solution – the groundbreaking concept of a monolithic IC where all circuit elements are integrated on a single piece (*mono*) of semiconductor substrate (*lithic*). In the following year, Robert Noyce would develop the silicon implementation of the IC.

1.1.2.2 Spring: Explosive Growth through Scaling

Around the same time in December 1959, American physicist Richard Feynman delivered a lecture entitled "There's Plenty of Room at the Bottom," where he discussed scientific manipulation at the atomic level. This could be considered the start of nanotechnology research, a field that fueled the advance in IC technology and hence its explosive growth.

In 1965, Gordon Moore of Fairchild Semiconductor (and later of Intel Corporation) published a paper entitled "Cramming More Components into Integrated Circuits," where he described the potential of the IC with its exponentially growing integration density. He observed that the number of components in an IC had doubled approximately every year, and speculated that such a growth rate would continue. This speculation would later become known as Moore's law and set the collective target for the advance of the semiconductor industry.

Figure 1.2 illustrates how the dominant computing device has evolved as transistor count increases following Moore's law, and the changing driving force behind the evolution. From around 1980 to 1995, transistor channel length shrank from 4 to 0.35 μm. Meanwhile, the number of transistors that can be integrated on an IC increased

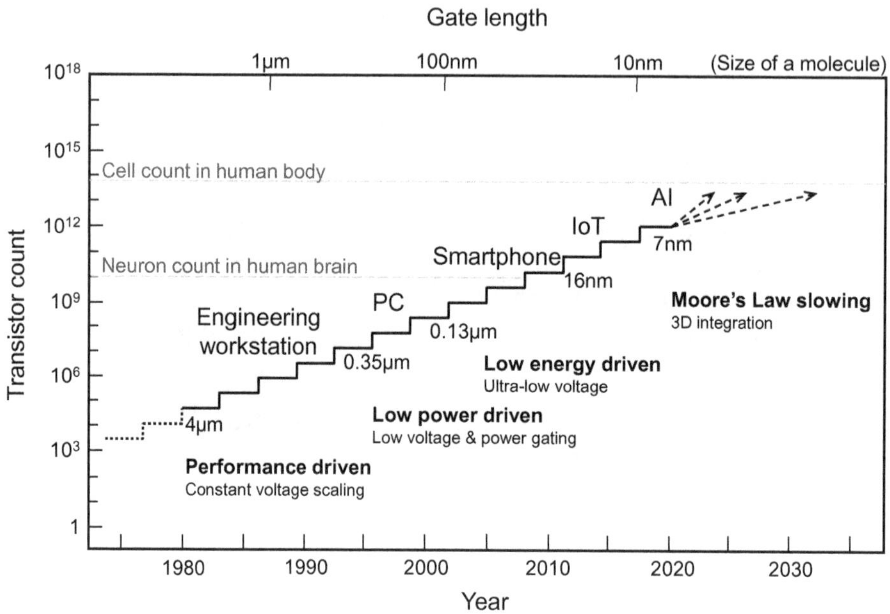

Figure 1.2 Evolution of the integrated circuit [1].

from 100,000 to 100,000,000. As a result of this increased chip-level integration, the computer evolved from being an engineering workstation to a personal computer (PC), and its adoption spread from being one per group to one per person. As the trend continued, the smartphone replaced the PC as the dominant computing device, the Internet of Things (IoT) became ubiquitous, and artificial intelligence (AI) computing finally entered into the mainstream.

In this manner, scaling fueled the explosive growth of the IC. Unfortunately, there was a trade-off with an undesirable consequence that lurked in the background as computational performance rose. Eventually, around 1995, it reared its ugly head in the form of a wall that threatened to stop further scaling.

1.1.2.3 Summer: The Power Wall

Around 1995, process scaling ran into a power wall. Scaling resulted in a continuous increase in power consumption, to the point that the IC released so much heat as to degrade its reliability, bringing further scaling to a halt. As an illustration, from 1980 to 1995, in 15 years, power consumption of an IC increased 1,000-fold. On a per-unit-area basis, an IC generated as much heat as an electric cooktop – you can fry an egg by powering an array of these chips under the fry pan! How did that happen?

In 1974, Robert Dennard, inventor of dynamic random-access memory (DRAM), formulated with his colleagues the scaling theory that describes an ideal scaling scenario where the electric field within the semiconductor device is kept constant. If you scale the supply voltage and physical dimensions of the device by the same factor,

Table 1.1. Transistor scaling scenarios

Scaling scenario		Constant electric field (ideal)	Constant voltage (actual, 1980–1995)
Scaling factor	Device size, x	½	½
	Gate oxide thickness, t_{ox}	½	½
	Voltage, V	½	1
Resultant change	Electric field, E	1	2
	Current, $I \propto V^2/t_{ox}$	½	2
	Resistance, $R \propto V/I$	1	½
	Capacitance, $C \propto x^2/t_{ox}$	½	½
	RC delay, $\tau \propto CR$	½	¼
	Power, $P \propto CV^2/\tau$	¼	2
	Power density, $p \propto P/x^2$	1	8

Source: © 1995 IEICE, [2] table 1.

the electric field, which is a function of the ratio of voltage to distance, will remain constant. As a result, transistors whose operation is driven by manipulation of the electric field will perform the same way after scaling. Under this scaling scenario, power density would remain unchanged as well. Therefore, for the same chip size, power remains the same after scaling, and hence no heat generation or dissipation problems should arise (see the column "Constant electric field" in Table 1.1). Meanwhile, as the device gets smaller, more devices can be integrated in the same area, leading to lower manufacturing costs. Furthermore, the circuit resistor–capacitor (RC) time constant is reduced, resulting in higher operating speed. It is indeed a beautiful scenario that, if realized, would have no undesirable consequences.

The reality was that scaling was implemented with no reduction in supply voltage. In other words, it was not a constant electric field but rather constant voltage scaling. In this case, the circuit ran even faster as RC delay was further reduced, IC computational performance went up further, and more chips were sold bringing in higher revenues. The flip side was a large increase in power (see the "Constant voltage" column in Table 1.1). Given that IC power consumption was small to begin with, its increase had initially only a secondary effect. Nevertheless, continuous scaling resulted in rapid increase in power consumption. Before long, it hit a painful limit. This power wall was therefore the result of an unintended side effect of nonideal scaling.

The power consumption increase with scaling was the result of the laws of physics. It is therefore not an easy problem to solve. Faced with the power wall, various schemes were devised to handle the extra power that came with performance increase. For instance, in power gating, circuit operations are diligently monitored. Whenever a circuit is not in use, its power is cut off. In another scheme, supply voltage is lowered whenever high circuit performance is not required. Such power-saving actions may sound obvious given that we practice similar power conservation in our daily life. But with hundreds of millions of transistors on a large-scale IC, uncovering waste is no simple task. For any waste we can find and eliminate, that much power can be exchanged for increased performance. In other words, "no gain in power efficiency, no gain in IC performance." This is a trade-off IC development has continued to face to this day.

1.1.2.4 Fall: The Leakage Wall

Around 2005, scaling hit yet another wall – the leakage wall. Unlike the power wall, the leakage wall was not a result of the progression of scaling; rather, it was the consequence of scaling reaching its limits. As transistor channel length was scaled down to 100 nm, gate oxide thickness was reduced to 1.2 nm, about the thickness of 4 molecules. If the oxide layer were thinned further, it would be subject to quantum effects under which tunneling current would flow. But without proportional reduction in gate oxide thickness, further transistor scaling would weaken ability of the gate to properly turn off the channel, resulting in leakage current. It is analogous to a faucet not properly turned off, allowing water to drip.

The options for fixing the "leaky faucet" include changing materials, modifying the manufacturing process, or adjusting how the device is constructed, as shown in Figure 1.3. For example, dielectric materials with higher dielectric constant were introduced to increase electric field strength without reducing thickness. In parallel, development of a manufacturing process was started to introduce distortions in the silicon structure (strained silicon) to stretch the silicon atoms to boost mobility and hence current when the transistor is on, without increasing leakage when the transistor

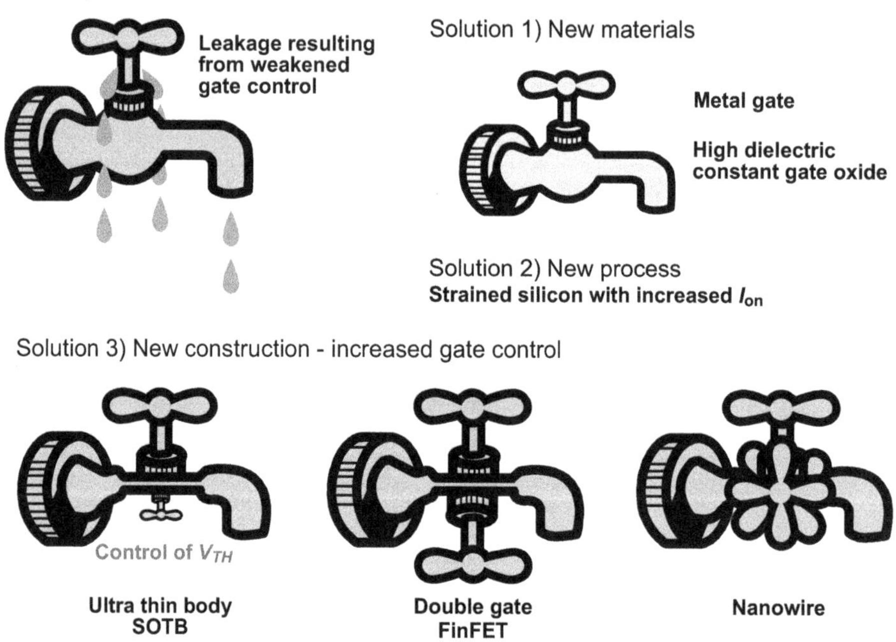

Figure 1.3 Low leakage device technologies [3].

is off. Furthermore, transistor gate construction was transformed from being planar to three-dimensional, such as in a fin field-effect transistor (FinFET), in order to strengthen gate control of the channel.

Nevertheless, all these are measures to provide temporary relief only; there are no good ways to counter quantum effects that result from the laws of physics.

1.1.2.5 Winter: The End of Scaling?

The cover of the April 2015 issue of *IEEE Spectrum* (Figure 1.4) depicts a flipped-over computer chip in the shape of a belly-up, dead bug, with the competing headlines of “Moore’s Law is dead” vs. “Long live Moore’s Law.” “Is Moore’s Law dead?” is the billion-dollar question faced by the IC industry: Is scaling, that has propelled the industry in the last half of a century, coming to an end? Has IC development entered its winter season?

Over the prior 40 years, the manufacturing cost per transistor had dropped by a factor of 4 million. However, as channel length shrank below 28 nm, this unit cost started to rise. The economic benefits from scaling finally started to disappear. IC manufacturing has become a privilege that only a few companies with deep pockets can afford due to the enormous investment required. Even the costs for making an R&D test chip have risen from only several hundred thousand dollars to several million dollars. Faced with such adverse economic reality, once thriving IC research in both the academia and industry is gradually losing momentum.

Figure 1.4 *IEEE Spectrum* cover, April 2015. © 2015 IEEE

1.1.2.6 The Second Spring

Nonetheless, innovation is not slowing; it is only taking a different form. Innovation in the semiconductor industry over the last 60 years has been mainly driven by Moore's law in process scaling that has fueled an exponential proliferation in IC quantity and made it ubiquitous in our society. With the slowing or even ending of scaling, innovation is shifting to enhancing quality, in terms of how we apply IC technology in various innovative ways to enrich our life. The excitement in dramatically improving the price performance of an IC has been replaced by excitement in research and development of novel applications, including automated driving, artificial intelligence, big data, and IoT. Each of these areas is exploring different innovative ways to realize the potential provided by the performance achieved in scaling – IC's first act. After experiencing the excitement of achieving a 100 million-fold improvement in the price performance of a computer, we are witnessing the opening of the curtain on IC's second act, where advance in IC technology enters its second spring to deliver enrichment across various facets of our life.

1.1.3 History of the Connector

Regardless of where we are on the transistor count growth curve, there is a limit on how many functions can be packed into an IC. Similarly, there is a limit on how big a package, a module, and a printed circuit board can practically and economically be manufactured. Furthermore, not all desired functions can be performed by transistors. For instance, many functions require the use of sensors, actuators, and other input/output devices. On the other hand, economics oftentimes requires sourcing components made with different processes by different suppliers, and economy of scale favors the use of standardized parts. As a result, a computing system is generally constructed by connecting separate parts together, which necessitates not only interchip integration but also module integration using connectors.

In contrast to the IC, the history of the connector over the last 70 years has been characterized more by advances in quality instead of quantity, by how connections are made instead of exponential increase in the number of connections. The major technological innovations during this time include the following:

- Crimped termination
- Insulation displacement contact
- Wire wrapping
- Press-fit
- Optical connector

In particular, the crimped termination represented a major paradigm shift since it gave rise to a solderless connection.

1.1.3.1 The Solderless Connection

The solderless electrical connection was invented by American engineer Uncas A. Whitaker in 1941 [4]. Known as a crimped termination, it is a connection formed by

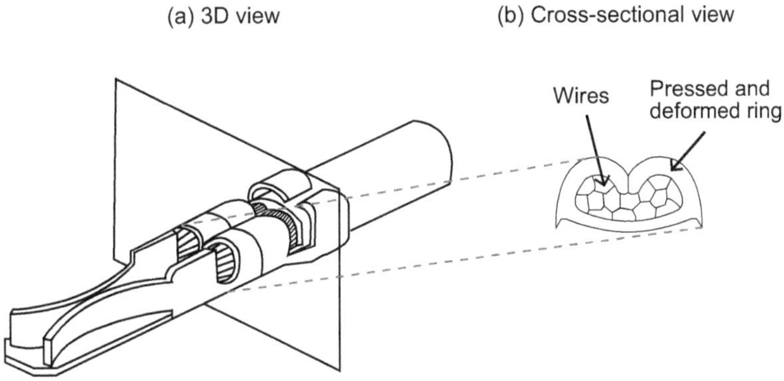

Figure 1.5 A crimped termination. © 2010 SMK Corporation [5]

inserting open wires on one end into a small metal tube with a ring on the other end, and pressing the ring against the wires using a specialized tool (Figure 1.5). Compared to a rigid, soldered connection, this revolutionary technology delivers a much faster way to make and break a connection in a smaller space. Furthermore, it offers higher reliability in the face of shock and vibration. Consequently, when the United States entered World War II, the technology was widely adopted in military applications since it maintained electrical connections even under harsh operating conditions. Furthermore, it significantly increased the speed of equipment repair, including that of fighter jets on aircraft carriers, which was a critical advantage in wartime. After the war ended, the application of crimped termination was extended to commercial applications, especially those that required operation in harsh environment, including automotive and appliances, such as power tools for the homebuilding boom that ensued after the war.

The solderless electrical connection opened up the possibility for an easily breakable connection and hence the idea of an electronic connector, which enabled modularization and flexible selection of different combinations of modules in building a large-scale electronic system, resulting in flexibility, expandability, and scalability. System assembly can be done by users in the field, and it is easy to replace or upgrade part of the system. Consequently, it represented an advance that is parallel to the IC in solving the connection problem of large-scale systems.

1.1.3.2 Recent Challenges for the Connector

The introduction of the iPhone in 2007 set in motion the rapid shift from the PC to the smartphone as the dominant form of electronic and computing system. Even the first-generation iPhone represented an order-of-magnitude shrink in form factor compared to the PC. The subsequent exponential growth in functionality and performance of the smartphone with continuous, albeit incremental, shrink in form factor in the last decade only added to the pressure to continuously shrink both the footprint and height of the components within. Nevertheless, scaling of connector dimensions is not keeping pace with that of the IC. This is illustrated by the historic trend in the scaling of flexible printed circuit (FPC) connector pitch. Because of its flexibility, low profile,

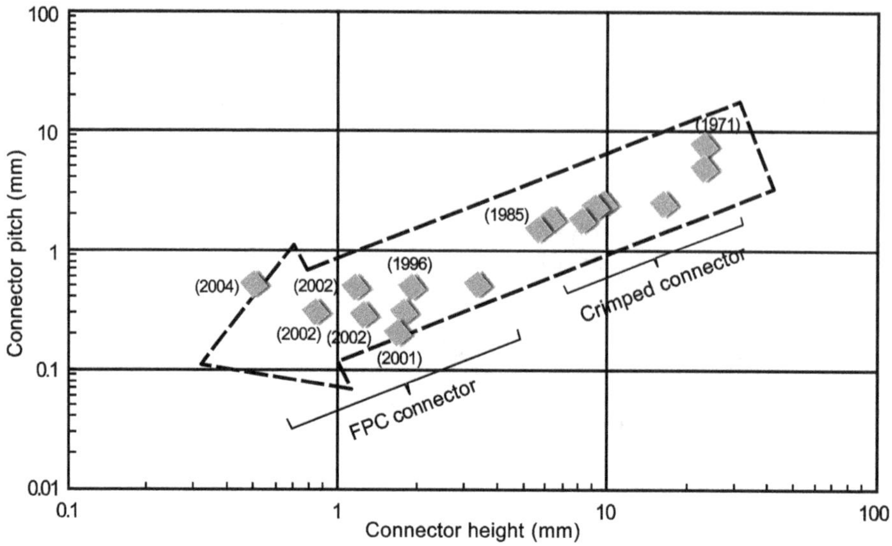

Figure 1.6 FPC connector dimensions scaling, 1996–2004. © 2010 SMK Corporation [5]

and small dimensions, FPC and its connector are commonly used in mobile devices. From 1996 to 2004, FPC connector pitch scaled by roughly 0.4× and its height by roughly 0.25× in eight years, corresponding to average scaling of 0.89× and 0.84× per year, respectively (Figure 1.6). By contrast, around the same time from 1995 to 2003, Intel process migrated from 350 to 90 nm, scaling gate length from 350 to 50 nm, or 0.14× in eight years [6]. More recently, based on the roadmap published by the Japan Electronics and Information Technology Industries Association (JEITA), mainstream pitch of FPC-to-PCB connectors used in small form-factor devices is projected to shift from 0.3–0.4 mm in 2016 to 0.2–0.3 mm in 2026 (Figure 1.7), which translates to total scaling of 0.714× over 10 years, and a modest average scaling of 0.967× per year. This represents a rapid deceleration in the scaling of FPC connector pitch.

The challenges in scaling connector dimensions are not surprising given the mechanical processes required in manufacturing connectors. Furthermore, it is difficult to significantly scale the dimensions of the connector housing that is needed to provide mechanical stability to ensure a reliable connection, especially when it is subject to shock and vibration, such as in an automotive or space application. This is on top of the challenge in maintaining signal integrity as the data rates of signals passing through connectors rise rapidly. The time is ripe for another revolutionary connector technology to enable a leap in miniaturization and performance of electronic systems.

1.1.4 Closing Thoughts

Figure 1.8 depicts the evolution of large-scale computer systems, their current challenges, and proposed solutions. The connection problem of the early large-scale

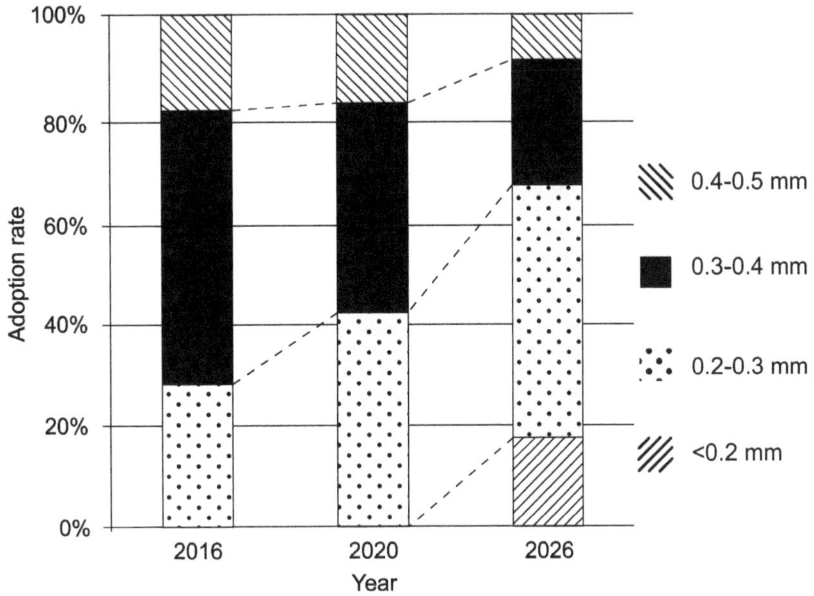

Figure 1.7 FPC connector pitch roadmap, 2016–2026. © 2017 JEITA [7]

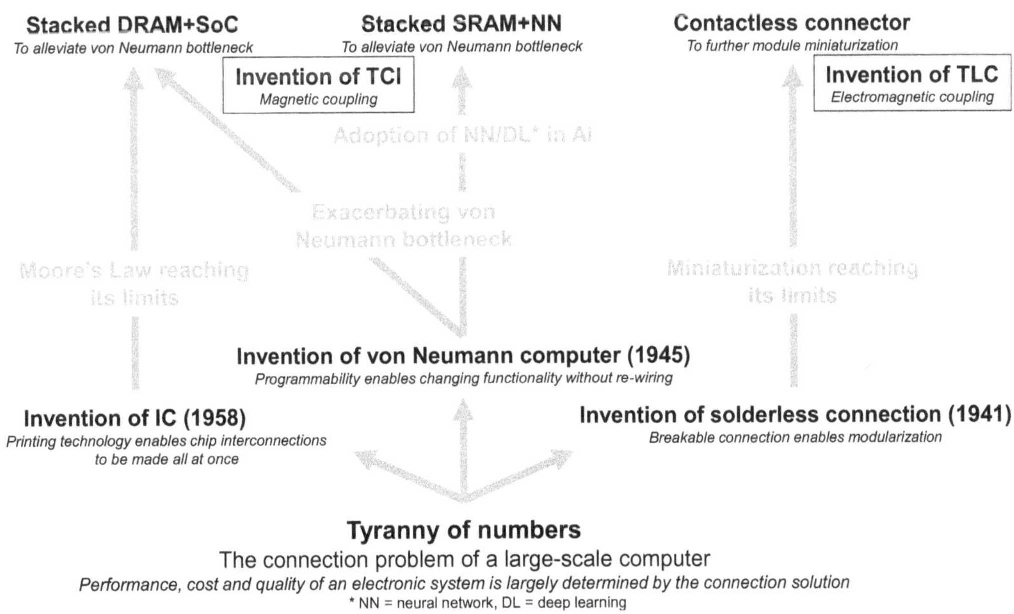

Figure 1.8 Evolution of solutions to the connection problem of a large-scale computer.

computers, the tyranny of numbers, was solved by the invention of the von Neumann computer, integrated circuit, and solderless connection. However, the advance that has been achieved in all these technologies has made them the victim of their own success, with Moore's law scaling and connector miniaturization each reaching its limits, and von Neumann bottleneck exacerbating. Furthermore, the adoption of neural network and deep learning in AI has once again brought to the forefront the challenges of scale and wiring in building hardwired computers. We believe the solutions can be found in our two wireless coupling technologies – ThruChip Interface (TCI) and Transmission Line Coupler (TLC). TCI, a magnetic coupling technology, enables stacking DRAMs with the system on a chip (SoC) to alleviate the von Neumann bottleneck. The same technology also enables stacking of static random-access memory (SRAM) with neural network chips implemented in field-progammable gate arrays (FPGAs) or reconfigurable processors to solve the challenges of scale and wiring in an AI computer. On the other hand, TLC, an electromagnetic coupling technology, enables a contactless connector that overcomes the scaling limits of its electromechanical counterpart. Each of these two wireless coupling technologies will be introduced in ensuing sections.

1.2 Energy-Efficient Computing

1.2.1 High-Performance IC

As noted in Figure 1.2 in Section 1.1, early IC development was performance driven, where process scaling delivered rapid performance increase as anticipated by Moore's law. As an illustration, comparing the Intel Pentium from 1995 to the 8086 processor from 1980, in the span of 15 years, transistor count increased from 30,000 to 3,000,000, clock frequency from 5 to 300 MHz, computing power from less than 1 to several hundred MIPS, in line with the target rate of a twofold increase every two years. Since minimum line width shrank from 3 to 0.35 μm, device size scaled by approximately 0.75 times every two years. Such exponential growth was fueled by smooth progress in process scaling, including advances in lithography, process technology, wafer size, and manufacturing yield.

As explained in Section 1.1.2.3, under an ideal constant electric field scaling scenario where both dimensions and voltage are halved, delay is halved and power reduced to 1/4, while power density remains unchanged. However, if voltage is kept unchanged, delay is further reduced by half. At a time when the engineering workstation was the technology driver, and hence performance the driving factor, and with the switch from n-channel metal–oxide–semiconductor (NMOS) to complementary metal–oxide–semiconductor (CMOS), which produced a three-orders-of-magnitude drop in power, constant voltage scaling became an easy choice over constant electric field scaling. But since constant voltage scaling results in a twofold increase in power and an eightfold increase in power density, from 1980 to around 1992, in 12 years IC power skyrocketed by increasing fourfold every three years (Figure 1.9).

Table 1.2. Constant electric field scaling changes from around 1995.

Time period		Pre-1995	Post-1995
Scaling factor	Device size, x	1/2	1/2
	Gate oxide thickness, t_{ox}	1/2	1/2
	Voltage, V	1/2	1/2
Resultant change	Electric field, E	1	1
	Current dependency on V, β	2 (ideal)	1.3
	Current, $I \propto V^{\beta}/t_{ox}$	1/2	1/1.23
	Resistance, $R \propto V/I$	1	1/1.62
	Capacitance, $C \propto x^2/t_{ox}$	1/2	1/2
	RC delay, $\tau \propto CR$	1/2	1/3.25
	Power, $P \propto CV^2/\tau$	1/4	1/2.46
	Power density, $p \propto P/x^2$	1	1.62

Source: © 2007 IEICE, [9], table 1.

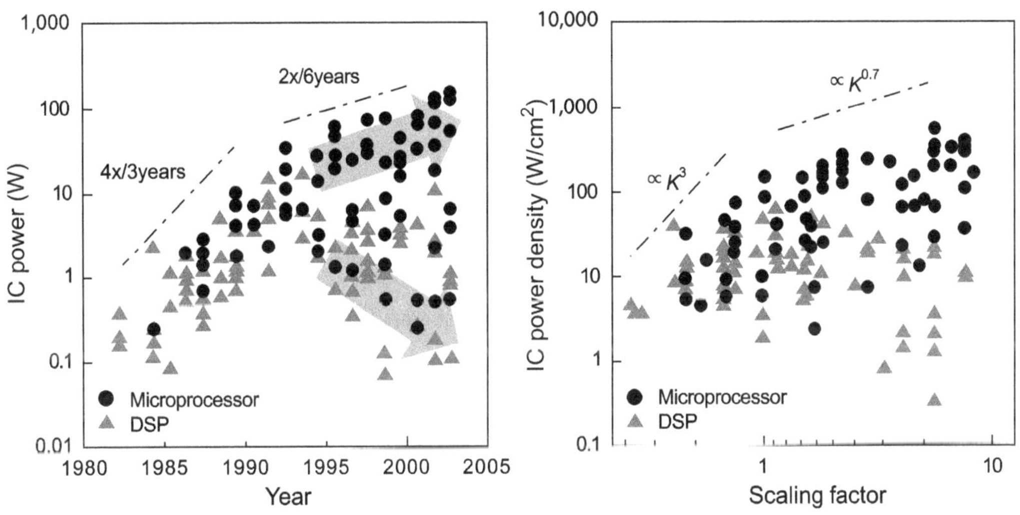

Figure 1.9 Evolution of IC power and power density. © 1999 IPSJ [8]

Eventually when IC power exceeded 10 W, the problem of increasing power could no longer be ignored. Consequently, constant electric field scaling was introduced as a countermeasure and voltage started to drop. Nevertheless, the electric field had risen to the point where charge carrier mobility had reached saturation. As a result, while ideally current scales as V^2, in reality it was scaling as $V^{1.3}$ (Table 1.2). Furthermore, leakage current prevented V from fully scaling proportionally with device dimensions. For example, while V scaled from 2.5 to 1.8 and then 1.3 V for the 250, 180, and 130 nm process nodes respectively, its scaling slowed to 1.2, 1.0, and 0.9 V as process scaled to 90, 65, and 45 nm. As a result, power increase from scaling was not suppressed as much as desired, and IC power continued to climb at a rate of twice every six years (Figure 1.9). Eventually, in the second half of 1990s it approached 100 W, and the IC heat generation problem became quite severe.

1.2.2 Low-Power IC

In the decade from around 1995 to 2005, low power became a key design goal in IC development, in addition to performance. Solutions were developed to scale the power wall, while process scaling continued to raise the transistor count. During this period, the technology driver shifted from the engineering workstation to personal computer, which further exacerbated the power problem due to the shrinking system form factor.

One low-power solution that drew a lot of attention was the optimization of drain voltage (V_{DD}) and transistor threshold voltage (V_{TH}). Figure 1.10 depicts how IC power and RC delay vary respectively as a function of V_{DD} and V_{TH}. When V_{DD} is lowered to reduce power, delay goes up. But if V_{TH} is lowered accordingly, it is possible to suppress the rise in delay while significantly reducing power. For instance, if V_{DD} and V_{TH} are lowered simultaneously to move from operating point A to B in Figure 1.10b along the equip–delay line of 100 ps, there is no change in delay, but total power is reduced as shown in Figure 1.10a.

The optimization of V_{DD} and V_{TH} for the best trade-off of power and delay can be performed both dynamically during chip operation and spatially based on the design requirements of individual circuit blocks. With close monitoring of chip activity, V_{DD} and V_{TH} can be adjusted dynamically in accordance with the speed requirement at any particular time.

Spatially, circuit blocks that demand high speed can be supplied with a high V_{DD} combined with a low V_{TH}, while blocks that can afford to run slower are supplied with a low V_{DD} combined with a high V_{TH}, trading off execution time for lower power. The

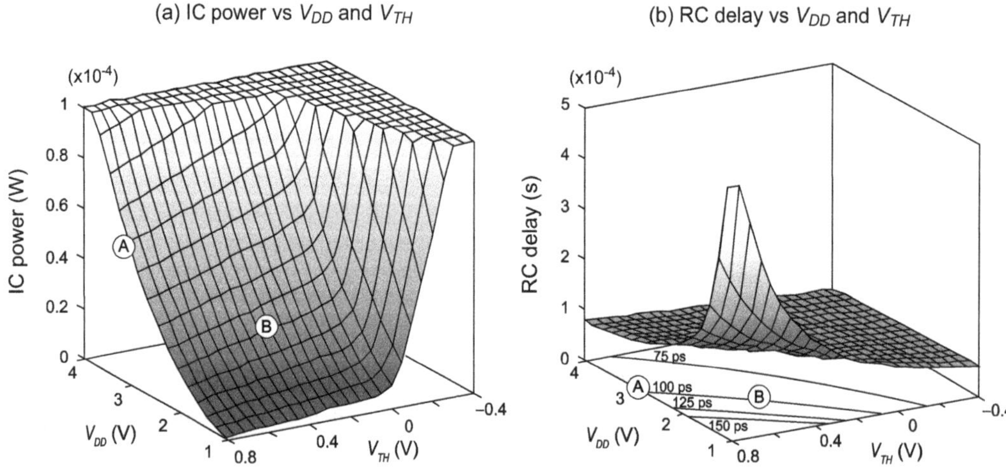

Figure 1.10 IC power and RC delay as a function of V_{DD} and V_{TH}. © 1995 IEICE, [2], figure 4

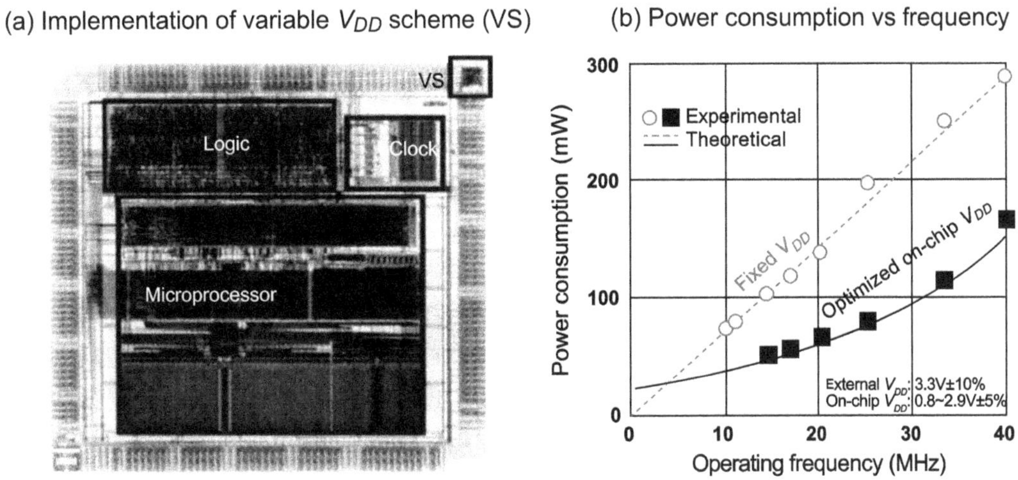

Figure 1.11 Power consumption reduction through V_{DD} optimization. © 1997 IEEE. Reprinted, with permission, from [10]

drawback is an increase in the complexity of the power supply, especially in the case of spatial optimization, where multiple voltages have to be generated at the same time.

By moving the on-board DC–DC converter that supplies V_{DD} to on-chip for internal adjustment, and by varying the substrate bias to adjust V_{TH}, it is possible to dynamically optimize these two parameters. Figure 1.11 depicts an example processor that implemented an on-chip, variable V_{DD} supply. By setting V_{DD} to its minimum acceptable value for correct operation at any particular frequency, power consumption was reduced by up to 50%.

Figure 1.12 depicts an example image processor that implemented an on-chip substrate bias circuit to vary V_{TH} to achieve significant reduction in leakage current.

Between 1996 and 1998, Toshiba in Japan successfully developed the world's first processors with controllable V_{DD} and V_{TH} (Figure 1.13, [10, 12–13]). The technology was soon adopted by Transmeta, followed by Intel and AMD, validating the need for processor developers to counter the pressing power problem.

Nonetheless, even though Intel had been aggressively driving up processor performance by raising clock frequency, by October 2004, the company decided to change course and shifted focus away from pushing clock frequency beyond 4 GHz. It was more a business than engineering decision since it was technically possible to further raise the clock frequency. But by doing so, the leakage and heat problem would become too expensive to solve. Instead, Intel shifted to boosting performance by increasing the number of processor cores.

1.2.3 Low-Power IC Interface

An IC must communicate with the outside world to perform useful functions. Therefore, in minimizing overall system power, in addition to minimizing power of

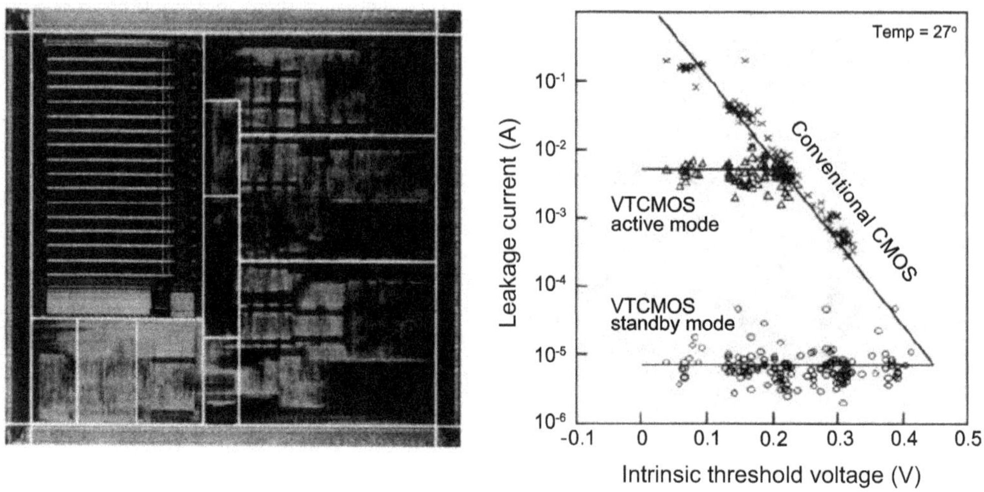

Figure 1.12 Leakage current reduction through V_{TH} optimization. (b) © 2002 IEEE. Reprinted, with permission, from [11]

the IC itself, IC interface power must also be taken into consideration. Conceptually, the IC interface can be divided into two components – the on-chip input/output (I/O) circuits (transmitters or drivers, and receivers) and the off-chip physical interconnections. Power optimization of the IC interface thus requires both low power I/O circuits and off-chip interconnect design.

Signal transmission in and out of the IC involves generating an analog waveform that represents the transmitted digital information, preserving the integrity (quality) of the waveform as it propagates, and receiving and correctly extracting the embedded digital information. It is therefore an operation in both the digital and analog domains. As a result, unlike their digital counterpart, I/O circuits' performance, including power consumption, does not scale proportionally with process. As transmission speed increases, circuit design must change to support the higher data rates, leading to rising power. For example, buffers must be inserted in the clock path for higher speed clock distribution to compensate for the higher attenuation at higher frequency. Furthermore, increased adoption of common mode logic (CML) circuits to support faster digital switching also contributes to higher power. As a result, while IC interface power roughly followed the scaling of CMOS gate power in earlier processes, from around the 130 nm node it reversed course and rose steadily thereafter (Figure 1.14). To counter this trend, an effective solution is to limit the increase in signal data rate and instead increase the number of parallel signals transmitted to increase total interface bandwidth. This was facilitated by the shift in the placement of I/O circuits from being along the perimeter to across the surface of the IC, leading to the shift from wirebond to flip-chip packaging.

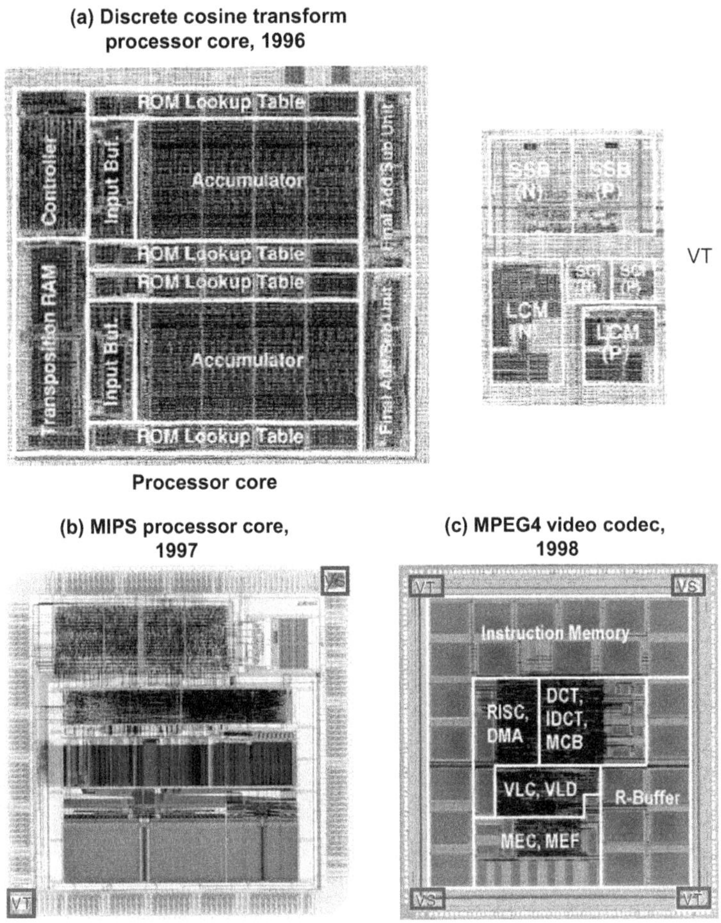

Figure 1.13 Example ICs with variable V_{DD} (VS) and V_{TH} (VT). © 1996, 1997, 1998 IEEE. Reprinted, with permission, from (a) [12]; (b) [10]; (c) [13]

The off-chip physical interconnections of an IC generally consist of bond wires or flip-chip bumps and solder balls, package traces, vias, PCB traces and connector pins, with total distance of more than an order-of-magnitude longer than the size of the IC (Figure 1.15). Consequently, off-chip interconnects have different electrical behavior than on-chip interconnects. As a rule of thumb, when an interconnect has delay that satisfies the following condition, it appears as a transmission line to signal propagation:

$$2t_d > 0.1t_r \tag{1.1}$$

where t_d is the propagation delay of the interconnect, and t_r is the edge rate of the signal. Delay of a typical PCB interconnect is around 6 ps/mm, while the typical edge rate of a 1 Gbps signal is around 200 ps. Therefore, a PCB interconnect only needs to be longer than 1.67 mm to appear as a transmission line to a 1 Gbps signal. Hence,

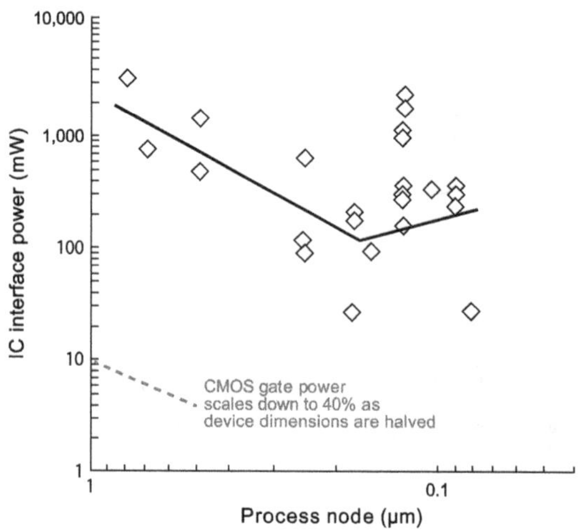

Figure 1.14 IC interface power scaling with process. © 2007 IEICE, [9], figure 4

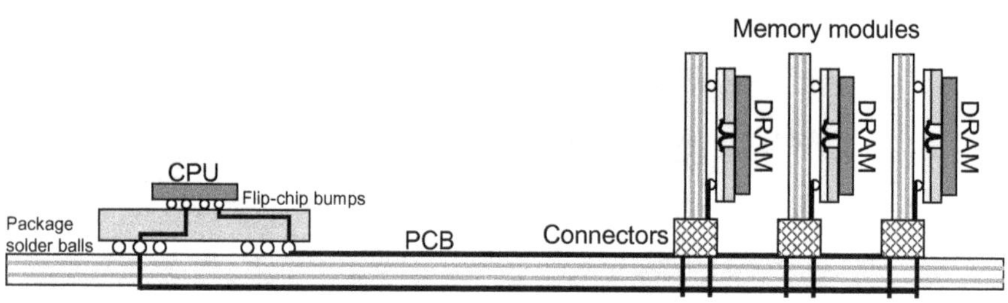

Figure 1.15 An off-chip interconnection in a typical PC memory system.

typical PCB interconnects in high-performance computing systems that are on the order of a centimeter or more in length behave as transmission lines. Furthermore, due to the combination of high conductivity and sufficiently large surface area, PCB interconnects are generally low-loss transmission lines with relatively low resistive loss and hence low power consumption.

The approximate electrical behavior of a low-loss transmission line is fully defined by its characteristic impedance Z_0 and propagation delay γ, which can be computed as follows:

$$Z_0 = \sqrt{\frac{L_0}{C_0}} \tag{1.2}$$

$$\gamma = \sqrt{L_0 C_0}, \tag{1.3}$$

where L_0 and C_0 are the per unit length inductance and capacitance respectively of the transmission line. In a low-loss transmission line, attenuation of signal amplitude is

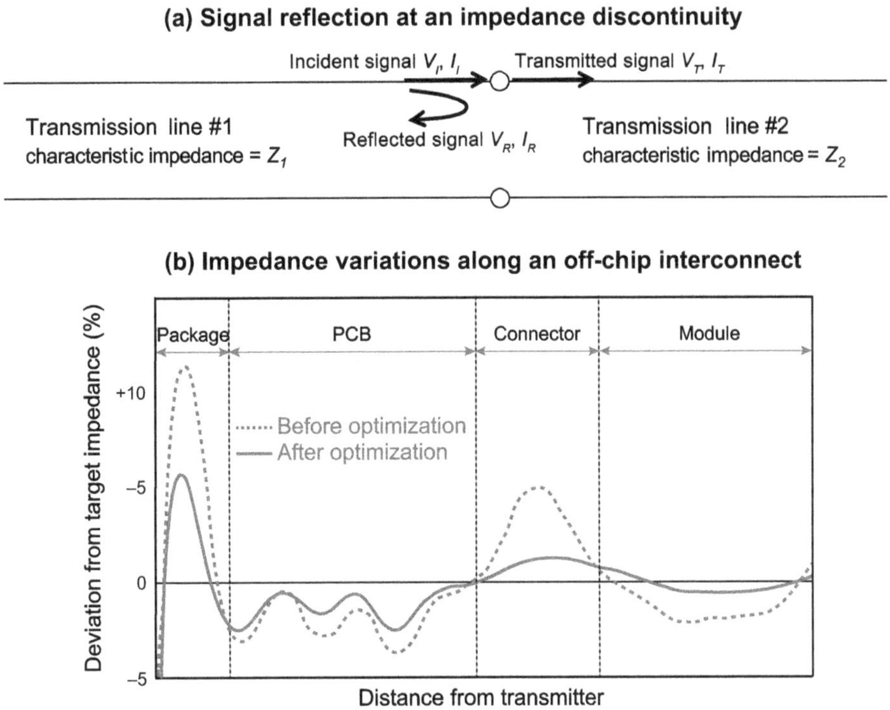

Figure 1.16 Impedance variations of off-chip interconnects.

small. However, if the transmission line Z_0 varies along its length, some of the signal energy will be reflected. When that happens, the shape of the received signal in a certain time slot depends not only on the shape of the transmitted signal in that time slot, but also the shape of reflection of the transmitted signal in prior time slots, resulting in what is known as intersymbol interference (ISI). Therefore, impedance variations along a transmission line result in ISI, which degrades integrity of the received signal.

When a voltage step V_I arrives at an impedance discontinuity, a voltage step V_R is reflected, resulting in a voltage step of V_T being transmitted (Figure 1.16a). The voltage steps are related through the following two equations:

$$\rho = \frac{V_R}{V_I} = \frac{Z_2 - Z_1}{Z_2 + Z_1} \tag{1.4}$$

$$\tau = \frac{V_T}{V_I} = (1 + \rho) = \frac{2Z_2}{Z_2 + Z_1}, \tag{1.5}$$

where ρ and τ are known as the reflection and transmissoin coefficients, and Z_1 and Z_2 the characteristic impedance of the transmission line before and after the discotninuity respectively. Therefore, the goal of off-chip interconnect design is to maintain a smooth characteristic impedance profile along its entire length to minimize reflection in order to preserve the integrity of the transmitted signal. Nevertheless, given the complexity of the off-chip interconnect, characteristic impedance variations along its length are unavoidable even with meticulous design optimization (Figure 1.16b).

To compensate for the loss in integrity of the transmitted signal due to impedance variations of off-chip interconnects, I/O circuits must be modified accordingly, for example by adding equalization circuits, resulting in increased complexity and power consumption. Furthermore, in general, the driver at the near end of the interconnect has lower impedance while the receiver at the far end higher impedance than the interconnect, representing additional discontinuities. To minimize reflection due to these discontinuities requires adding resistors at the driver and/or receiver to alter their impedances, resulting in additional area, cost, as well as power consumption. Consequently, a more effective way to preserve signal integrity of a low-power IC interface is to minimize the communication distance to limit transmission line effects in order to reduce I/O circuit complexity and power consumption.

Putting the entire system on a single chip in the form of an SoC will result in the shortest communication distances. However, in practice, there is a limit on how much of the system can be brought on-chip. In addition to the high manufacturing costs of a bigger die, a larger and more complex SoC increases development costs and time to market. Furthermore, an SoC precludes tailoring the manufacturing process toward individual functions to minimize manufacturing costs. For instance, logic and memory chips are manufactured in substantially different processes for economic reasons. Therefore, putting the two functions on the same chip means the manufacturing process would be suboptimal for one or both functions, leading to suboptimal costs and performance. An alternate solution to minimize communication distance that is compatible with a multichip system is a system-in-package (SiP). The best example of an SiP is the memory subsystem of a smartphone implemented in a package-on-package (PoP), where a packaged memory chip is stacked on top of the application processor (AP) mounted on a second package substrate, thus eliminating the PCB connection that exists when the two chips are mounted separately on the board (Figure 1.17).

In the smartphone PoP, the AP and memory chips are connected indirectly through bond wires, solder bumps, and package substrate traces. In a later section, we will introduce a wireless chip interface that allows direct connection between multiple chips, thereby reducing the communication distance to a minimum.

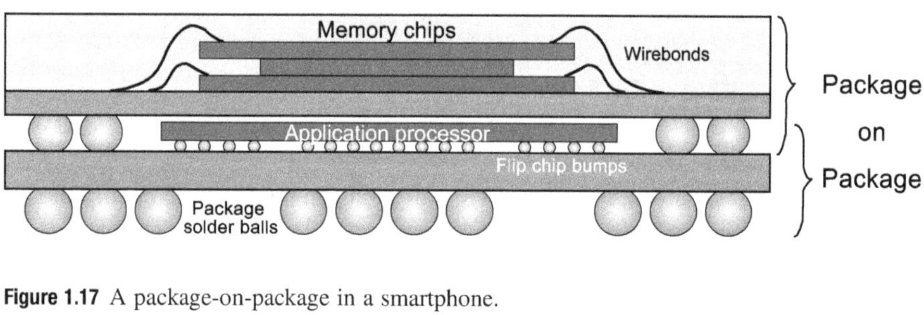

Figure 1.17 A package-on-package in a smartphone.

1.2.4 Energy-Efficient IC

Around 2005, the technology driver for the IC industry changed again, from the PC to the smartphone, which demands continuous, aggressive downsizing. Minimum silicon line width has continued to shrink from 0.13 μm to 28 nm, and down to 7 nm, for example, for the A12 application processor in the iPhone XS released in 2018. Meanwhile, battery has become a severe performance limiter, since it occupies a large percentage of both the weight and size of a smartphone. Furthermore, the perceived performance of the device is greatly affected by how often the battery needs to be recharged, and hence how much energy its components consume. As a result, low-energy has become a key IC design objective.

With the smartphone market growth slowing, new technology drivers are emerging. Two of them are undoubtedly data center and IoT, each of which presents different design challenges than the smartphone. Servers and supercomputers for the data center drive the advance in power efficiency. Furthermore, given their tremendous power consumption, the heat they generate becomes a limiting factor on performance, while cooling becomes another large adder to operating costs. On the other hand, IoT, which includes wearables and implantables together with smartphones, demands aggressive reduction in energy consumption. Some of their applications require very small form factor, which significantly limits battery capacity.

Power and energy consumption are interrelated. But while power $P = \alpha CV_{DD}^2 f$ is a linear function of switching frequency f and capacitance C, and a quadratic function of voltage V_{DD}, energy $E = CV_{DD}^2$ varies only with C and V_{DD}. (α is the switching probability). Reducing V_{DD} is therefore an effective way to reduce both power and energy consumption. In practice, when V_{DD} is initially lowered, power and energy will indeed decrease proportionally. But when V_{DD} drops below a certain value, power and energy will reverse course and rise (Figure 1.18). This is due to rising leakage current. When V_{DD} is lowered, V_{TH} needs to be lowered accordingly to avoid performance loss. But when V_{TH} is lowered, leakage current rises exponentially. Consequently, leakage current must be taken into account when considering overall power and energy efficiency. In the case of a metal–oxide–semiconductor field-effect transistor (MOSFET), transistor switching is controlled by the electric field between the gate and the substrate. This electric field is like the packing attached to the handle of a water faucet that controls the water flow. If this packing is weakened, the faucet will leak. In the MOS transistor, when the gate length is reduced without a corresponding reduction in oxide thickness, the electric field controlling switching is weakened, resulting in increased leakage. Therefore, as we lower V_{DD} to reduce power and energy consumption, we must implement countermeasures to rising leakage, such as the material and structural changes described in Section 1.1.2.4 and Figure 1.3.

Given that the overall power consumption of a chip is the sum of its active power P_{active} and leakage power P_{leak}, it is a function of how active the chip is. A logic chip that spends a lot of time performing active computation has a larger proportion of active power than a memory chip that is active only during a memory access in a localized area and during management operations. As a result, the optimal V_{DD} and

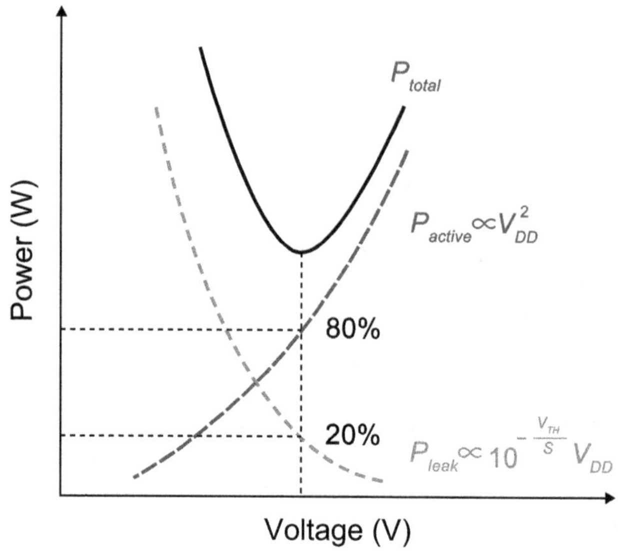

Figure 1.18 IC power as a function of V_{DD} [3].

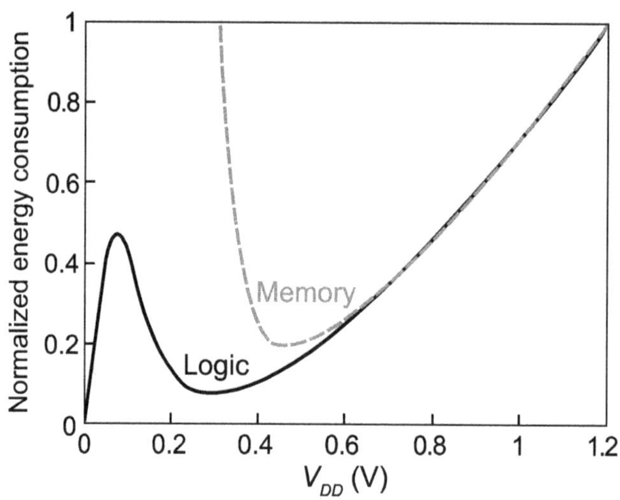

Figure 1.19 Logic and memory energy consumption as a function of supply voltage. © 2011 IEEE. Reprinted, with permission, from [14]

V_{TH} for the lowest power and energy are different between logic and memory chips (Figure 1.19). It is therefore necessary to understand the operation of a chip when we design for low power and low energy.

As an illustration, from experience, lowest overall power is achieved when active and leakage power are around 80% and 20% respectively of overall processor power. For example, both Intel's Pentium4 and IBM's Power5 had about 20% leakage power.

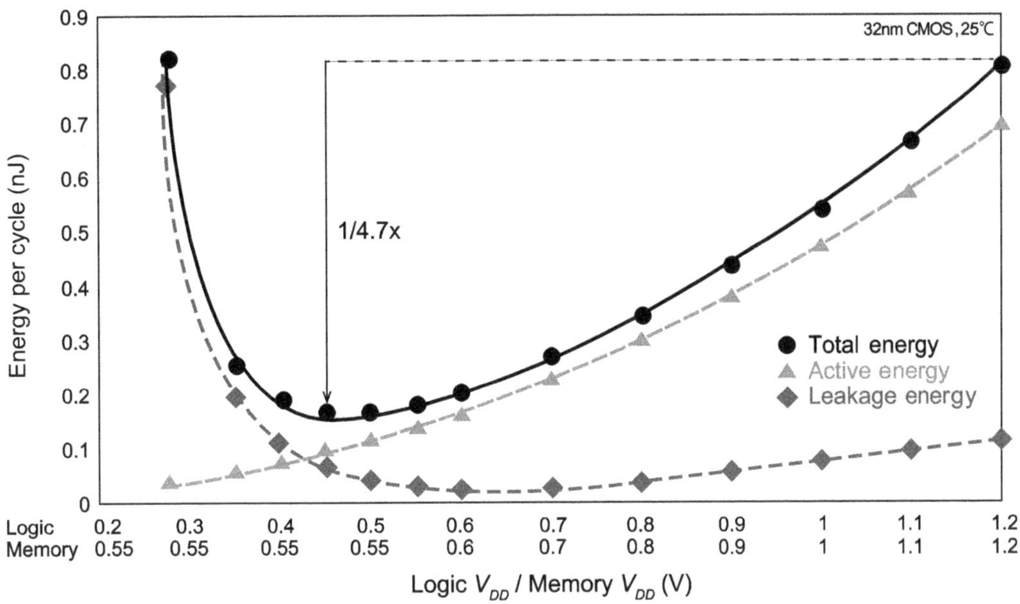

Figure 1.20 Optimized V_{DD} for minimum total energy. © 2014 IEEE. Reprinted, with permission, from [15]

Figure 1.20 depicts the energy consumption as a function of logic and memory V_{DD} for an example processor in a 32nm CMOS process [15]. Based on these data, overall energy is minimized at a logic V_{DD} of around 0.45 V. The Low-Power Electronics Association & Project (LEAP) in Japan has successfully developed a new device known as silicon on thin buried oxide (SOTB) that can operate down to 0.4 V [16], which makes it possible to operate the processor at its optimal V_{DD} for low power.

1.2.5 Energy-Efficient Computing Trends

With SOTB, a low voltage of 0.4 V has been achieved. Yet, there is headroom for further reduction. For logic circuits to function correctly, the output of a CMOS invertor must be able to achieve full swing up to its supply voltage of V_{DD} to drive the invertor in the next stage. Theoretically, V_{DD} can be lowered to 0.036 V before the invertor gain drops below 1. Therefore, V_{DD} can be lowered by another order of magnitude. Since energy is a quadratic function of voltage, there is room to reduce energy further by two orders of magnitude.

Nevertheless, in practice, a lower V_{DD} is difficult to achieve. When V_{DD} is lowered further, gate operation becomes unstable. For example, since the on and off (leakage) currents of the gate inside a flip-flop go against each other, when the on-to-off current ratio becomes too small, the circuit will stop functioning. It is therefore necessary to maintain a certain balance between on and off currents to ensure proper operation. Furthermore, increase in leakage current resulting from lower voltage as discussed in

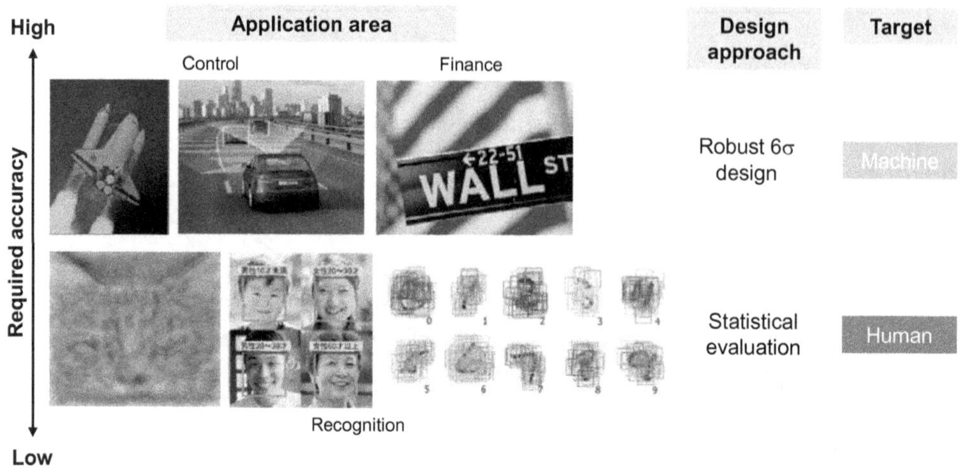

Figure 1.21 Statistical system design approach and its application.

the previous section will need to be addressed. As a result, various research efforts are under way to develop steep slope devices with more pronounced subthreshold characteristics [17].

In addition, as voltage is lowered further, manufacturing variations will surface as another challenge. When V_{DD} approaches the threshold voltage, small changes in V_{TH} will lead to large changes in current, resulting in large variations in circuit speed that cannot be ignored. As a result, synchronous circuit designs may not be possible anymore [18–19]. One solution is to allow certain amount of functional errors, as long as the probability of an error is sufficiently low. In other words, the chip will be functionally correct most of the time, but not 100% of the time. It is a statistical approach.

Of course, this is not acceptable for mission critical applications such as airplane and automotive navigation, as well as financial transaction computation. On the other hand, there are applications where small errors do not alter the outcome, such as image and face recognition (Figure 1.21).

The statistical system design approach is compatible with some of the trending new applications such as image recognition and artificial intelligence. A hot area of technology development in image recognition and artificial intelligence is machine learning, where a deep neural network (DNN) is used to create an artificial intelligence system. The excitement in DNN research started when one such network achieved a breakthrough in dramatically reducing the error rate from 25% to 16% in image recognition in the 2012 ImageNet Large Scale Visual Recognition Challenge. Technology bellwethers such as Google, Microsoft, Amazon, and Baidu are all competing to deliver the next breakthrough in machine learning, in addition to startups funded by billions of dollars in venture capital [20]. The technology is rooted in mimicking the algorithms humans use for image recognition. Since human actions are based on statistical and probability analysis, a statistical design approach can be a good fit for developing energy-efficient machine learning systems.

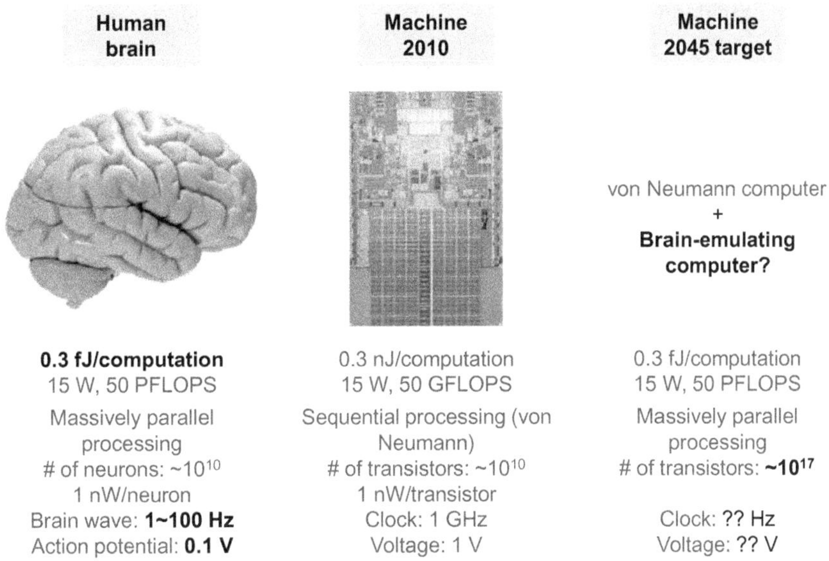

Figure 1.22 Power efficiency comparison: machine vs. human.

To see how much we need to improve power efficiency for machine learning applications, the natural benchmark for comparison is the human brain. The human brain consumes about 0.3 fJ of energy per computation, corresponding to a power efficiency of 15 W/50 PFLOPS. By contrast, Intel processors from 2010 consume about 0.3 nJ of energy per computation, resulting in an efficiency of 15 W/50 GFLOPS (Figure 1.22). In other words, the human brain is six orders of magnitude more power efficient. It will take a lot of effort, including introducing the statistical system design approach, running the machine nonstop to gain an advantage over the human brain, which needs to sleep every day, and so forth, in order to close this gap.

Another trend in energy-efficient computing is distributed computing. At the system level, the trend is to distribute processing over many physically small processing elements that are connected through short-range wireless communication so no one element has to work very hard. Early development efforts include Ubiquitous Computing (PARC/Weiser), Things That Think (MIT), Disappearing Computer (EU), Invisible Computing (Microsoft), Pervasive Computing (IBM), Paintable Computing (MIT), and Pushpin Computing (MIT). The proliferation of IoT can be considered an implementation of distributed computing at a more global level.

Figure 1.23 depicts an example of distributed computing and its benefits. It illustrates how the problem of searching for a parking space can be solved efficiently with short-range wireless communication and small capacity memory. First, a parking space broadcasts its availability to a passing car (Car_3). Then the information is passed from this car to another (Car_2) when they cross each other, and the process continues. A car (Car_1) searching for a space can capture this information as it crosses Car_2

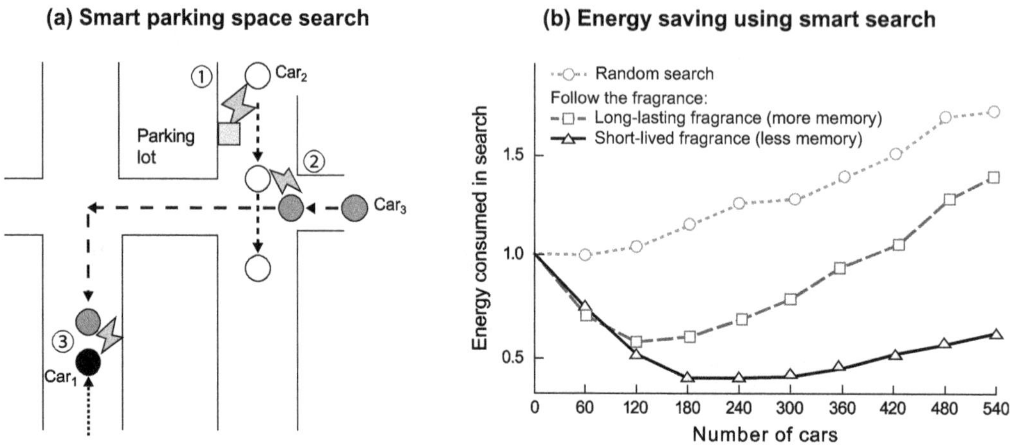

Car$_1$ is searching for a parking space. A space is available in a parking lot at a distance.
① The parking space broadcasts its availability to a passing car (Car$_2$).
② The information is passed from Car$_2$ to Car$_3$ as they cross each other.
③ Car$_1$ captures the information as it crosses Car$_3$, without being in the vicinity of the parking space.
This scheme enables the use of wireless communication for short distance only, minimizing transmission power. Furthermore, by discarding information as soon as updated information arrives (short-lived fragrance), memory requirement can be kept to a minimum. Together they propagate and populate space with information to create a smart environment in an energy-efficient manner.

Figure 1.23 Using "fragrance" to create a smart environment. © 2007 IEICE, [9], figure 5

without being in the vicinity of the parking space to directly receive the broadcast information. This is akin to the spread of fragrance, where locating the parking space is like tracing the fragrance back to its origin. The "fragrance" together with the group of cars that spread it makes the environment intelligent, or smart, to use the buzzword of the IoT era. With this smart environment, only short-range communication is required. Therefore, the amount of energy needed to locate the parking space can be dramatically reduced compared to random search using the traditional, centralized processing approach, as illustrated in Figure 1.23b.

1.2.6 Closing Thoughts

Low-power and energy-efficient designs have allowed IC to continue to scale. Using a 10 nm process, it is now possible to put 100 million transistors on 1 mm^2 of silicon [21], resulting in 0.1 trillion transistors on a 1,000 mm^2 chip. While still two orders of magnitude away, the transistor count is starting to approach the 60 trillion cell count of the human body (Figure 1.24). When the chip transistor count finally crosses over the human body cell count, IC development may become dominated by innovative application of the large quantity of transistors made available through advance in power and energy efficiency to achieve the next level of enrichment of our life.

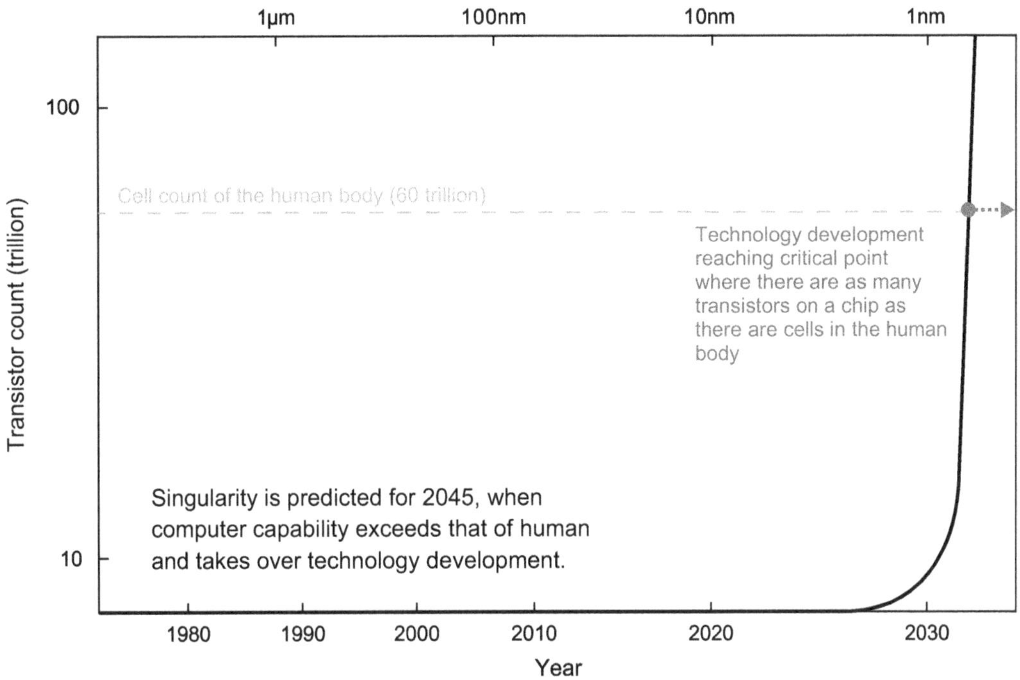

Figure 1.24 Critical point in IC technology development [1].

1.3 Evolution from 2D to 3D Integration

1.3.1 Motivation for 3D Integration

IC chips and other semiconductors are generally packaged and mounted individually on a PCB to build a computer system. Since the individual components are mounted on a 2D surface and connected primarily through traces running in the *x*- and *y*-direction (plus vertical vias), this is known as 2D integration. However, as discussed in Section 1.2.3, this complicated off-chip interconnect structure can result in signal integrity (SI) problems due to transmission line effects, which lead to more complex I/O circuit design and higher power. To circumvent these SI problems, 3D integration is adopted where chips are placed on top of each other vertically and connected using primarily interconnects in the *z*-direction to minimize the communication distance between chips (Figure 1.25). Moreover, stacking chips on top of each other instead of spreading them across a 2D surface minimizes package footprint, which is critical for applications with limited board area such as mobile. This is exemplified by the use of PoP in a smartphone as introduced in Section 1.2.3. But in recent years, a new driver of 3D integration has emerged in the form of More than Moore. As the name suggests, this is an effort to offer more performance than what Moore's law delivers.

As explained in Section 1.1, we can no longer achieve IC performance increase through Moore's law scaling without having to improve power efficiency.

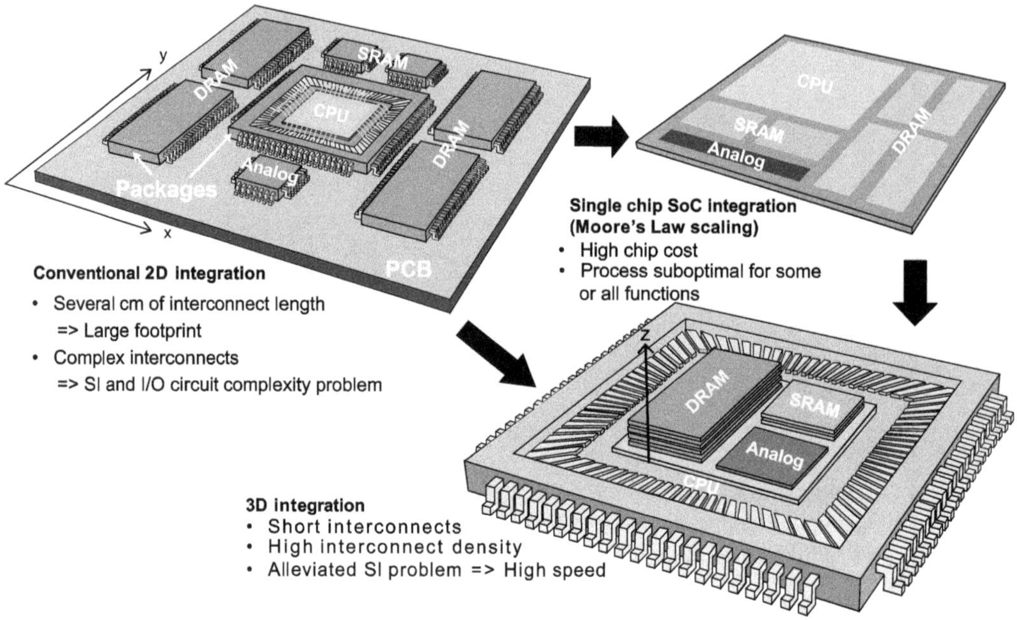

Figure 1.25 From 2D to 3D integration.

Furthermore, around the 28 nm process node, manufacturing cost per transistor started to rise. In addition, nonrecurring engineering (NRE) and fixed costs including process development cost, IC design cost, mask cost, and manufacturing equipment cost are all going up exponentially with advanced process scaling. For example, it was predicted in 2018 that 3 nm would cost $4 billion to $5 billion in process development, while total IC design cost including software development and validation was estimated to almost triple from 16 to 7 nm [22]. On the other hand, it was estimated that by 2014 extreme ultraviolet (EUV) lithography manufacturing equipment development would have already cost $14 billion [23]; yet it required another four years of development before its initial production use was announced in 2018 [24]. It is therefore imperative for the IC industry to explore alternatives to Moore's law scaling to continue to improve IC cost performance. Since Moore's law has been about scaling of 2D device structures, it is only natural to explore using the third dimension as an alternative or complement, hence the emergence of 3D solutions.

1.3.2 Monolithic 3D IC

The most advanced IC processes today employ a FinFET, which is a 3D device. This is because in FinFET the transistor gate is no longer two-dimensional but instead is wrapped around the channel utilizing the *z*-direction as well (Figure 1.26a). Another way to extend the 2D IC to 3D is a monolithic IC where multiple device layers are integrated in the vertical direction within the IC. A 3D NAND flash memory chip is

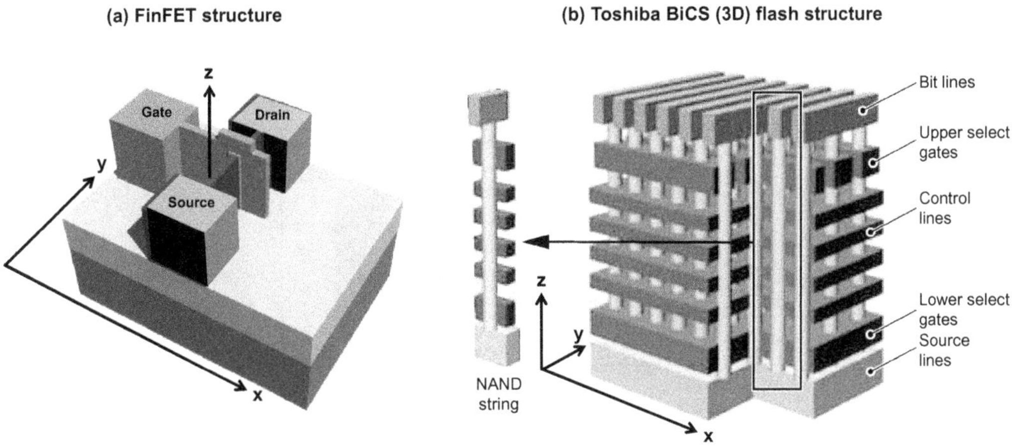

Figure 1.26 Monolithic 3D ICs – FinFET and 3D NAND. (b) © 2011 Toshiba Corporation. Reprinted, with permission, from [25]

one such example where multiple layers of memory cells are stacked on top of each other (Figure 1.26b). Scaling is achieved by increasing the number of layers. One merit of this solution is that since memory cell layers are measured in nanometer in thickness, stacking does not appreciably increase the overall thickness of the IC. In 2018, the most advanced 3D NAND flash in production had a total of more than 90 cell layers.

However, since IC manufacturing cost and yield are highly dependent on the number of processing steps and processing complexity, monolithic 3D IC has significant cost and yield disadvantage, such that its adoption for mass production has been limited to NAND flash. Instead, 3D integration of IC chips is the leading 3D solution for logic and DRAM memory chips.

1.3.3 Conventional 3D Integration Solutions

In addition to lowering IC manufacturing cost, 3D integration of IC offers two advantages over monolithic 3D IC and Moore's law scaling of 2D IC. First, by keeping digital, analog, and memory functions implemented as separate chips, the manufacturing process for each function can be independently optimized to achieve better overall performance. Second, 3D integration enables reuse of intellectual properties (IP) by mixing and matching different chips to develop different systems without IC redesign, significantly reducing development cost in terms of both engineering cost and time to market.

Due to these benefits, 3D integration of IC chips based on conventional packaging technologies have been in use for more than two decades. Figure 1.27 (from [26] with added PoP data point based on [27]) depicts the transition of IC packaging solution from 2D to 3D integration based on conventional packaging technologies in the late

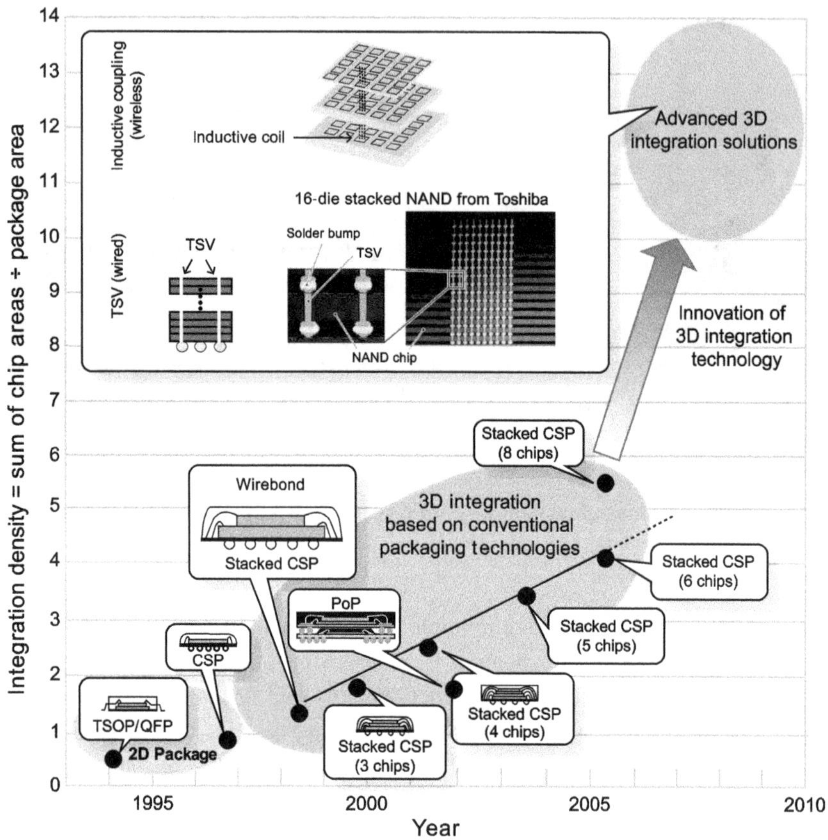

Figure 1.27 Evolution from 2 to 3D IC integration. © 2005 Nikkei BP [26]; photo: © 2016 Toshiba Corporation [28]

1990s and the evolution of 3D integration technology thereafter. The initial solution was a stacked IC wirebond (WB) package where multiple chips are stacked face-up and wirebonded to a ball grid array (BGA) or chip scale package (CSP) substrate. The chips can be stacked directly on top of each other to minimize overall package height, or with spacers in between to create vertical clearance for wirebonding when the stacking chips are square and of the same size (Figure 1.28a and b). Then in early 2000, PoP was introduced, which stacks single- or multichip BGA packages on top of each other (Figure 1.28c).

Table 1.3 compares the stacked IC WB package with PoP. While innovations are necessary to reduce wire length and loop height in the stacked IC WB package to increase the number of stacked chips, wirebonding has been in mass production for years and is therefore low cost and reliable. However, since bare dice are difficult to test, it is challenging to provide known good dice. As a result, the packaged part may have a yield problem, especially as the number of chips increases. By contrast, individual packages in a PoP are tested by their respective suppliers before they are

Table 1.3. Comparison of conventional 3D IC integration solutions.

	Stacked IC WB package	PoP
Cost	+	–
Yield (known good die)	–	+
Electrical performance	–	+
Interconnection density	–	Upper package: – Lower package: +
System flexibility (IC mix-and-match)	–	+
Package height	+	–
Application	High-capacity memory	SoC memory SiP

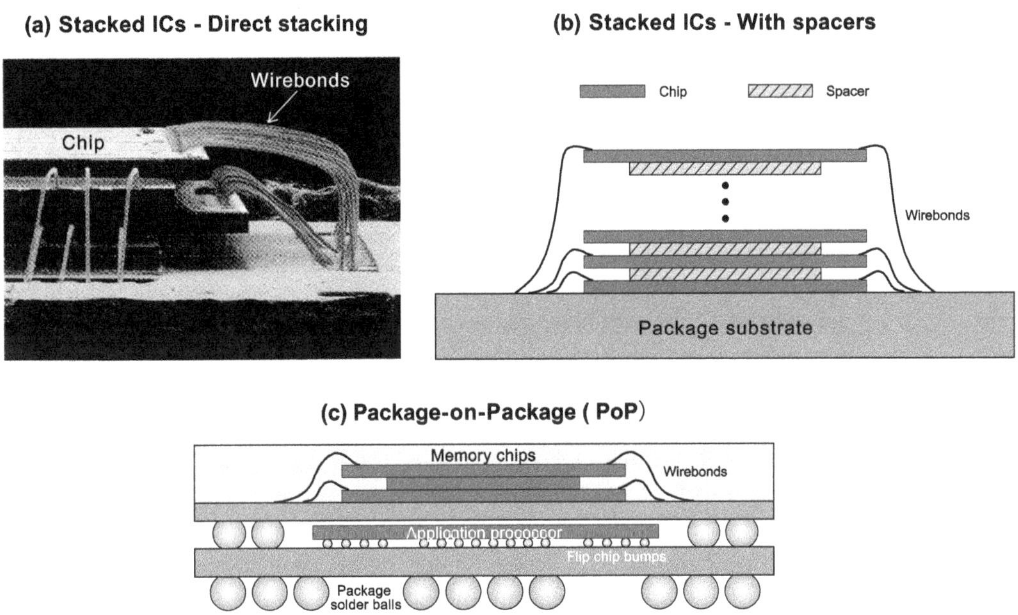

Figure 1.28 Conventional 3D IC integration solutions. (a) © 2007 IEEE. Reprinted, with permission, from [29]

stacked. Furthermore, the PoP assembly is done by the system user who is free to mix and match the individual packages. This provides flexibility not only in changing the content of each packaged part, such as changing the mix of memory types or capacity, but also in multisourcing. Meanwhile, since the interconnection between individual packages consists of solder balls and substrate traces, its electrical performance is generally better than wirebond interconnection, which is long and narrow. The lower package, with a full ball grid array, has high interconnection density. However, the upper package, with depopulation of the ball grid array in the center to accommodate the die on the lower package, may have limited interconnection density, especially when the lower package has a large die. Individual packages can use either wirebonding or flip-chip for die attachment. In fact, as previously noted, PoP is commonly used to build a memory SiP in a smartphone where the SoC is housed in a high-performance flip-chip package at the bottom, while memory dice in the upper package are connected to the package substrate with bond wires. Nevertheless, PoP has a disadvantage in package height since individual packages, with a substrate, solder balls, bond wires, and molding compound, are significantly thicker than individual chips.

Rather than stacking heterogeneous chips to create an SiP, the stacked IC WB package, given its drawbacks, is commonly used for stacking homogeneous chips to create a high-capacity memory solution. This is because stacking of low-cost memory chips requires a low-cost solution. Furthermore, being low cost and having high yield means known good die is not as important a requirement. Finally, many chips need to be stacked for capacity, making a stacked IC WB package a better fit than PoP. For instance, [30] reports a nine-stack, 32 Gb micro-SD flash memory card from Sandisk, as well as a 16-stack, 64 Gb NAND part in an iPhone from Samsung.

Meanwhile, despite the advantages of PoP over the stacked IC WB package, its height disadvantage limits its applicability. When Sharp introduced its PoP technology in 2002, it showed a three-stack example [27], so the technology supports stacking of more than two packages. However, the total stack thickness increases rapidly as the number of packages increases, since package thickness is on the order of 500 μm (without including the solder ball standoff) as opposed to die thickness that is on the order of 50 μm after thinning. As a result, direct die stacking is the preferred solution for stacking many ICs together.

Nevertheless, the use of wirebonding technology in the stacked IC WB package makes it difficult to overcome its disadvantages. In particular, the poor electrical characteristics of bond wires, which result in poor signal and power integrity, prevents it from being used for high-performance chips such as processors, especially as wire length increases with the number of chips being stacked. Furthermore, since wirebonding is a peripheral interconnect technology where bond pads are placed along the perimeter of the IC, most commonly in one or two staggered rows, it is difficult to significantly increase the interconnection density to increase bandwidth. Consequently, an area array interconnect technology with better electrical characteristics and shorter interconnects is needed for heterogeneous 3D chip integration for high-performance systems.

1.3.4 Advanced 3D Integration Solutions

To address the limitations of conventional 3D IC integration solutions, various advanced solutions have been in development to dramatically increase interconnection density and/or the number of ICs that can be integrated. These solutions can generally be classified as wired vs. wireless, and $n_{st} = 2$ (face-to-face) vs. $n_{st} \geq 2$ (free orientation) solutions, where n_{st} is the number of stacked chips (Figure 1.29, [31]–[34]). All four classes of solutions offer significant improvement in interconnection

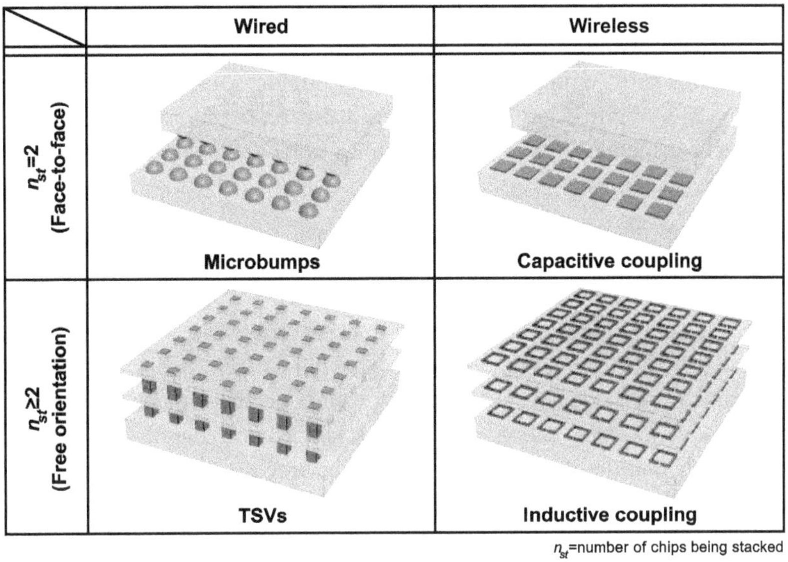

Figure 1.29 Advanced 3D IC integration solutions.

density and communication distance compared to conventional solutions. Since connection terminals are placed in an area instead of peripheral array in all solutions, the number of interconnections can be increased in a quadratic fashion. On the other hand, because adjacent chips communicate with each other directly instead of through a package substrate, communication distance and hence interconnect lengths are greatly reduced.

Two variants of wired 3D IC integration solutions using microbumps [31], [35] and through-silicon-vias (TSVs) [32] respectively have entered production. Microbump technology employs small solder bumps formed on the chip surface to vertically connect two chips together. Since connection terminals are on the front side of the chips, the two chips must be oriented face to face for interconnection, which precludes integration of more than two chips. Hence, $n_{st} = 2$. On the other hand, with TSV, as the name suggests, via holes through the silicon substrate are created to add connection terminals to the backside of the chip as well. Interconnection between chips may still be created using microbumps. But since interconnection can be established using both the front and backside of the chip, integration of more than two chips ($n_{st} \geq 2$) becomes possible. TSV therefore has broader applicability than microbump alone.

For wireless 3D IC integration, the initial solution proposed uses capacitive coupling [33, 36–37]. It establishes an electrical connection by aligning connection terminals in the form of metal pads on the surface of the mating chips across a dielectric gap to form a capacitor. Since the impedance of a capacitor is inversely proportional to frequency, this is an alternating current (AC) only connection with no direct physical contact – information is passed through the gap between mating chips by varying the electric field associated with the capacitor. Given that electric field cannot effectively penetrate a silicon substrate, this capacitive coupling solution can only be applied to two chips oriented face to face. Hence, $n_{st} = 2$. To overcome this and other shortcomings of the capacitive coupling solution, an alternate solution using inductive coupling has been developed that replaces the matching metal pads with matching inductive coils and communicates by changing the magnetic field linked to the coils [34], [38]–[41].

1.3.4.1 Nature of a TSV

Since electronic circuits are formed on the top surface of an IC, the straightforward way to connect to them is from the front side. Therefore, the natural way to stack more than two dice over each other is to stack all of them facing up and connect the top surface of individual dices to a common point to create interconnections, which is how it is done in a conventional wirebond stack. To enable a drastically different integration scheme, the die must be additionally processed in such a way as to enable its electronic circuits to be connected from the back side as well. Such is the concept of a TSV. It is an electrical connection that traverses the silicon substrate, thereby enabling connection to active circuits of the chip from its back side as well.

Figure 1.30 depicts the basic structure of a TSV fabricated using one of two commonly employed processes known as a via-middle process [42]. It connects the metal layers on the front side of the die to its back side. A microbump on each of the front and back sides allows the die to connect with an adjacent die above and below respectively. While the die in a single-chip package is generally 200–500 μm thick, for TSV 3D integration it is thinned to about 50 μm. A thinner die helps to keep the stack height low, even if many of them are stacked. But more importantly, it helps to keep the TSV short and hence its diameter small, while maintaining a certain via aspect ratio for manufacturability. A shorter TSV in turn results in a shorter fabrication time.

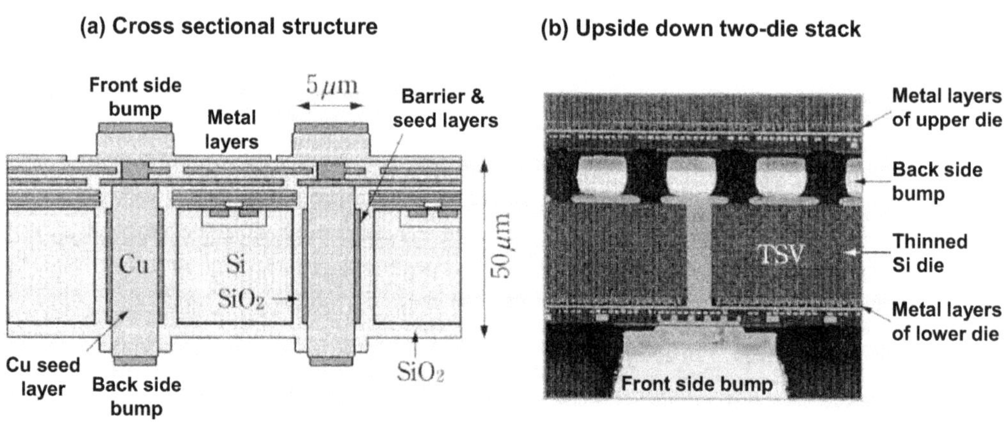

Figure 1.30 Structure of a TSV. Reproduced by permission from Sei-ichi Denda, 半導体の高次元化技術 [Enabling Technology for Higher Dimensional Semiconductors], Japan: Tokyo Denki University Press, 2015. © Denda Sei-ichi 2015

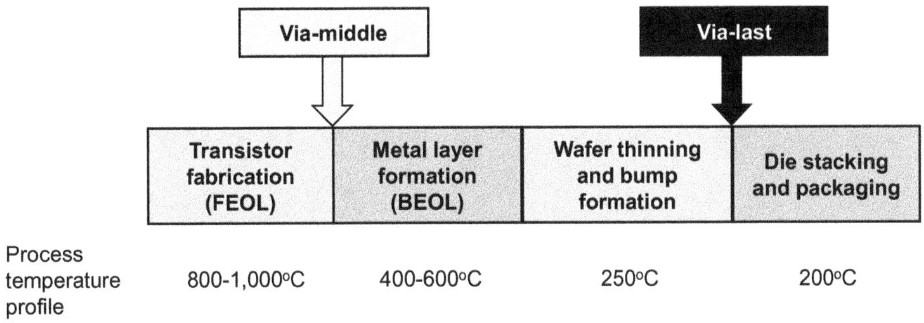

Figure 1.31 TSV formation within the IC fabrication process flow. Reproduced by permission from Sei-ichi Denda, 半導体の高次元化技術 [Enabling Technology for Higher Dimensional Semiconductors], Japan: Tokyo Denki University Press, 2015. © 2015 Denda Sei-ichi

A 50 μm die thickness results in a via diameter between 3 μm and 20 μm, with 5 μm being a typical value. The via wall consist of a SiO_2 insulating layer, a barrier layer that is, for example, made of tantalum (Ta), titanium oxide (TiO_2), or titanium nitride (TiN) to prevent copper from diffusing into the silicon substrate, as well as a copper seed layer. The copper seed layer serves as an electrode for plating to fill the via with copper, thereby completing the electrical connection from the front to back side of the die. Even with this simplified picture, it is clear that the structure of TSV consists of multiple structural components made of different materials. This has implications in processing complexity as well as yield and reliability due to thermomechanical stresses.

The via-middle process is so named because the TSV formation occurs in the middle of the IC fabrication process, after the front end of line (FEOL) process to create the transistors, but before the back end of line (BEOL) process to form the metal layers. A second common process known as via-last creates the TSVs at the end of the IC fabrication process, after wafer thinning and bump formation. Figure 1.31 depicts where the TSV formation occurs within the IC fabrication flow, including the temperature profiles for each of the fabrication steps. The two TSV processes are different not only because of the different positions within the IC fabrication process flow, but also because of the very different process temperature profiles they are exposed to, resulting in different process complexities and cost and yield equations. Furthermore, for outsourced manufacturing, while the via-middle TSV process falls naturally within the flow of the foundry, the via-last process may occur either at the foundry or outsourced semiconductor assembly and test (OSAT) vendor. As a result, the choice of the TSV fabrication process has supply chain implications as well.

Figure 1.32 contrasts the use of TSV versus wirebond in 3D IC integration. The following key differences can be observed:

- TSV enables a smaller form factor: the wirebond stack requires additional area and height to accommodate the wire loops as well as bond pads on the package substrate. Besides, while the example in the figure uses 90 degree rotation of every

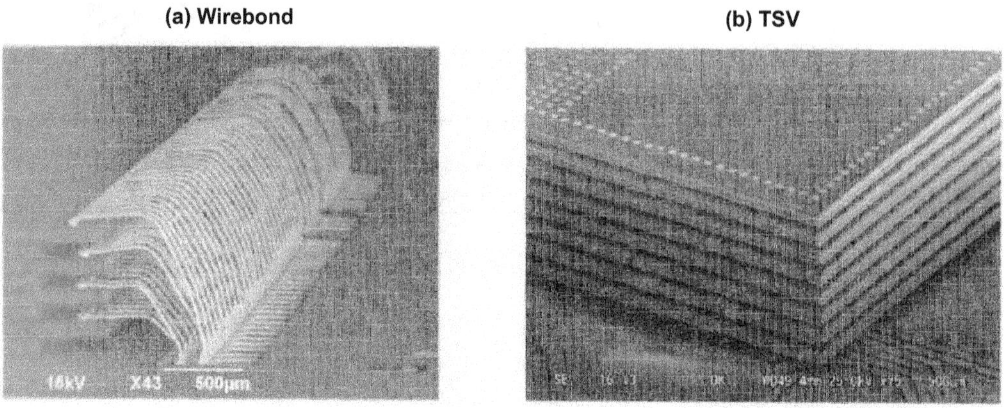

Figure 1.32 3D IC integration technology comparison: wirebond vs. TSV. Reproduced by permission from Sei-ichi Denda, 半導体の高次元化技術 [Enabling Technology for Higher Dimensional Semiconductors], Japan: Tokyo Denki University Press, 2015. © 2015 Denda Sei-ichi

other die to create clearance for the wire loops, when the dice are square and of the same size, spacers must be inserted between adjacent dice to create such clearance, which increases the stack height further.

- TSV enables more robust interconnection: the visual comparison in Figure 1.32 illustrates the contrast in interconnection complexity between the two solutions. While the interconnects in the TSV stack are short and tucked away, those in the wirebond stack are long and susceptible to mechanical force that can tangle them and cause electrical shorts.
- TSV enables a larger number of dice to be stacked: in the wirebond stack, the number of wires and their maximum length increase with the number of dice. Therefore, the maximum number of dice in the stack is limited by manufacturability and electrical requirements. Longer wires are more susceptible to sweeping force that can cause electrical shorts. Furthermore, longer wires have higher inductance that can lead to unacceptably large power supply noise and signal degradation, especially at high frequency.
- The TSV stack enables a higher interconnect density: while both TSV and wirebond pitch are on the order of 50 μm, TSVs are placed across the surface of the die while wirebonds are limited to its perimeter. Consequently, TSV increases interconnect density in a quadratic manner.

TSV is therefore an advanced technology offering a small form factor, robust, high-capacity stacking, and high interconnect density solution for 3D IC integration.

1.3.4.2 TSV for Wired IC Integration

The first industrywide effort to commercialize the application of TSV to wired 3D IC integration was driven by the Joint Electron Device Engineering Council (JEDEC)

Table 1.4. Evolution of mobile DRAM energy efficiency.

	Specification release	Data rate (Mb/s)	Bandwidth (Gb/s)	Relative I/O energy per bit (%)	Relative total I/O power @ min BW (%)
LPDDR	2007	400	1.6	100	100
LPDDR2	2009	800–1,066	3.2–4.3	50.3	101
LPDDR3	2012	1,600–1,866	6.4–7.5	32.7	131
Wide I/O	2011	200	12.8	4.5	36

through its release of the Wide I/O specification in 2011. At the time, the mobile DRAM industry was looking for a solution to deliver a quantum leap in energy efficiency. The transition from the first- to second-generation low-power DRAM technology, LPDDR to LPDDR2, achieved an almost 50% reduction in I/O energy per bit in conjunction with a doubling of bandwidth, resulting in the same total power. However, the next transition achieved only another 35% reduction in energy per bit while the bandwidth again doubled, resulting in a 31% increase in total power. This is summarized in Table 1.4, which is based on data published in [42] and [43]. The continuous increase in data rate demanded by the industry was anticipated to make improvement in energy efficiency more and more difficult as LPDDR evolves. In particular, for the memory interface, higher data rates exacerbate signal integrity problems, including signal reflection, ISI, and crosstalk. Furthermore, skew in delay across LPDDR's parallel data bus becomes a larger and larger percentage of bit time, especially as the bus width increases to achieve higher bandwidth. Meanwhile, a higher-frequency clock is harder to generate and distribute due to higher jitter, higher sensitivity to noise, and higher attenuation. All these problems require architectural and circuit improvements to solve, such as the use of equalization circuits, transmission line terminations, signal phase alignment circuits, phase locked loop (PLL), delay locked loop (DLL), on-chip power supply bypassing, and insertion of clock buffers that lead to increased power. As a result, the Wide I/O solution was proposed to take low-power memory development in a different direction to enable continuous bandwidth scaling while drastically improving energy efficiency.

Figure 1.33 depicts the construction of a Wide I/O memory system. It consists of one or more Wide I/O DRAM dice stacked on top of a processor die. The 3D die stack is interconnected with TSVs and microbumps, while the bottom processor die is flip-chip attached to the package substrate.

As the name suggests, Wide I/O adopts a very wide data bus to achieve high bandwidth while keeping the data rate and hence I/O power low. Despite running at ¼ the data rate, it delivers four times the bandwidth of LPDDR2. This is made possible by the use of TSV to enable the 512-bit wide data bus, which is 16 times that of LPDDR2. The use of TSV instead of the conventional PCB interconnect results in drastic reduction in both the interconnect area (5 μm TSV diameter vs. 100 μm PCB trace width) and length (50 μm TSV length vs. tens of mm of PCB trace length), thereby avoiding an order-of-magnitude increase in interface area while significantly

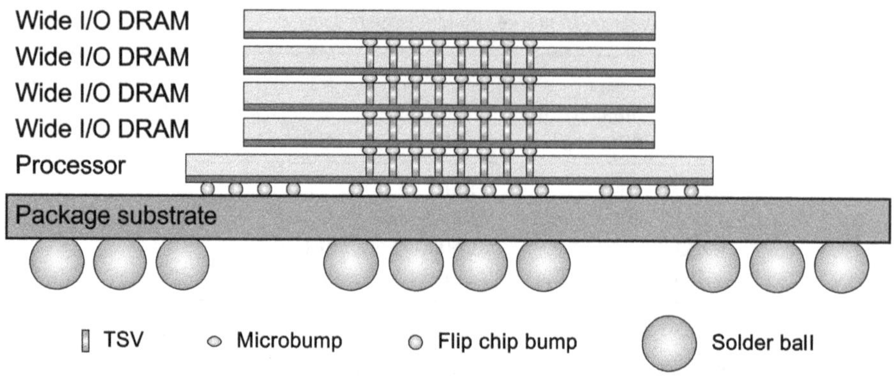

Figure 1.33 Construction of a Wide I/O memory system.

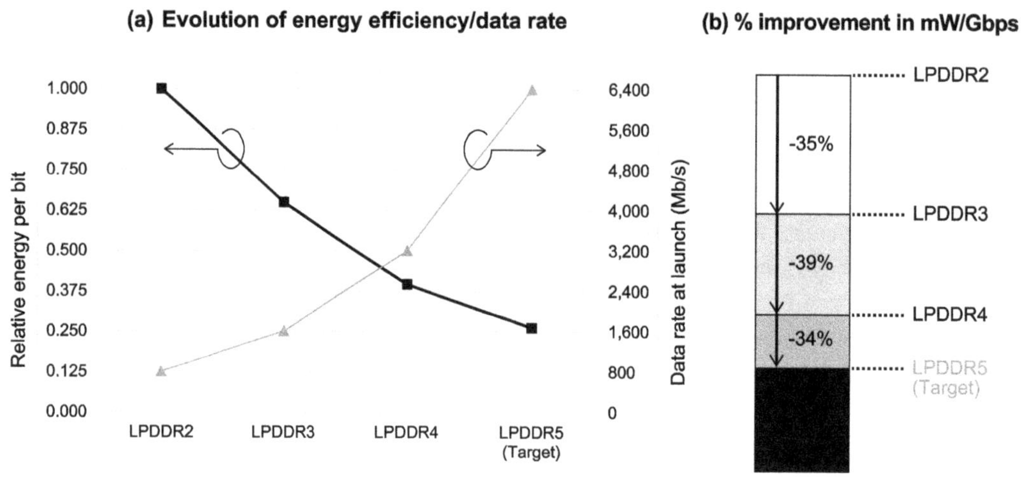

Figure 1.34 Evolution of LPDDR bandwidth and energy efficiency.

reducing interconnect capacitance and transmission line effects. Furthermore, since the interface bypasses the package, the bus width is not limited by the pin count and cost of the package. As a result of the lower data rate and simpler interconnects, the energy per bit is reduced by an order of magnitude while bandwidth is more than tripled compared to LPDDR2 (Table 1.4).

1.3.4.3 2.5D IC Integration Using TSV

Despite the promise of dramatically improved energy efficiency, Wide I/O (both versions 1 and 2 released in 2011 and 2014 respectively) was not able to replace LPDDR as the low-power memory solution for mobile applications. After the release of the Wide I/O specification, LPDDR continued to evolve from LPDDR3 in 2012 to LPDDR5 in 2019, doubling the data rate in every generation, as shown in Figure 1.34

based on [43]. Meanwhile, energy efficiency improved by about 35% in every generation. This is not sufficient to fully offset the increase in power due to doubling of the data rate, so total memory I/O power goes up at maximum bandwidth. Nevertheless, the industry has so far decided to stay with this evolutionary path, due to challenges in applying TSV to 3D integration of heterogeneous dice.

The challenges in applying TSV to 3D integration of heterogeneous dice include added manufacturing cost, yield loss, thermomechanical reliability problems, and device performance degradation. The added manufacturing cost is the result of both added processing for TSV formation and die area overhead. Meanwhile, the complexity of the TSV structure and the multitude of materials used that have diverse coefficients of thermal expansion (CTEs) lead to creation of thermomechanical stresses when the die stack is subject to thermal cycling during both manufacturing and normal operation. Such stresses result in lower manufacturing yield, lower reliability, and semiconductor device performance degradation [44]. Furthermore, thermomechanical stress also occurs at the flip-chip bump interface between the bottom die in the stack and the package substrate below, due to CTE mismatch between silicon and the package substrate material, such as epoxy resin. Such stress exists in single-die packages as well. But the much thinner die in a 3D stack means it is more susceptible to warpage.

There are solutions to overcome some of these challenges, such as adding redundant TSVs and enlarging the TSV keep-out zone to keep active circuits away. But such solutions increase the already large area overhead. Figure 1.35 depicts a relative size comparison between TSVs and NAND gates. The study in [45] from 2016 estimated a die area overhead of almost 40% for a TSV diameter of 5 μm in a 14 nm FinFET process, while noting that practical TSV diameters could be up to 10 μm large.

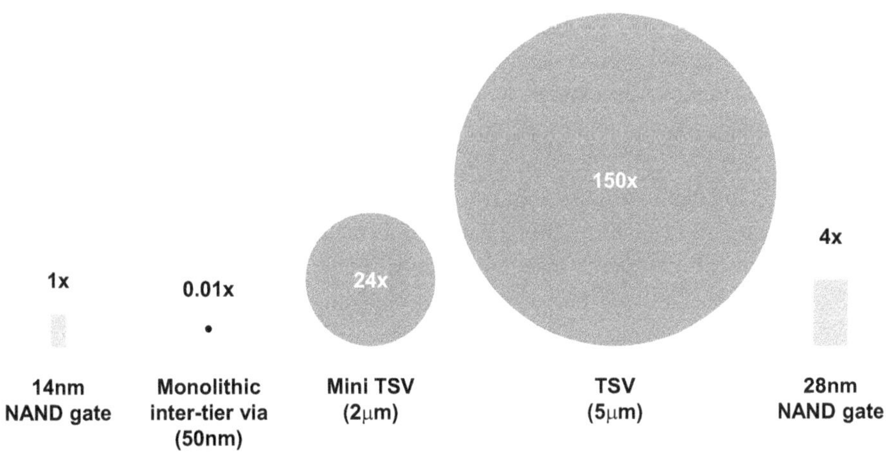

Figure 1.35 Relative size comparison: TSV vs. NAND gate. © 2016 IEEE. Reprinted, with permission, from [45]

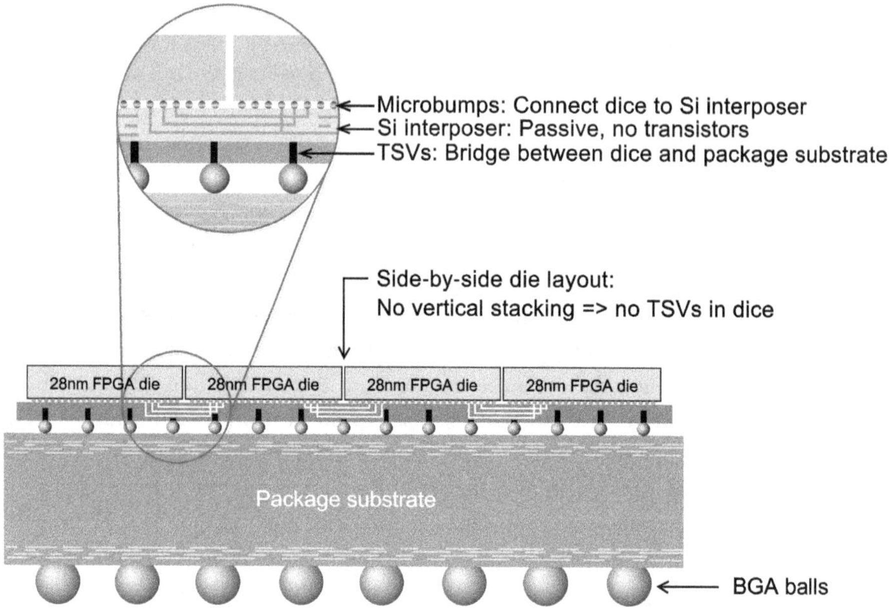

Figure 1.36 Cross section of a 2.5D integrated FPGA. © 2012 IEEE. Reprinted, with permission, from [47]

One additional challenge for TSV 3D integration of heterogeneous dice arises when a high-power chip is placed toward the bottom of the stack, which is the case with Wide I/O, where the processor chip sits at the bottom. In this case, heat removal from the high-power chip can be a challenge even if there is a heat sink at the top, especially if there are many chips in the stack.

To alleviate these problems, an intermediate solution known as 2.5D IC integration has emerged that has achieved commercial success especially in the high-end FPGA and graphics processing unit (GPU) markets.

Figure 1.36 depicts the cross section of the first 2.5D integrated FPGA commercial product announced by Xilinx in late 2011 [46]. It illustrates the basic construction of a 2.5D IC using TSV. First, the ICs are not stacked and hence have no TSVs. This facilitates heat removal from the individual ICs and eliminates IC manufacturing and reliability problems related to the use of TSV. Second, there is a passive silicon interposer inserted between the ICs and package substrate. The interposer serves the sole purpose of interconnecting the ICs and the package substrate and hence contains no active circuits. Interchip connections consist of microbumps and metal traces only without using TSVs. TSVs in the interposer are only used for power, ground, and external I/Os. Consequently, the TSV density is much lower than in 3D IC integration. As a result, the interposer can be fabricated in a lower-performance, more mature, and hence less expensive process. Furthermore, since the use of TSV is limited to the passive interposer, there is no TSV-induced device performance degradation.

Table 1.5. Pros and cons of 2.5D IC integration.

	2.5 vs. 2D	2.5 vs. 3D
Pros	– Higher interconnection density – Shorter IC interconnections ◦ Lower power, higher speed – Smaller footprint	– No TSV cost adder to ICs – No TSV-related yield and reliability problems for ICs – No TSV-induced device performance degradation – Easy heat removal
Cons	– Added interposer cost including TSVs – Added microbump assembly complexity	– Larger footprint – Longer IC interconnections

Figure 1.36 shows an example of homogeneous 2.5D integration where all the chips are of the same type. The same technology can be applied to integration of heterogeneous chips. For instance, in Virtex-7 HT, which is also introduced in [46], three FPGA chips are integrated with two 28G SerDes transceiver chips in a 2.5D package. Furthermore, 2.5D can be combined with 3D integration where one of the chips mounted on the silicon interposer is a 3D integrated IC. A representative example is high bandwidth memory (HBM), which is introduced later in this section.

As elaborated in [46], compared to 3D, 2.5D integration represents an evolutionary solution in terms of design flow, test, thermal, and reliability. Furthermore, a 2.5D integrated FPGA offers high logic cell capacity at much lower power. As an illustration, the Virtex-7 2000T FPGA from Xilinx is a homogeneous 2.5D integration of four identical 28 nm FPGA chips on top of a 65 nm silicon interposer. It contains a total of 2 million logic cells and about 10,000 connections between adjacent chips and consumes 19 W of power. Using conventional FPGAs, four individually packaged chips will be required to match the capacity, while consuming almost six times the power, with 28.6% consumed by the interconnections.

Without vertically stacking the ICs, 2.5D integration requires a larger area than 3D. But it offers similar interconnection density as 3D integration, as well as shorter interconnections and higher IC packing density and hence a smaller footprint than 2D integration, while overcoming some key challenges faced by 3D integration. The pros and cons of 2.5D integration compared to 2 and 3D are summarized in Table 1.5.

Given the smaller footprint, 3D is still the preferred solution over 2.5D when many dice need to be integrated, and when the application is less cost sensitive. A good example can be found in high-end GPU systems that have adopted the HBM DRAM solution to meet their large memory capacity and high-speed requirements. HBM utilizes a combination of 2.5D and 3D integration to deliver the highest memory performance. HBM DRAM is a 3D stack of DRAM core chips on top of a logic chip. This memory stack is then placed side by side with a high-performance processor on a passive silicon interposer to form a 2.5D integrated memory system. It thus combines the best of both worlds. Three-dimensional integration allows the DRAM footprint to

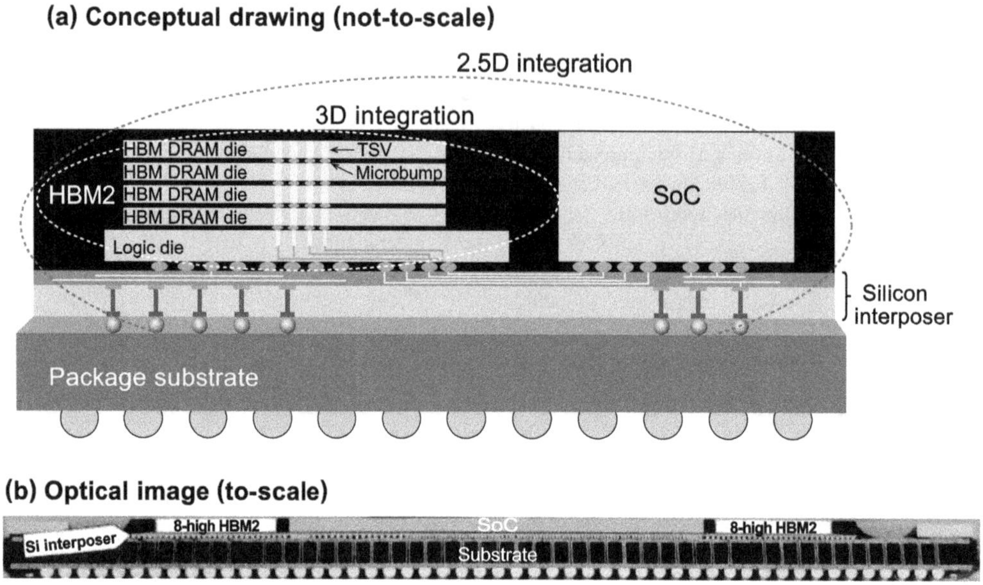

Figure 1.37 Cross section of an HBM memory subsystem. © 2017 IEEE. Reprinted, with permission, from [48]

remain small while providing large capacity. Meanwhile, 2.5D integration of the processor keeps TSVs out of the large and expensive die of the system, while enabling effective removal of the large amount of heat it generates.

Figure 1.37 shows both a conceptual drawing and an optical image of the cross section of an HBM solution from [48]. The optical image illustrates the relative dimensions of the different parts of the memory system. Note that the use of an eight-stack HBM memory component incurs no height penalty since it stays within the height of the SoC. However, the interposer does add to the overall thickness of the package even though it is relatively thin.

1.3.4.4 Wireless 3D Integration

Wireless communication, as opposed to wired communication, does not require physical contact, e.g., by the use of a wire, to establish the communication channel. As any user of WiFi knows, wireless communication frees you from being tethered and hence offers you great mobility and convenience. However, since the receiver can be anywhere within the communication range, the transmitted signal must likewise be available everywhere. As a result, wireless communication is generally higher cost and offers lower bandwidth (throughput) than wired communication. By contrast, due to its very short communication distance of <100 μm, wireless 3D IC integration can be optimized to compete with its wired counterpart in cost and performance, while offering other benefits.

First, the connection terminals in the form of either metal pads or inductive coils for wireless integration are manufactured as part of the standard IC metal layer fabrication process. As a result, there is minimal added processing cost. By contrast, TSVs require complicated mechanical and wafer-level processing, adding significant manufacturing cost. Second, since wireless integration uses AC coupling, it can be applied to chips supplied with different voltage levels without any level shifters. Third, because there is no physical connection, wirelessly integrated chips are detachable – it is as easy to disconnect as it is to connect them together. Likewise, individual chips can easily be wirelessly connected to and disconnected from probe cards, making them easy to test and without concern for damages that a physically connected probe can inflict.

Performance of wireless integration in terms of communication data rate and reliability, power consumption, and chip area can also be optimized to rival that of wired integration, for three reasons. First, unlike WiFi, the very short communication distance makes it possible to create large capacitive and inductive coupling coefficients. As a result, high communication reliability can be achieved without consuming a lot of power. Second, since the connection terminals of the integrated chips do not require a DC path to the external world, they can be insulated to avoid electrostatic discharge (ESD) so as to eliminate ESD protection circuits, thereby reducing capacitive loading on the interconnection. This enables both high data rate and low power consumption. Circuit area is reduced as well. Third, in the case of inductive coupling, since the magnetic field can penetrate through circuits on the silicon substrate, the inductive coils can be placed above active circuits with no keep-out area. (Keep-out area is, however, needed to suppress crosstalk to sensitive circuits.) Consequently, die area overhead is small. Contrast this with the use of TSVs, where each via requires a keep-out area, thus adding significant area overhead, especially when they are numbered in the thousands. To experimentally confirm these and other merits of the inductive coupling solution and to advance its development, an inductively coupled wireless interface was prototyped in several test chips between 2006 and 2008 (Figure 1.38).

Figure 1.39 depicts the schematics of the transmitter and receiver circuits, and simulated and measured waveforms from an inductive coupling interface fabricated in a 180 nm CMOS process in 2006 (top-left of Figure 1.38 and [39]). A data rate of 1 Gbps at a bit error rate (BER) $<10^{-13}$ was achieved, delivering reliable communication comparable to that of a wired interface. To minimize chip area, the inductive coils were placed above the transmitter and receiver circuits. Yet, no appreciable degradation in communication reliability was detected. Furthermore, by implementing a four-phase time-multiplexing scheme to skew the timing of data switching in adjacent communication channels (interconnections), total crosstalk from neighboring channels was significantly reduced. This enabled operating 1,024 parallel channels at a 30 μm pitch simultaneously at a per-channel data rate of 1 Gbps at BER $<10^{-13}$, resulting in a total interface throughput of 1 Tb/s.

Figure 1.40 compares the test interface from 2006 with alternate solutions from around the same time frame, in both total channel throughput and area/throughput, confirming the merits of the inductive coupling solution.

The inductive coupling interface was further refined in a test chip in 2007 (middle of Figure 1.38 and [40]) that achieved an energy efficiency comparable to

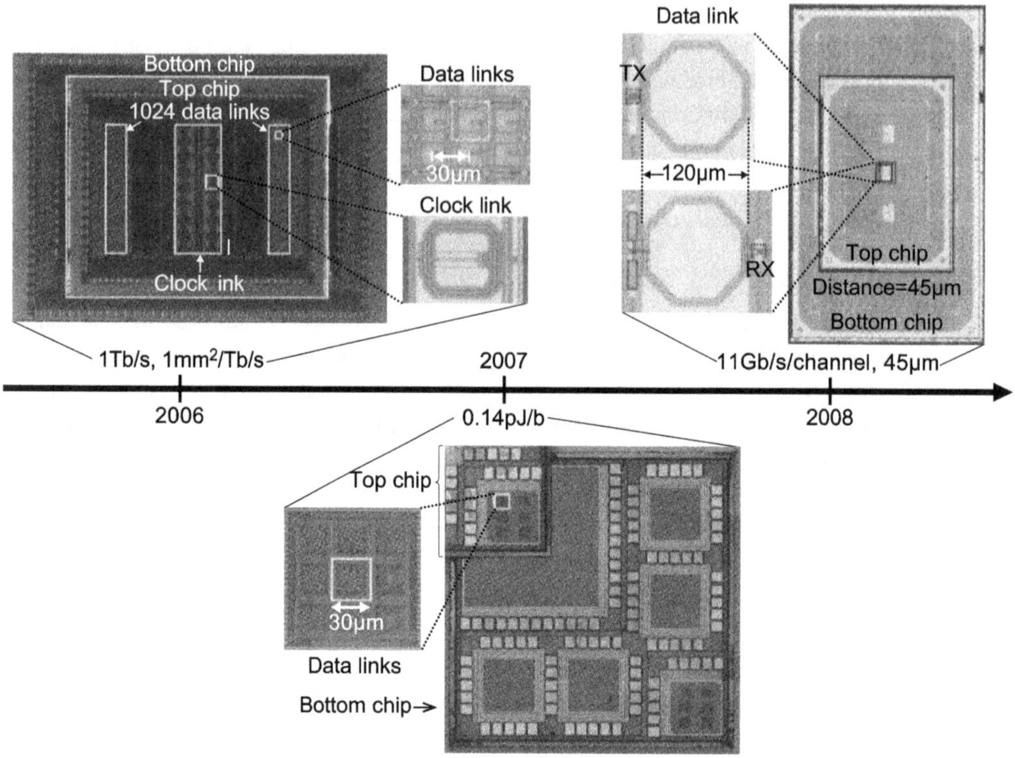

Figure 1.38 Inductive coupling interface test chips from 2006–2008. © 2008 The Japan Institute of Electronics Packaging. Reprinted, with permission, from [49]

that of wired communication. The energy efficiency of the 2006 test chip interface was dominated by the transmitter circuits, which consumed 2.2 pJ/b out of a total energy consumption of 2.8 pJ/b. By optimizing the shape of the transmitted pulse and reducing its width, in addition to migrating to a 90 nm process and reducing V_{DD}, the (transmitter + receiver) energy consumption was reduced by a factor of 1/20 to achieve 0.14 pJ/b, making it comparable to alternate wired solutions (Figure 1.41, [40]).

1.3.4.5 Inductive vs. Capacitive Coupling for 3D Integration

The key advantage of inductive coupling over capacitive coupling for 3D IC integration is twofold. First, signal loss through silicon substrate is small. While capacitive coupling relies on the vertical electric field between the two coupled terminals, inductive coupling utilizes the magnetic field linking the two terminals instead. Figure 1.42 depicts the simulated S_{21} of the communication channel as a function of substrate resistivity when the substrate is in the transmission path, comparing inductive against capacitive coupling. S_{21}, which is defined as the ratio of output to input signal, is a parameter that quantifies signal transmission efficiency in the frequency

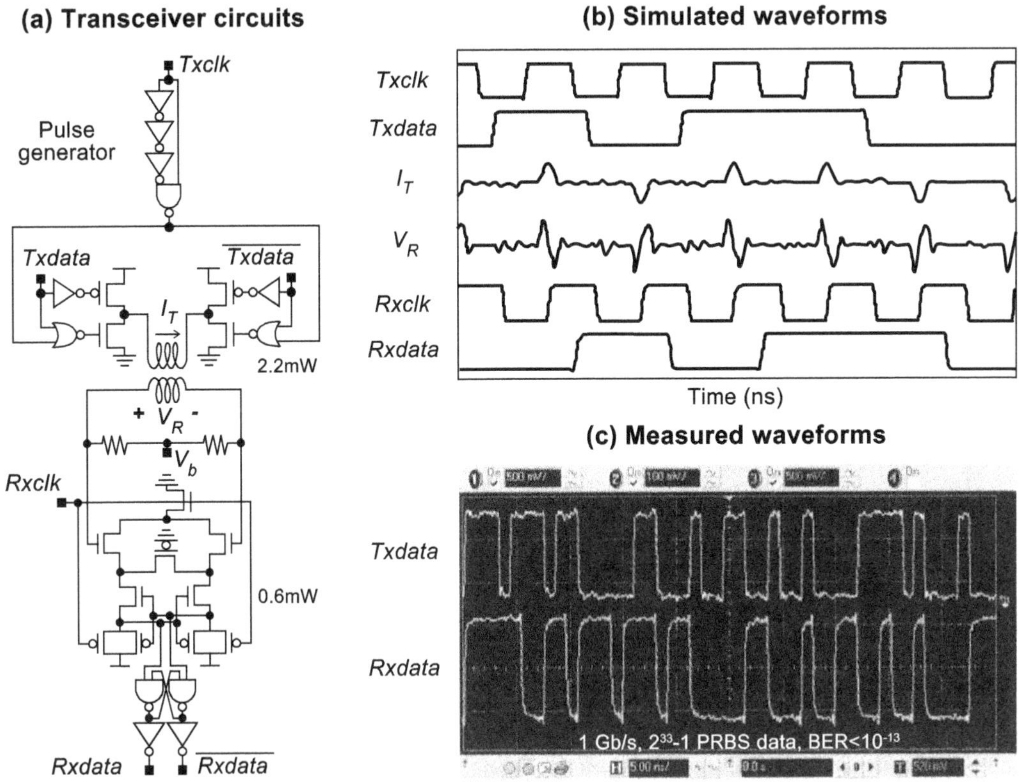

Figure 1.39 Inductive coupling test chip, 2006. © 2008 The Japan Institute of Electronics Packaging. Reprinted, with permission, from [49]

domain. The results reveal that electric field is significantly more susceptible to loss to substrate resistivity than the magnetic field used in inductive coupling. When resistivity reaches around the 1~0.1 Ωcm value of p+ Si, capacitive coupling suffers a dramatic increase in loss. As a result, capacitive coupling cannot be used to communicate through a silicon substrate effectively. Instead, it is confined to face-to-face IC integration. The magnetic field does suffer loss due to eddy current induced in the substrate. However, the loss is small enough that the magnetic field can effectively penetrate p+ Si substrate, making it possible to apply inductive coupling to both face-up and back-to-back chip stacking.

A second advantage of inductive over capacitive coupling is its support for a longer communication distance. The equations defining the coupling and the transceiver circuits are compared in Figure 1.43. The voltage of the received signal through capacitive coupling is defined by the following equation:

$$V_R = \frac{V_T C_C}{C_C + C_{SUB}}, \tag{1.6}$$

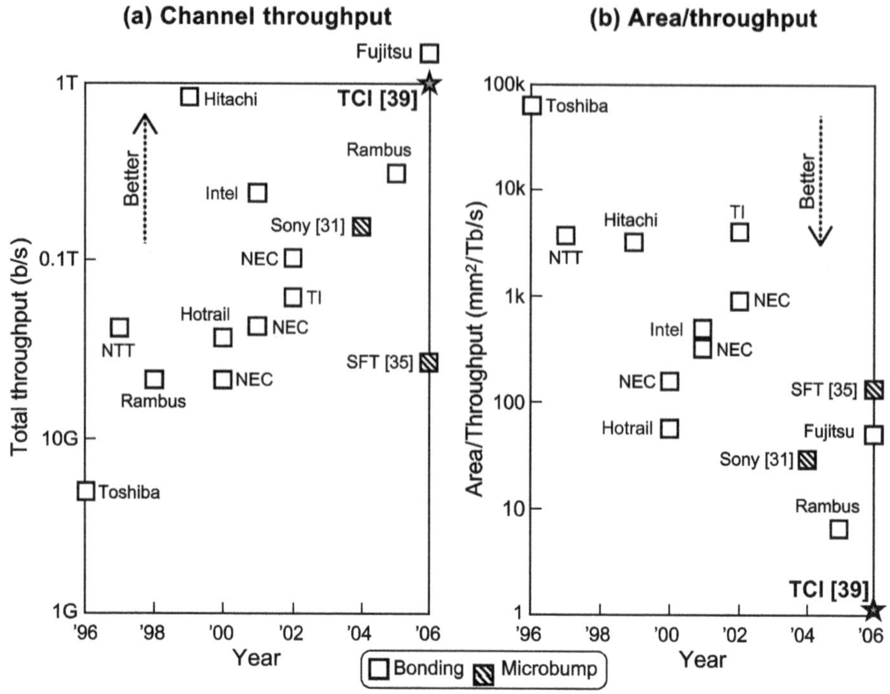

Figure 1.40 Benchmarking the 2006 inductive coupling test chip. © 2008 The Japan Institute of Electronics Packaging. Reprinted, with permission, from [49]

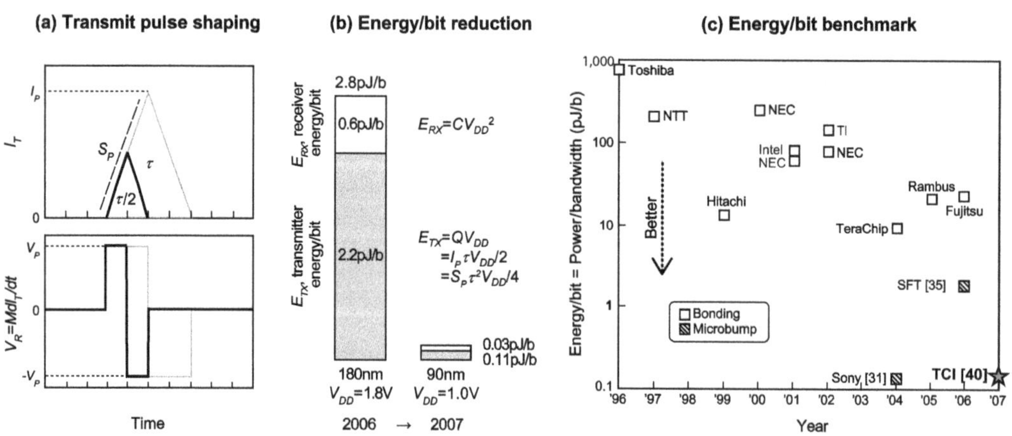

Figure 1.41 Inductive coupling test chip, 2007. © 2008 The Japan Institute of Electronics Packaging. Reprinted, with permission, from [49]

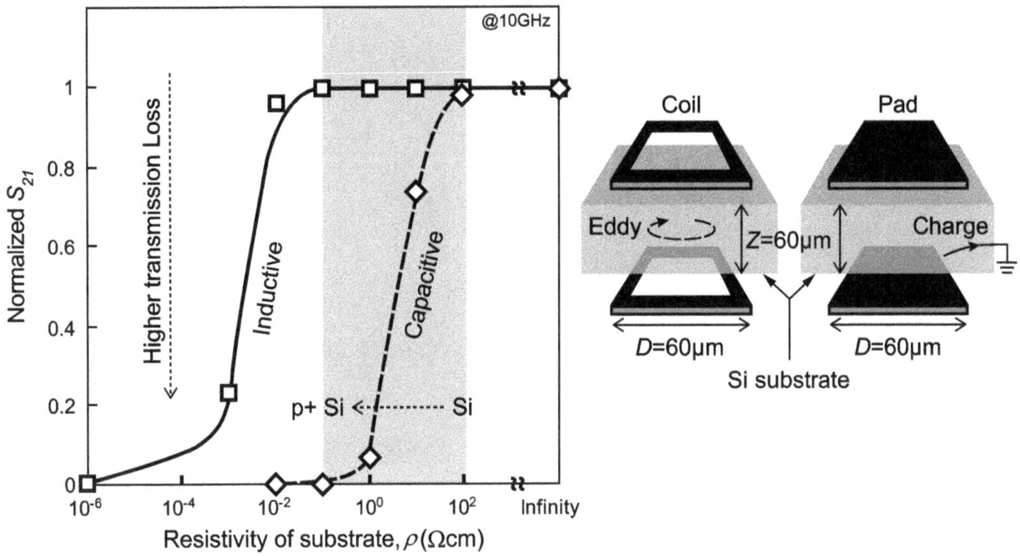

Figure 1.42 Simulated S_{21} dependency on substrate resistivity. © 2008 The Japan Institute of Electronics Packaging. Reprinted, with permission, from [49]

where V_T is the transmitted voltage, C_C the capacitance between the coupled terminals, and C_{SUB} the capacitance between the receiver terminal and the underlying substrate (Figure 1.43a). As the communication distance and hence the distance between the coupled terminals increases, C_C and hence V_R decreases. This loss in V_R cannot be compensated by increasing the area of the coupled terminals since larger area increases both C_C and C_{SUB} proportionally. By contrast, the voltage of the received signal through inductive coupling is defined by the following equation:

$$V_R = M\frac{dI_T}{dt}, \tag{1.7}$$

where I_T is the transmitted current and M the mutual inductance between the coupled terminals (Figure 1.43b). The received voltage is proportional to M, which can be increased by increasing the area of the coupled coils. Therefore, within practical limits, longer communication distance can be supported by enlarging the inductive coils.

A test chip from 2008 (top-right of Figure 1.38 and [41]) confirmed the scalability of communication distance of inductive coupling. Figure 1.44 compares the performance of the test chip with an alternate capacitive coupling interface from 2007. The same data rate of 11 Gb/s at comparable BER was realized using a smaller area for five times the communication distance. Furthermore, communication distance could be extended from 15 to 45 µm with only a 23% drop in data rate to 8.5 Gb/s at the same BER.

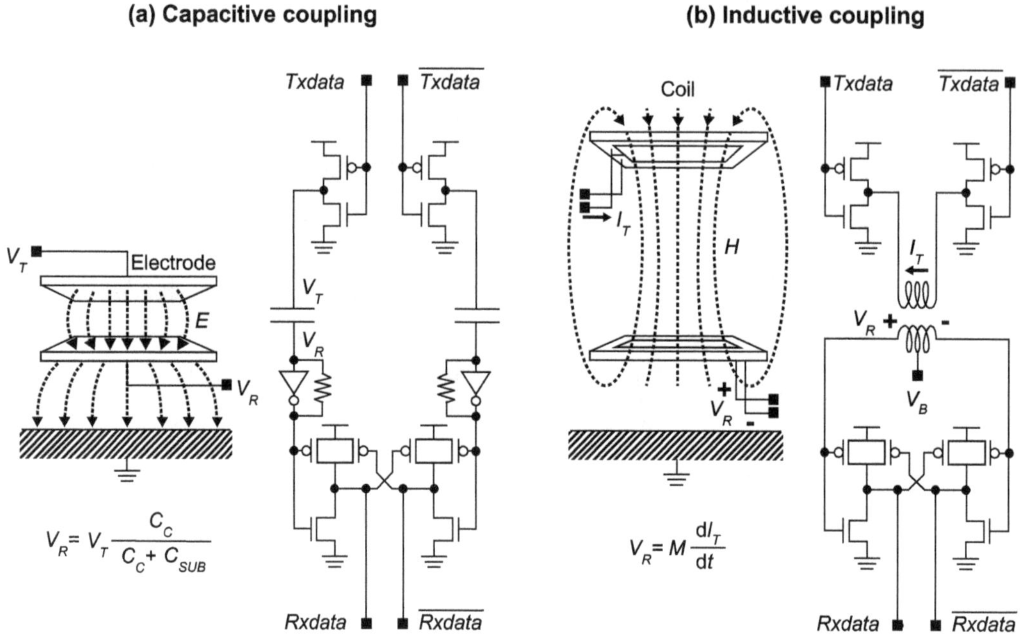

Figure 1.43 Mechanism of capacitive and inductive coupling. © 2008 The Japan Institute of Electronics Packaging. Reprinted, with permission, from [49]

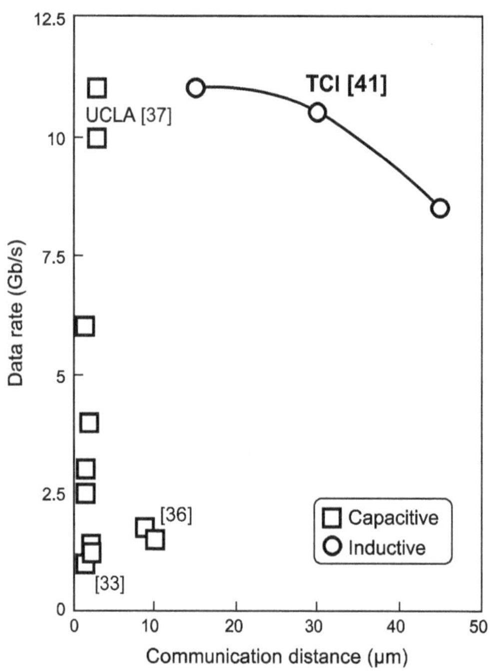

	TCI [41]		UCLA [37]
Channel	Inductive		Capacitive
Distance	15μm	45μm	3μm
Data rate	11Gb/s	8.5Gb/s	11Gb/s
Area	0.015mm²		0.016mm²
BER	$<10^{-14}$		1.02×10^{-14}
Signaling	Pulse		25GHz ASK
Process	180nm stamdard CMOS V_{DD}=1.8V		180nm 3D CMOS V_{DD}=1.0V

Figure 1.44 Performance comparison between inductive and capacitive coupling. © 2008 The Japan Institute of Electronics Packaging. Reprinted, with permission, from [49]

1.3.4.6 Wireless <3D Integration

While 2.5D integration using TSVs as introduced in Section 1.3.4.3 offers a lower-cost alternative to TSV-based 3D integration, the introduction of TSVs and a silicon interposer, albeit fabricated in a more mature process, still represents a significant cost adder. Furthermore, the integrated package contains three levels of bumps and solder balls – microbumps between chips and interposer, C4 or flip-chip bumps between interposer and package substrate, and solder balls connecting package substrate to the PCB (Figure 1.45a). As a result, thermomechanical stress can occur in three physical interfaces, degrading reliability of the integrated part. The problem is exacerbated by the increasing silicon interposer size to accommodate larger processor chips and/or more integrated chips placed side by side. For instance, the CoWoS-2 integration technology from TSMC supports a large interposer of 1,200 mm^2 in size. To alleviate the cost and reliability problems of conventional 2.5D integration, inductive coupling (wireless, $n_{st} \geq 2$) has been explored for enhancing <3D integration. In all three solutions of TCI <3D integration in Figure 1.45, both TSVs and microbumps are eliminated. Furthermore, in TCI 2.5D integration, the large silicon interposer in conventional 2.5D integration is replaced by a small piece of silicon interposer between each pair of adjacent chips. Together they add up to attractive cost saving and higher reliability to enhance <3D integration solutions. These solutions will be further discussed in Section 2.3.

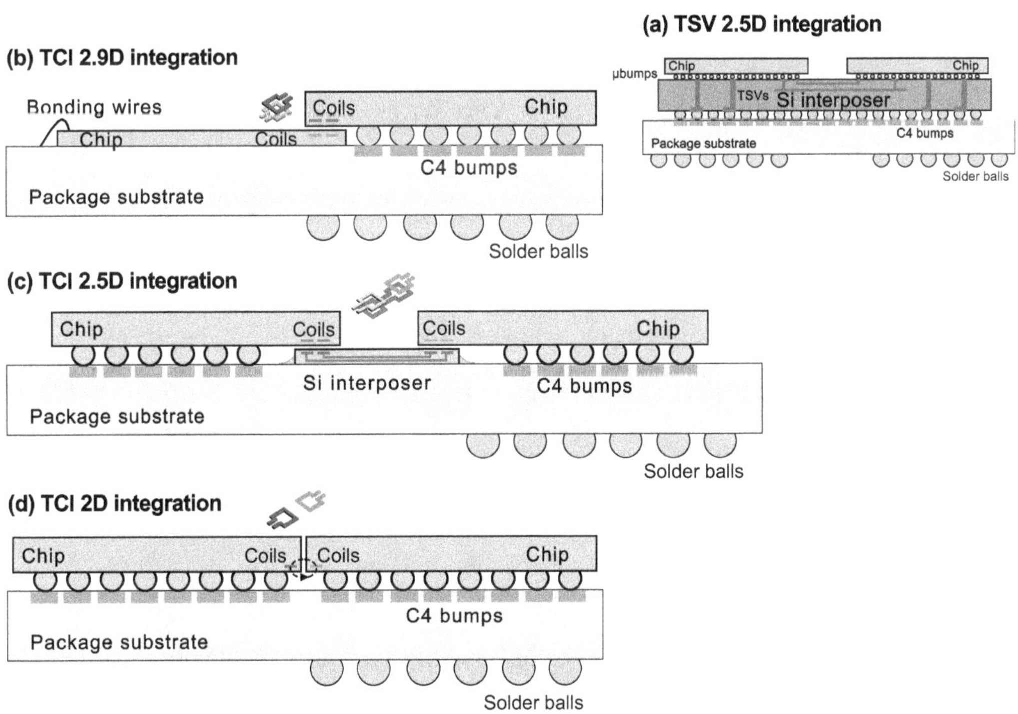

Figure 1.45 TCI <3D integration solutions.

1.3.5 Closing Thoughts

The slowing of Moore's law scaling has spawned a multitude of More than Moore innovations to provide additional improvement over what Moore's law can deliver. They include monolithic 3D IC such as 3D NAND flash memory, as well as 3D and 2.5D IC integration. Among 3D IC integration solutions, there are wired vs. wireless, $n_{st} = 2$ vs. $n_{st} \geq 2$ solutions utilizing microbumps, TSVs, and capacitive and inductive coupling. TSVs are also utilized in 2.5D integration, which offers better cost with some disadvantages, such as a larger footprint than its 3D counterpart. Meanwhile, inductive coupling can also be used for various <3D More than Moore solutions. To determine the optimal solution requires careful comparison of a host of factors.

As we learned from looking back at the history of the IC, the industry has evolved from designing for performance only, to designing for performance, power, and energy efficiency. A frequently used IC evaluation metric is PPA, as in performance, power, and area. But even that is insufficient for evaluating an integrated solution. For instance, since the early days of integrating multiple chips in one package, known good die (KGD) has been a major technical challenge. The overall yield of a multichip package is the product of the yields of individual chips and the yield of the integration process. As a result, the yield of the integrated subsystem falls off rapidly as the number of integrated chips increases, unless the chips are known to be good before integration. Unfortunately, the mainstream IC test infrastructure has been set up to test packaged chips instead of bare chips. Therefore, 2.5 and 3D integration of bare IC chips has the additional challenge of KGD. The point of this discussion is that, given the complex performance and cost equations of an integrated IC subsystem, many factors must be considered in comparing different More than Moore solutions. Some key factors are the following:

- PPA
- Fabrication costs of individual chips, including overhead for integration
- Packaging or integration cost
- Chip yield (KGD)
- Circuit design complexity and power and signal integrity
- Development cost and time to market
- Thermal management especially heat removal
- Form factor (footprint and height) of integrated subsystem

Careful trade-offs between these factors must be made based on your specific product requirements in order to choose the optimal solution. One size does not fit all.

1.4 Near-Field Coupling Interconnect Technology

1.4.1 Motivation for Using Near-Field Coupling for Interconnection

As enumerated in Section 1.3.4.4, wireless technology has numerous merits for IC integration compared to wired technology. These merits include flexibility in the placement of receiver relative to transmitter, elimination of level shifters between chips supplied with different voltage levels, and removal of ESD protection circuits, which reduces capacitive loading on the interconnection and hence enables a higher data rate, lower power consumption, and smaller circuit area. In addition, chips can be tested wirelessly without concern for damages from a physically connected probe.

However, wireless is actually a catch-all term that includes a multitude of technologies with very different characteristics. In particular, one useful way to classify wireless technologies is to divide them into far-field and near-field. As the names suggest, far-field technology is utilized for communicating through long distances, while near-field short distances, using changing electric and/or magnetic field. Figure 1.46 from [50] illustrates the comparison between near- and far-field. The two wireless technologies widely adopted in smartphones – Wi-Fi for Internet connection and near-field connection (NFC) for mobile payment – are example of far-field and near-field technologies respectively (Figure 1.46a). This agrees with our experience – Wi-Fi supports more than 10 meters in communication distance, while NFC only a few centimeters. Nevertheless, the technical distinction is more complicated since the boundary between far-field and near-field is not absolute, and it depends on factors such as the size of the antenna and wavelength of the signal being transmitted. As a rule of thumb, near-field communication occurs when the communication distance d meets the following criterion:

$$d < \frac{\lambda}{2\pi}, \tag{1.8}$$

where λ is the wavelength of the signal. Therefore, for a 50 GHz signal in free space, a communication distance of <1 mm is within the near-field region.

Another important distinction between the two modes of wireless communication lies in how the field strength varies over distance. While the field strength degrades inversely with distance from the source in far-field communication, it degrades significantly faster in near-field communication, following an inverse-square law for electric field and inverse-cubed law for magnetic field (Figure 1.46b). As a result, the transmitted energy in near-field communication is concentrated in a small space, leading to more efficient use of energy (Figure 1.46c). Furthermore, parallel communication channels can be placed close to each other without excessive crosstalk, enabling a large number of interconnections and hence high integration and aggregate bandwidth.

Because of these merits of near-field communication, we have adopted it in our TCI, using inductive coupling for wireless IC integration to deliver More than Moore (Figure 1.47a). In addition, we have used it to implement our TLC using electromagnetic coupling to create a contactless connector to enable LEGO-like modular construction of electronic systems (Figure 1.47b).

1.4.2 Near-Field IC Interconnect Technology

1.4.2.1 Motivation for a Near-Field IC Interface

ThruChip Interface (TCI) is a near-field IC interconnect technology that has been developed to address two major challenges facing the IC industry. The first challenge

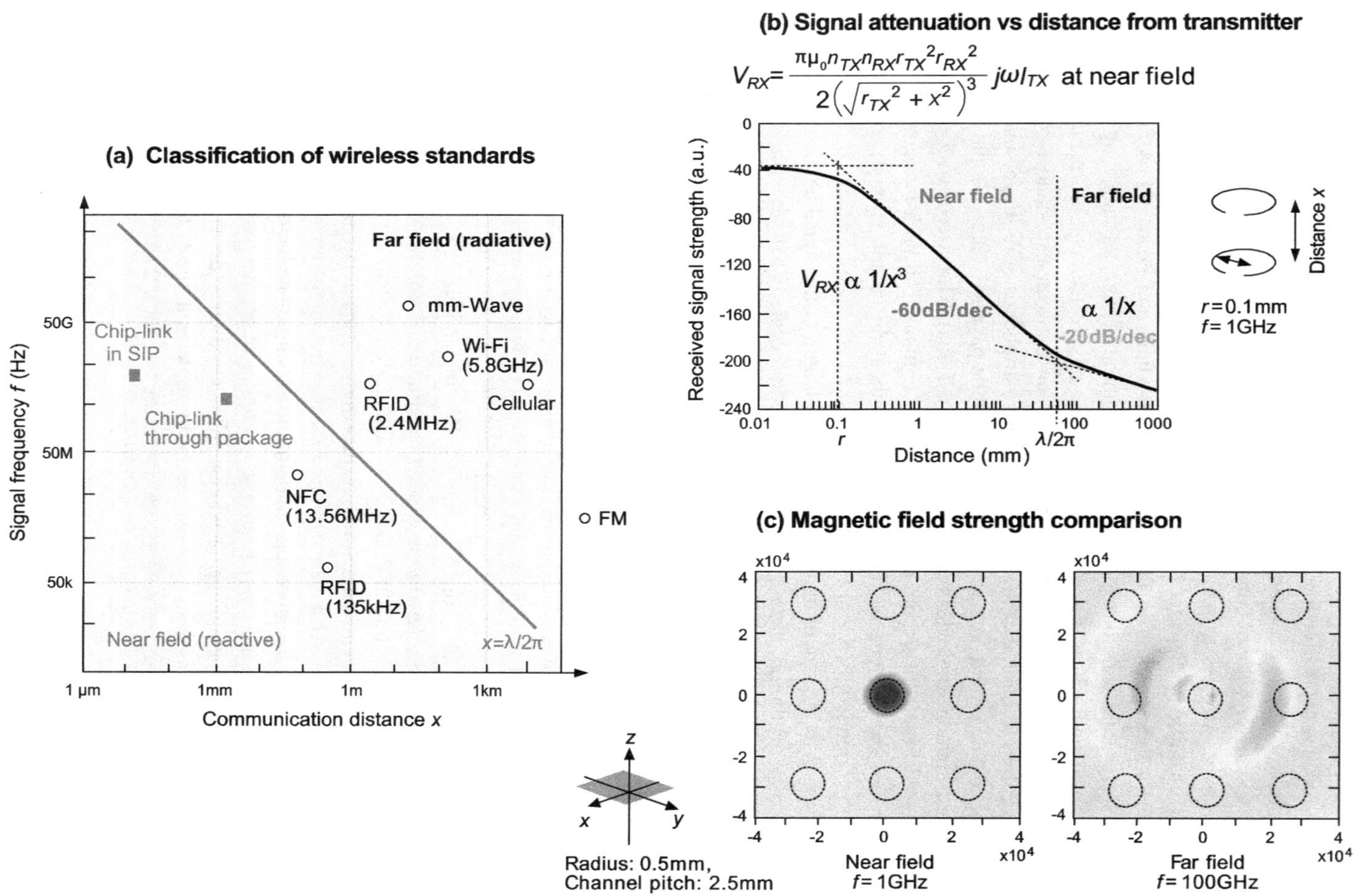

Figure 1.46 Comparison between near- and far-field. © 2010 IEEE. Reprinted, with permission, from [50]

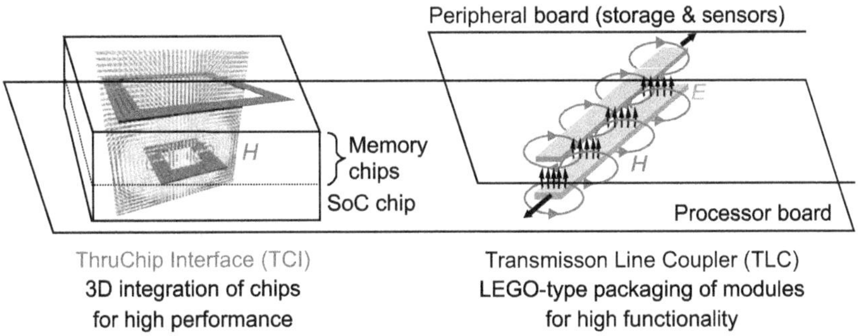

Figure 1.47 Near-field coupling interconnect technologies. © 2017 IEEE. Reprinted, with permission, from [51]

is the slowing of Moore's law, which calls for innovative More than Moore solutions for 3D IC integration as discussed in Section 1.3. Using inductive coupling, TCI offers a better $n_{st} \geq 2$ IC integration solution than TSV, enabling integration of more than two chips at lower cost while delivering high data rate at low power.

The second challenge addressed by TCI is the performance gap between chip core (computational performance) and I/O (interface communication bandwidth). Enabled by Moore's law, performance of digital circuits in the chip core has been increasing by about 71% every year. This is the result of an annual increase in transistor speed of about 15% coupled with annual increase in chip core functional density of about 49% ($1.15 * 1.49 = 1.71$). Based on Rent's rule, to take full advantage of this increased computational performance, the I/O bandwidth should increase at an annual rate of 46% ($1.71^{0.7} = 1.46$). In reality, however, process scaling only drives an annual I/O bandwidth increase of 28%. While flip-chip packaging is commonplace for high-performance processors today, for many years wirebond was the dominant packaging technology. Since wirebond packaging uses peripheral pads for I/O interconnection, chip I/O functional density improves only linearly with process scaling, by about 11% annually. Together with the annual increase of 15% in transistor speed, total I/O bandwidth improvement from scaling has been around 28% ($1.15 * 1.11 = 1.28$). To make up for the deficit, circuit innovations have been employed to boost the data rate to sustain a total annual bandwidth increase of about 46% (Figure 1.48).

Unfortunately, continuous increase in data rate is not sustainable, because of the high circuit power and off-chip interconnect signal integrity problems that result. For instance, to enable high data rate, a high-speed clock must be distributed across the chip, which requires insertion of buffers to compensate for loss, significantly increasing power consumption of the clock tree. In some cases, I/O power has risen to be more than 30% of total chip power. Therefore, in order to deliver increasing I/O bandwidth without breaking the power budget, instead of disproportionally boosting data rate, a better approach is to increase the number of parallel interconnections. The first step is to shift from using a peripheral to an area array of I/O connection

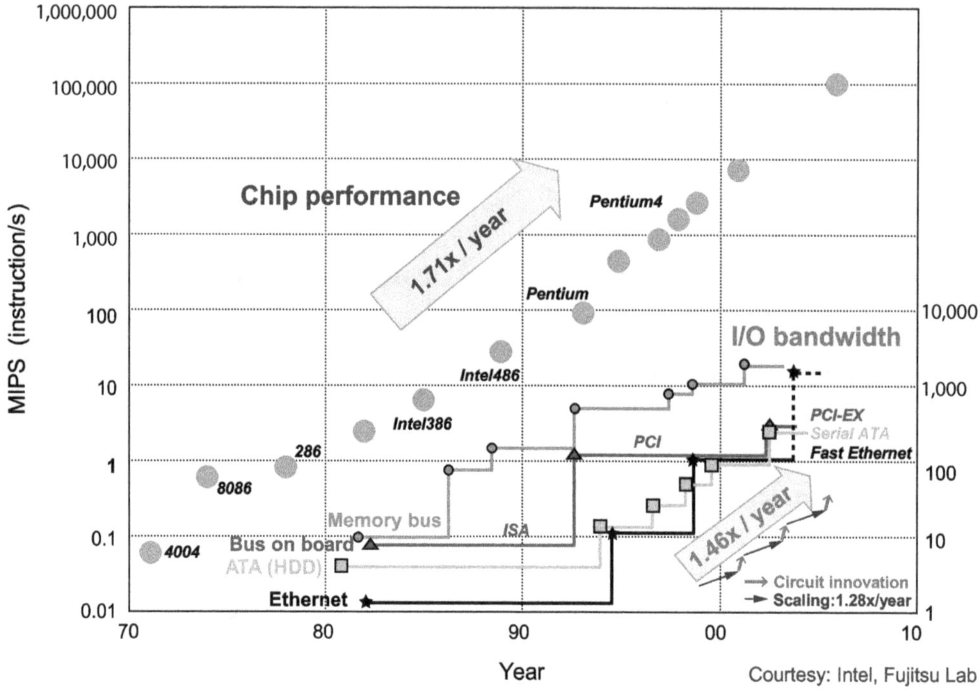

Figure 1.48 Disparity in performance scaling: chip vs. I/O.

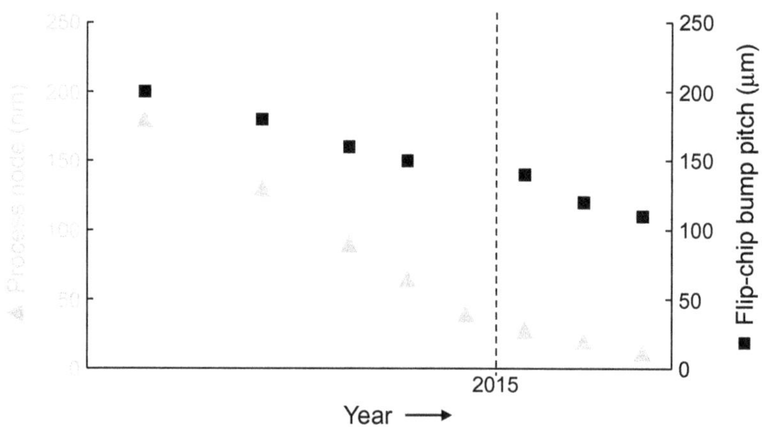

Figure 1.49 Scaling comparison: flip-chip bump pitch vs. process.

terminals. This has resulted in the wide adoption of flip-chip packaging for high-performance IC. Nevertheless, since the flip-chip bumping process has not been scaling as fast as the IC fabrication process, flip-chip bump pitch and hence I/O interconnection density scaling has not been keeping pace with IC performance increase in terms of process node scaling, as shown in Figure 1.49, based on the industry roadmap published in 2015 [52]. Consequently, an alternate area array interconnect technology that follows Moore's law scaling is desired.

The Wide I/O Mobile DRAM solution introduced in Section 1.3.4.2 was a solution that used TSV to create a large number of parallel interconnections to deliver high I/O bandwidth. But as previously noted, TSV carries a significant cost premium both because of the added processing cost and the area overhead resulting from its large keep-out zone. For a memory chip, the added manufacturing cost of TSV can amount to 10% of the total cost. Furthermore, thermomechanical stresses can lead to loss in reliability. The problem is magnified by the tens of thousands of TSVs and microbumps utilized by each chip. Therefore, a lower fabrication cost, lower area overhead, and higher reliability solution is warranted.

1.4.2.2 Characteristics and Merits of TCI

TCI is an inductive near-field coupling, wireless interconnect technology offering an alternate solution to TSV for 3D IC integration and for delivering high interface bandwidth at low power. Figure 1.50 depicts the basic structure of TCI. Specifically, it consists of a pair of coupled coils that communicate by varying the magnetic field

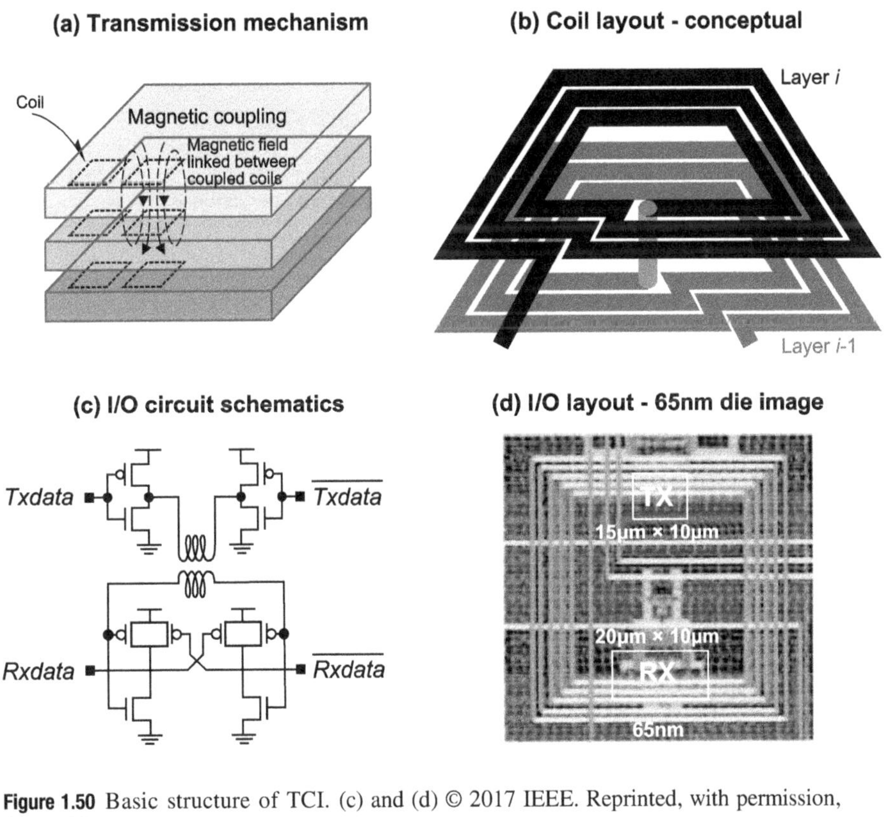

Figure 1.50 Basic structure of TCI. (c) and (d) © 2017 IEEE. Reprinted, with permission, from [51]

linked between them (Figure 1.50a). Each coil is implemented as a multiturn loop (n = number of turns) across multiple interconnect layers (m = number of layers) as part of the BEOL IC fabrication process (Figure 1.50b). The basic transceiver consists of simple digital CMOS circuits – the transmitter is an H-bridge circuit that switches current in the transmitter coil at the transition of *Txdata*, while the receiver is a sense-amplifier flip-flop that detects the resultant small voltage magnetically induced in the receiver coil to create *Rxdata* (Figure 1.50c). With proper mitigation measures, electromagnetic interference (EMI) from the coils can be made negligible, allowing transmitter and receiver circuits to be placed directly underneath the coils with no keep-out zone to form a compact I/O design (Figure 1.50d). In fact, test chip results have shown that TCI coils can be placed directly on top of active circuits and memory cells, including SRAM, NAND, and DRAM, with no appreciable degradation.

Table 1.6 enumerates the general merits of TCI given its characteristics, and the value delivered to 3D IC integration and to a low-power, high-bandwidth chip

Table 1.6. Merits of TCI interconnect technology.

		Value delivered	
TCI characteristic	**Merit**	**3D IC integration**	**Low-power, high-BW I/F**
Wireless	Contactless => no ESD	Low power, small area, high speed	Low power, small area, high BW
	Contactless => no level shifter	Integration flexibility	
	Contactless probing	KGD	Easy probing
	High placement tolerance	High integration yield	
	No thermal stress in interface	High reliability	
Near-field coupling	Concentration of transmitted power	Low transmission power	
	Small crosstalk ⇒ high interconnection density	Small form factor	High BW
	Short communication distance	Small form factor	No transmission line effects ⇒ high BW
Inductive	Communication through silicon substrate	Integration of more than 2 chips	
Coils fabricated in BEOL	No mechanical processing ⇒ high yield	Low cost	
	No fabrication cost-adder	Low cost	
	Benefit from process scaling	Headroom for PPA improvement	
Transceiver uses digital CMOS circuits	Benefit from process scaling	Headroom for PPA improvement	
Low EMI	Circuits under coils ⇒ small area overhead	Low cost	
	No keep-out zone ⇒ high interconnection density	High integration	High BW

Table 1.7. Comparison of TCI and TSV.

	TSV	TCI
Connection mechanism	Mechanical	Electrical
Fabrication process	Additional steps required	Standard CMOS
Dimensional scaling	Difficult	Easy
Yield	Low, difficult to improve	High ($\approx$100%)
Ecosystem	New model required	Conventional
Cost adder	>40%	A few %
Placement	Dedicated area with keep-out	Flexible
Bandwidth	<512 Gb/s	>512 Gb/s
ESD protection	Required	Not required
Power consumption	High	Low

Source: © 2017, IEEE. Reprinted, with permission, from [51].

interface respectively. Meanwhile, Table 1.7 highlights how TCI compares favorably against TSV. In contrast to TSV, which requires mechanical processing, TCI is an electrical and digital solution fabricated as part of the standard wafer process. As a result, TCI incurs much lower manufacturing cost. Because OSAT is not involved in implementing the interconnection, TCI can utilize the existing manufacturing ecosystem. Furthermore, unlike TSV, TCI can be placed directly over circuits, resulting in minimal area overhead. Additionally, being a contactless interconnection, TCI eliminates the need for ESD protection circuits, resulting in much smaller parasitic capacitance and hence significantly lower power and higher speed.

In summary, TCI is a low-cost, low-power, high-bandwidth technology with high interconnection density and reliability, which enables flexible and high-level integration of ICs in small form factor. Furthermore, its PPA will continue to improve as process scales under Moore's law, enabling it to advance together with IC performance.

The table in Figure 1.51 from [50] illustrates how TCI performance scales with process. Under this scaling scenario, coupling coefficient and crosstalk and hence signal-to-noise ratio will remain unchanged, while both aggregated data rate/area and energy/bit will improve drastically in a cubic fashion. This represents an optimistic scenario – in reality, supply voltage and chip thickness may not scale as fast as transistor size. Nevertheless, it illustrates how TCI performance will benefit from scaling in transistor size, power supply voltage, and chip thickness simultaneously. Hence, there is plenty of headroom for its performance to improve as process scaling continues.

The potential of TCI has been demonstrated in multiple test chips. In 2010, a data rate of 30 Gb/s/channel and 2.2 Tb/s/mm^2 was reported in [53], an energy efficiency of 0.01 pJ/b in [54], and successful random memory access to a stack of 128 NAND flash chips in [55].

1.4.2.3 Application of TCI

Given the myriad of benefits delivered by TCI, it is expected to offer improvement to a large variety of applications. Figure 1.52 illustrates some of these applications.

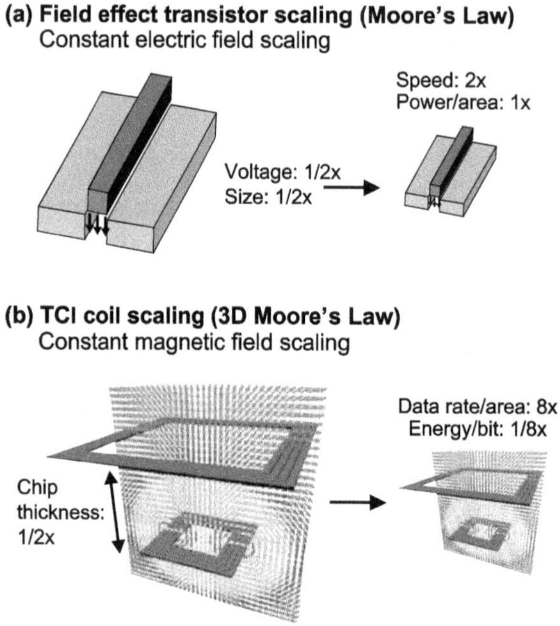

(c) TCI performance scaling

Transistor size	[x]	$1/\alpha$
Power supply voltage	[V]	$1/\alpha$
Chip thickness	[Z]	$1/\alpha$
Coil turn number (# of layers)	[n]	$\alpha^{0.8}$
Current	[I]	$1/\alpha$
Capacitance	[C]	$1/\alpha$
Circuit delay time	[t]~[CV/I]	$1/\alpha$
Coil diameter	[D]~[Z]	$1/\alpha$
Self inductance	[L]~[$n^2D^{1.6}$]	1
Magnetic coupling coefficient	[k]	1
Received signal	[V_R]~[$kn^2D^{1.6}(I/t)$]	1
Crosstalk	[V_{RS}/V_{RN}]	1
Data rate / channel	[$1/f$]	α
Channel number / area	[$1/D^2$]	α^2
Aggregated data rate / area	[$1/tD^2$]	α^3
Energy / bit	[ItV]	$1/\alpha^3$

Figure 1.51 TCI performance scaling scenario. © 2010 IEEE. Reprinted, with permission, from [50]

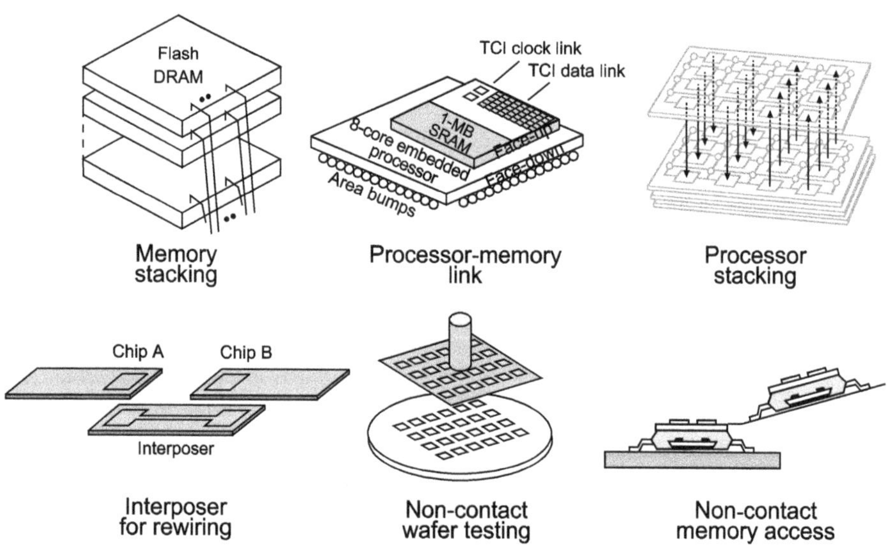

Figure 1.52 Example TCI applications. © 2010 IEEE. Reprinted, with permission, from [50]

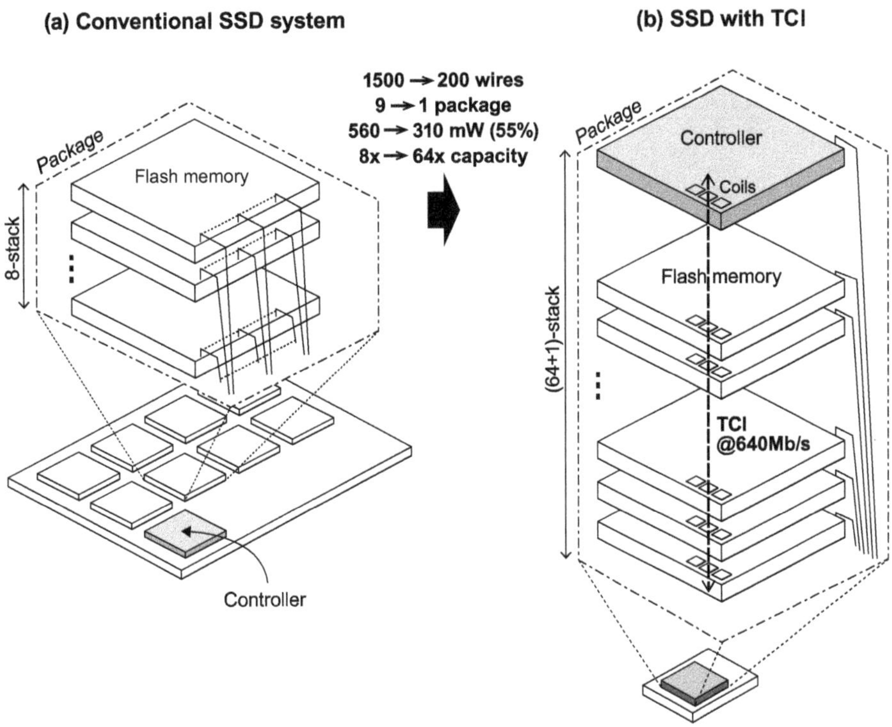

Figure 1.53 High-capacity SSD enabled with TCI. © 2010 IEEE. Reprinted, with permission, from [56]

For instance, a conventional SSD with a stack of eight wirebonded NAND flash memory chips can be replaced with a stack of 64 TCI-equipped chips. Even with eight times the capacity, the TCI-enabled SSD would use fewer packages (9→1) and bond wires (1500→200), hence be lower cost with lower assembly complexity in a smaller form factor, at almost half the power (Figure 1.53). Additional details are provided in Section 2.7.

On the other hand, using TCI, a processor can be directly connected to a stack of SRAM chips fabricated in a different process optimized for SRAM and supplied with a different voltage (Figure 1.54, [57–58]). This would offer an alternate memory solution to DRAM with better energy efficiency and bandwidth per unit area. Additional details are provided in Sections 2.3.1 and 4.1.

1.4.3 Near-Field Module Connector Technology

1.4.3.1 Motivation for a Near-Field Coupled Module Connector

TCI interconnect technology offers a versatile low-power, high-performance solution for connecting multiple IC chips directly together. However, it is not always possible to integrate all the IC chips that need to communicate with each other into a single

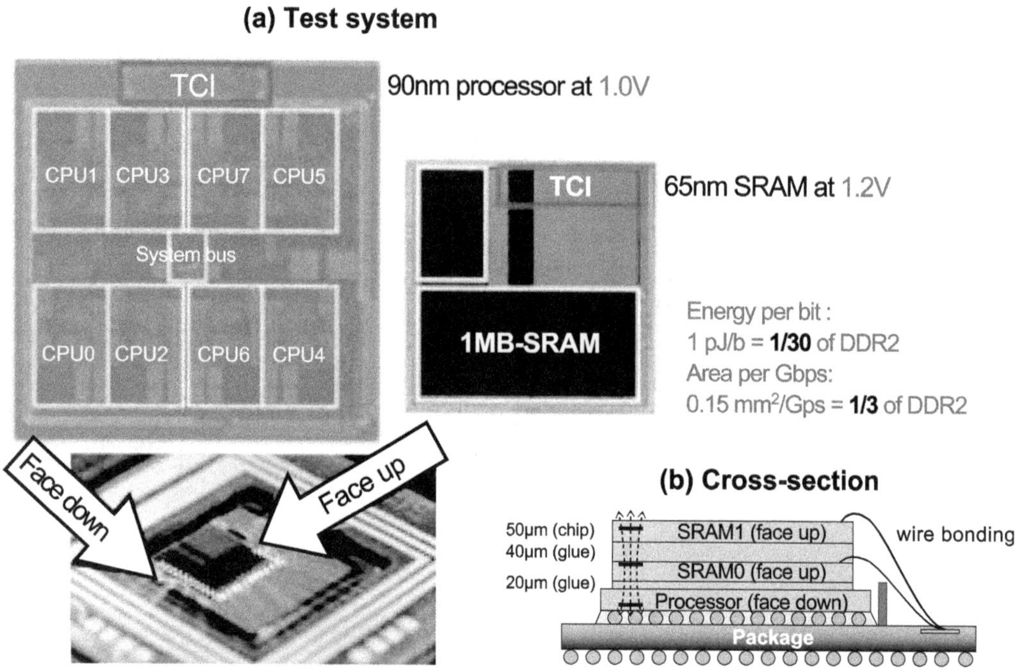

Figure 1.54 Processor and SRAM stack integrated with TCI. (a) © 2009 IEEE. Reprinted, with permission, from [57]; (b) Copyright 2009 The Japan Society of Applied Physics [58]

package. For instance, to build a large-scale computer system such as a supercomputer, it is necessary to modularize smaller building blocks such as a graphic or memory card. This is done for multiple reasons, including cost, design simplicity, ease of customization, serviceability (ease of replacing defective parts), and scalability. As a result, while it is best to have minimal communication distance to suppress signal integrity problems and interface power, it is also necessary to develop interface technology that supports long communication distance, traversing multiple printed circuit boards and connectors.

Interface technology that supports long communication distance must address similar challenges of delivering high bandwidth at low power as is described in Section 1.4.1. As an illustration, the communication data rate between the application processor and display in a smartphone continues to climb exponentially as new display format emerges (Figure 1.55, [59]). Full HD format, which has enjoyed widespread adoption starting around 2016, requires a date rate of 3 Gb/s. To support the next upgrade in resolution to 4K/8K, the maximum data rate will need to increase to 50 Gb/s. Even higher data rates will be necessary to support richer color and higher refresh rate. With these trends as the backdrop, the MIPI Alliance responsible for setting standards for mobile applications has created its high-performance interface physical layer M-PHY specification to support 10 Gb/s-class internal data communication.

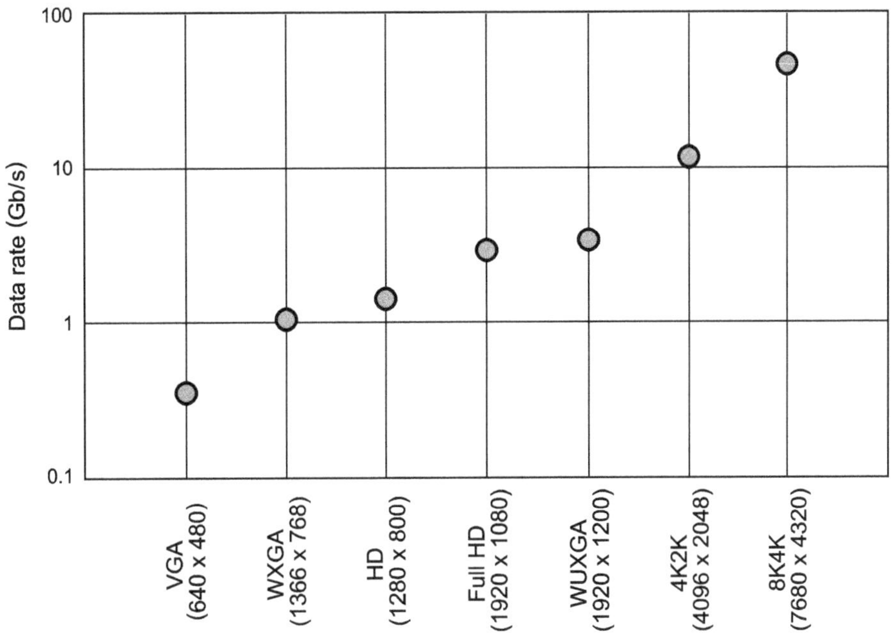

Figure 1.55 Data rate as a function of display format. © 2016 Atsutake Kosuge [59]

Meanwhile, the widely adopted DDR external memory interface, USB and PCIe peripheral interfaces, automotive LAN interface, as well as satellite processor-memory interfaces are all experiencing similar increases in data rate (Figure 1.56, [59]).

With the exception of automotive LAN interface, all interfaces have a data rate exceeding 1 Gbps/lane. Given that communication distance in these interfaces is longer than a centimeter, at such high data rates, signal transmission is subject to transmission line effects. (Remember that a PCB interconnect only needs to be longer than 1.67 mm to appear as a transmission line to a 1 Gbps signal.) As a result, signal transmission is subject to reflection and hence distortion at any discontinuity in characteristic impedance Z_0 along the interface (Section 1.2.3).

Figure 1.57 depicts a typical interconnect structure for some of these interfaces constructed with modules. To minimize signal distortion, the signal must see Z_0 close to being constant as it traverses this interconnect structure. For practical reasons, the default Z_0 for most applications is 50 Ω. Therefore, the target Z_0 for module PCB1, connector1, main PCB, connector2, and module PCB2 should all be 50 Ω. As data rate goes up such that smaller and smaller interconnect features begin to exhibit transmission line effects, designing for constant Z_0 becomes increasingly challenging. Z_0, which governs how an electromagnetic signal propagates, is determined by the electric and magnetic field patterns around the transmission line. It is therefore highly dependent on any large metallic component of the interconnect structure. Conversely, it is easier to control Z_0 of an interconnect when there is a large inherent metallic

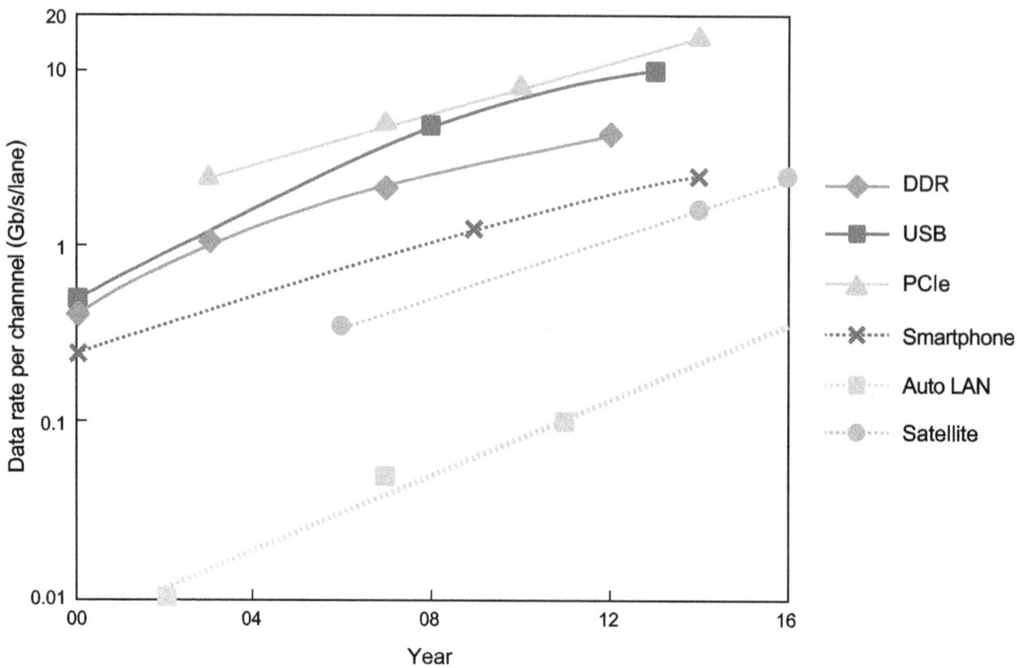

Figure 1.56 Data rate trends of widely adopted interfaces. © 2016 Atsutake Kosuge [59]

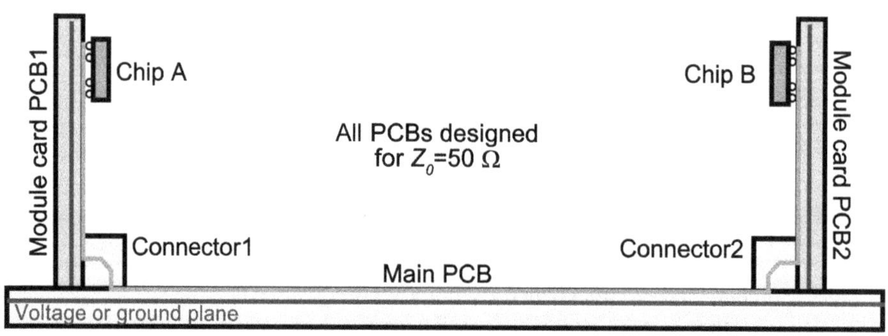

Figure 1.57 A typical interconnect structure in a modular system. © 2016 Atsutake Kosuge [59]

component. As a result, it is easier to control Z_0 of a PCB interconnect than a connector since the former generally contains large voltage and ground planes running parallel to signal paths while the latter does not (Figures 1.57 and 1.58). Figure 1.59 shows the simulated signal degradation due to an interconnect consisting of a backplane connector plus a 5 cm PCB trace. The significant degradation due to the connector is evident in both the frequency and time domain.

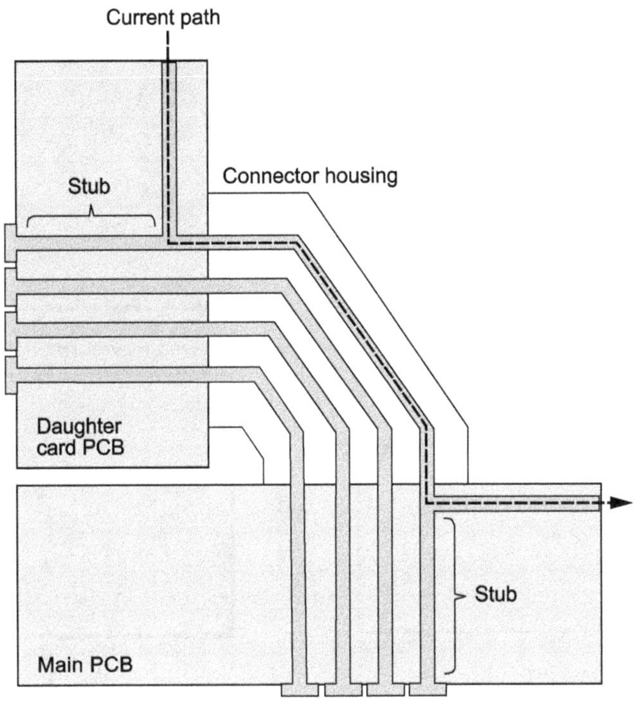

Figure 1.58 Internal structure of a backplane connector. © 2016 Atsutake Kosuge [59]

To compensate for the severe degradation due to transmission line effects, complicated equalization circuits are commonplace in the implementation of high-speed interfaces. Common equalization circuit solutions include TX (transmitter) deemphasis to boost high-frequency components while attenuating low-frequency components of the transmitted signal (Figure 1.60a, [60]), continuous time linear equalizer (CTLE) to boost high-frequency components of the received signal (Figure 1.60b, [61]), and decision feedback equalizer (DFE), which determines how to compensate each received bit based on the preceding bit pattern (Figure 1.60c, [62]). As the name implies, equalization circuits serve to equalize the transmission gain of an interconnect across different frequencies. For proper data transmission, such circuits are necessary to compensate for the wild variation in gain over frequency, as shown in Figure 1.59a. While CTLE is relatively simple, TX deemphasis and DFE circuits grow in complexity as the interconnect complexity increases. Equalization circuits can therefore consume a lot of both power and chip area, significantly increasing interface power and area overhead.

Another transmission line effect degrading signal integrity is crosstalk. In addition to controlling characteristic impedance, large voltage and ground planes also serve to suppress crosstalk since they limit the spreading of electric and magnetic fields from one transmission line to another. Conversely, the lack of such planes in a connector contributes to not only poorly controlled characteristic impedance, but also significant

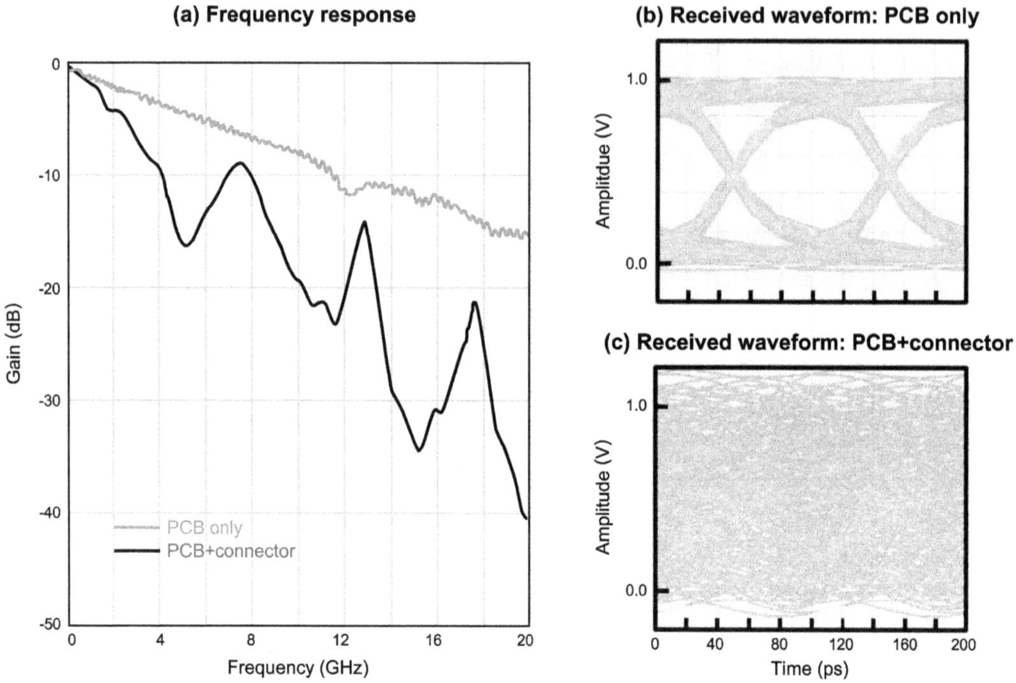

Figure 1.59 Simulated signal degradation of a backplane interconnect. © 2016 Atsutake Kosuge [59]

crosstalk. As a result, from an electrical signal integrity standpoint, a connector is generally the weakest link in an interconnect structure for a high-speed interface.

From a mechanical standpoint too, the connector has weaknesses in terms of environmental robustness given that a connection is made through physical contact. This can be problematic when the connector must operate under harsh environmental conditions, such as when it is used in consumer applications and subject to user mishandling, or in an automotive and subject to heat and vibration as well as high density of contaminants, or in a satellite and subject to stressful conditions during launch and a harsh environment in space. In particular, to allow for physical contact to make a connection, connector pins must be exposed and hence subject to deterioration due to environmental conditions. Furthermore, strong vibration can weaken the connection, especially when it is solderless. On the other hand, water can create a short, while contaminants can create a short or increase contact resistance. Additionally, for applications where the connector must endure multiple cycles of mating and disengaging, the physical connection can be weakened through wear and tear.

The ever-shrinking form factor demanded by applications such as mobile and IoT poses additional challenges to the conventional connector. Smaller physical dimensions further degrade the environmental robustness of the connector. Smaller pins and

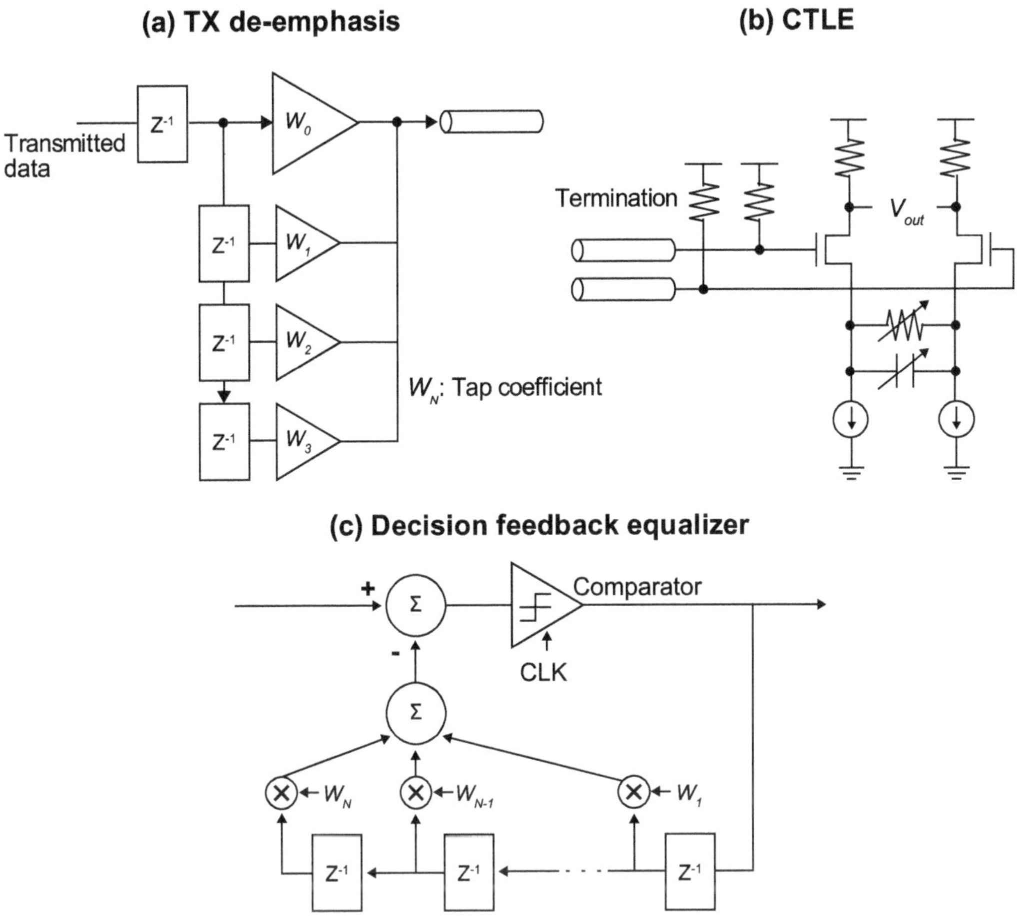

Figure 1.60 Common equalization circuit solutions. © 2016 Atsutake Kosuge [59]

pin pitch are more susceptible to the effects of contaminants and shorting, and wear and tear. Furthermore, smaller pins result in a smaller contact force making it easier to break a connection through vibration. On the other hand, since connector manufacturing involves mechanical processing, it is difficult to scale down its dimensions, making it challenging to support not only a shrinking form factor but also increasing interconnection density.

As a result, an alternate connector technology is necessary to enable better characteristic impedance and crosstalk control for high-speed interfaces, better robustness for operating under harsh environment, and better scaling to support increasing pin count in shrinking form factor.

While TCI can address a lot of these challenges, it suffers from two limitations when applied to a connector. First, since the coil in the coupler is inductive, its impedance varies with frequency. As a result, when it is used to connect two

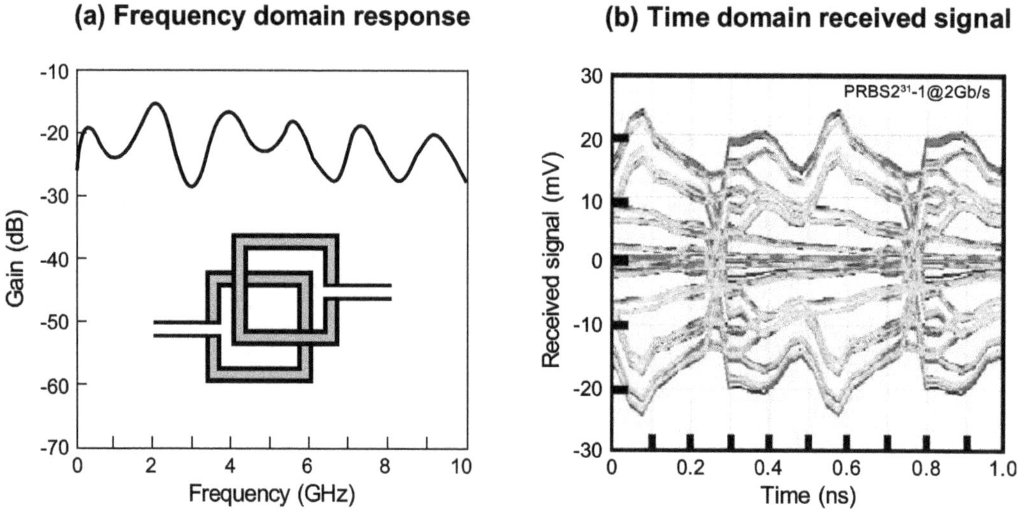

Figure 1.61 Simulated received signal through a TCI connector and transmission line. © 2016 IEEE. Reprinted, with permission, from [63]

transmission lines, it creates a discontinuity in the characteristic impedance of the interconnect. This is demonstrated by the simulated received signal through a TCI connector attached to a 5 cm, 50 Ω transmission line in Figure 1.61. The impedance discontinuity results in large ripples in the frequency domain gain, and closes the data eye in time domain. Second, the coil inductance together with its parasitic capacitance create a resonance circuit. Since coil size varies linearly with communication distance, the large communication distance required in connector applications means that the coil size must grow, simultaneously increasing its inductance and capacitance, resulting in a lower resonant frequency. Consequently, a TCI connector has a limited bandwidth.

1.4.3.2 Characteristics and Merits of TLC Technology

TLC is an electromagnetic near-field coupling, wireless interconnect technology developed to address the challenges faced by conventional connectors enumerated in the previous section. Figure 1.62a depicts the basic structure of TLC, which consists of two pairs of differential transmission lines aligned face to face with each other and can be implemented in a PCB manufacturing process. Each pair is designed to have well-controlled characteristic impedance and minimal crosstalk by virtue of a proximal ground plane (Figure 1.62b). This is evident from the simulated transmission gain in the frequency domain of an example TLC connector in Figure 1.63, which exhibits relative flatness over a wide frequency range. Note that since TLC is a wireless technology, there is no DC gain and hence the curve falls off at low frequency.

Figure 1.62c shows the basic structure of the TLC transceiver. It consists of a simple 50 Ω terminated current mode logic driver and a hysteresis comparator in the

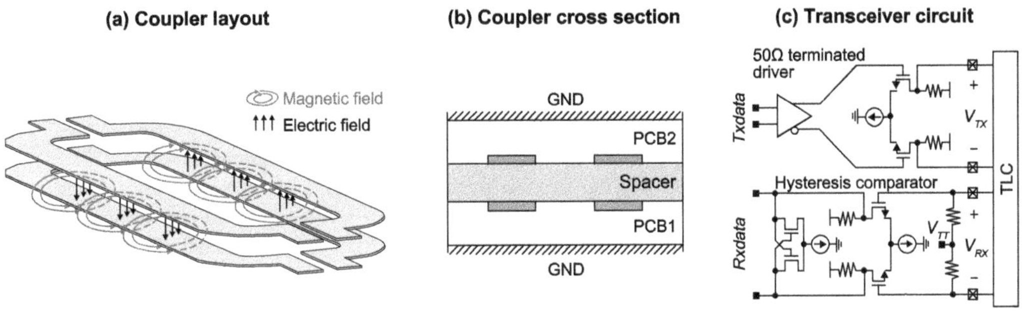

Figure 1.62 Basic structure of TLC. (a) and (b) © 2016 Atsutake Kosuge [59]; (c) © 2015 IEEE. Reprinted, with permission, from [64]

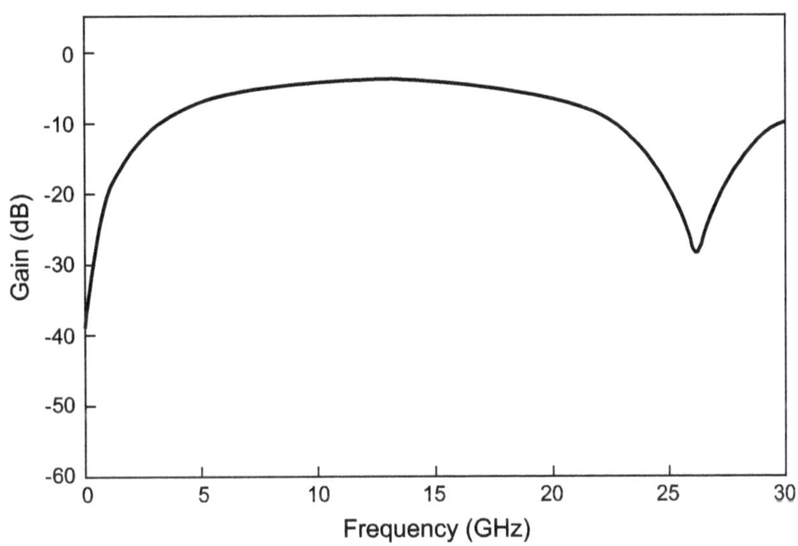

Figure 1.63 Transmission gain of an example TLC connector. © 2015 IEEE. Reprinted, with permission, from [65]

receiver. In contrast to conventional connectors, the well-controlled characteristic impedance of a TLC connector significantly reduces the need for complex equalization circuits.

Table 1.8 enumerates the general merits of TLC given its characteristics and how they address the challenges of implementing a high-speed interface that includes connectors. It lists the characteristics that strengthen the environmental robustness of the connector. Furthermore, it explains how it enables a low-power and high-bandwidth interface at a lower cost than conventional mechanical connectors. While the PCB manufacturing process does not scale as fast as the IC fabrication process, it enjoys better scaling than the mechanical processes used to manufacture conventional

Table 1.8. Merits of TLC connector technology.

Characteristic of TLC	Merit	Challenge addressed
Wireless	Contactless ⇒ connection terminals can be sealed	No shorting by water or contaminants No deterioration from direct contact with environment
	Contactless ⇒ no level shifter	Integration flexibility
	No physical interconnection	No wear and tear Vibration tolerance
Near-field coupling	Small crosstalk ⇒ high interconnection density	Small form factor High bandwidth
	Short communication distance	Small form factor
Fabricated in PCB process	No mechanical processing ⇒ high yield	Low cost
	No fabrication cost adder	Low cost
	Benefit from PCB process scaling	Continuous improvement of PPA
Transceiver using only digital CMOS circuits	Benefit from IC process scaling	Continuous improvement of PPA
Implemented with transmission lines	Well-controlled characteristic impedance and low crosstalk	High bandwidth
	No complicated equalization circuits	Low power, small area, simple transceiver design

connectors. As a result, there is room for scaling the TLC connector form factor. Besides, using digital CMOS circuits for its transceiver allows its PPA to improve as IC process scales. Finally, not requiring level shifters can be a real benefit when more and more heterogeneous subsystems are connected together as the semiconductor continues to proliferate, especially in IoT and automotive applications.

1.4.3.3 Application of TLC

Given the myriad of benefits delivered by TLC, it is expected to offer improvement to a large variety of applications. Figure 1.64 from [51] illustrates some of these applications, which range from consumer applications, including memory cards, displays, smartphones, and PC memory DIMM modules, to automotive and space applications.

For instance, TLC can enable a low-cost, low-power, high-bandwidth solution to build a modular smartphone with various heterogeneous subsystems, with low profile and small footprint (Figure 1.65, [68]), while offering headroom for future PPA improvement. Further discussion of TLC application in smartphones is provided in Section 3.3.

1.4.4 Closing Thoughts

A key merit of wireless interconnect technology can be appreciated by examining the conceptual structure of an interconnect. It is basically a perpendicular structure that

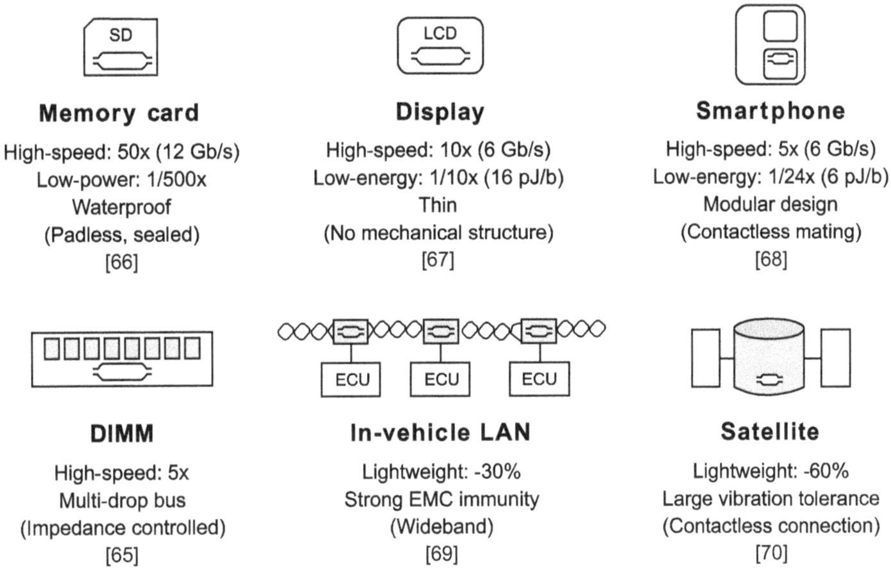

Figure 1.64 Example TLC applications. © 2017 IEEE. Reprinted, with permission, from [51]

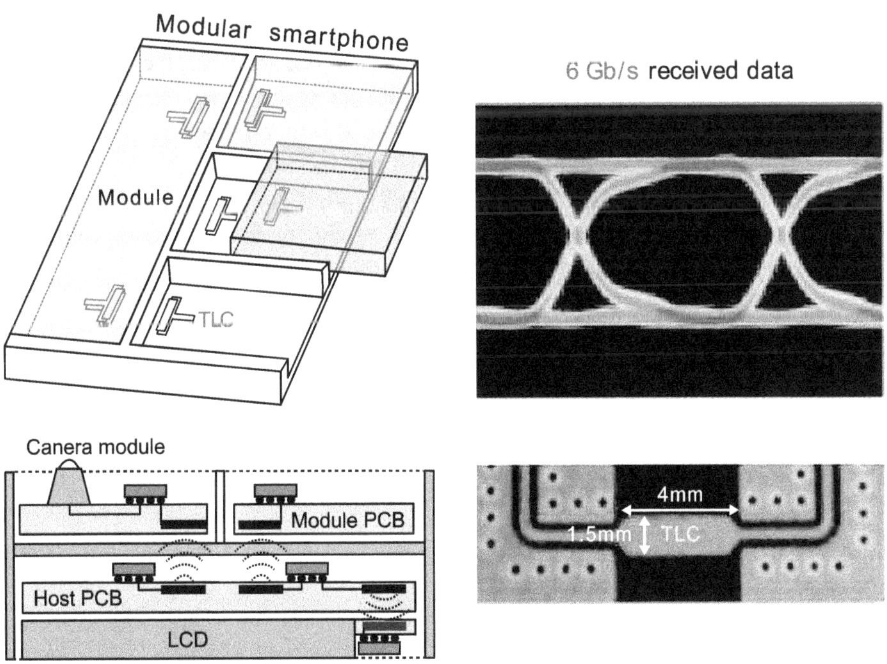

Figure 1.65 A modular smartphone design. © 2015 IEEE. Reprinted, with permission, from [68]

connects two parts fabricated in a layering process – IC process for chips and PCB process for boards attached with a connector. While the layering process can be scaled over time to downsize the parts, the perpendicular structure is not as scalable. Furthermore, relative movement between the two parts due to thermal expansion or vibration can induce stress in the perpendicular structure, thereby reducing its reliability. By using a wireless interconnect technology, this perpendicular structure is eliminated, resulting in a more scalable and reliable solution. Furthermore, eliminating the physical contact helps improve the environmental robustness of the solution, and facilitates interconnection of two heterogeneous parts, such as parts with different supply voltages.

Within the wireless solution space, using near-field coupling suppresses crosstalk and focuses the transmitted energy within the short communication distance, resulting in higher bandwidth, lower power, and higher interconnection density for smaller form factor.

In TCI, near-field coupling is realized through inductive coupling between coils implemented using BEOL of the IC fabrication process. This enables very small coils and very small communication distance, resulting in simple transceiver design in a low power solution for 3D IC integration and for delivering high interface bandwidth that scales with the IC fabrication process.

Meanwhile in TLC, near-field coupling is realized through electromagnetic coupling between transmission lines implemented using the PCB manufacturing process. This enables an interconnect with well-controlled characteristic impedance and crosstalk and hence simple transceiver design, resulting in a low-power and high-bandwidth connector solution that scales with the PCB manufacturing process.

With the myriad of benefits offered by TCI and TLC, they are expected to open up possibilities for performance improvement in various applications. For instance, although the concept of a modular smartphone design has not taken root, the LEGO-like modularization enabled by TLC should open up possibilities for system designers to create more customized products in different application areas using the same set of building block modules, driving rapid and further proliferation of IC and electronics in different facets of our life.

References

[1] T. Kuroda. "Semiconductor industry in 2025," Panel Discussion (presented but not published). *2010 IEEE International Solid-State Circuits Conference,* Feb. 2010. Updated Nov. 2019.

[2] T. Kuroda and T. Sakurai. (1995, Apr.). "Overview of low-power ULSI circuit techniques." *IEICE Transactions on Electronics. E78-C*(4), pp. 334–344.

[3] T. Kuroda. "Will SOI ever become a mainstream technology?" Panel Discussion (presented but not published), *2002 IEEE International Electron Devices Meeting,* Dec. 2002.

[4] TE Connectivity Ltd. Retrieved on Jun. 11, 2019. Available: www.te.com/usa-en/products/brands/amp.html?tab=pgp-story

[5] SMK Corporation, "電線対基板圧着コネクタの動向 [Crimped connector trends]," in Japanese. Retrieved on Jun. 11, 2019. Available: www.smk.co.jp/products/connectors/technology/1002dempaCSbtow/

[6] Process technology history – Intel. *WikiChip*. Retrieved on Jun. 11, 2019. Available: https://en.wikichip.org/wiki/intel/process

[7] Japan Electronics and Information Technology Industries Association (JEITA). (2017, Mar.). "2026 年までの電子部品技術ロードマップ [Roadmap of electronic component technology up to year 2026]." In Japanese. Retrieved on Jun. 11, 2019. Available: www.jeita.or.jp/japanese/assets/pdf/letter/vol21/15.pdf

[8] T. Kuroda. (1999, Apr.). "ディープサブミクロン時代の半導体集積回路の技術課題とEDAへの期待 [Technological challenges of IC in the deep submicron age and expectations for EDA]." *情報処理学会論文誌 [IPSJ Journal]*, *40*(4), pp. 1500–1506.

[9] T. Kuroda. (2007, Nov.). "システムLSIの低電力技術 [Low-power technology for system LSI]." *電子情報通信学会誌 [Journal of Institute of Electronics, Information and Communication Engineers]*, *90*(11), pp. 977–981.

[10] K. Suzuki, S. Mita, T. Fujita, F. Yamane, F. Sano, A. Chiba, Y. Watanabe, K. Matsuda, T. Maeda, and T. Kuroda. "A 300MIPS/W RISC core processor with variable supply-voltage scheme in variable threshold-voltage CMOS." *1997 IEEE Custom Integrated Circuits Conference*, pp. 587–590, May 1997.

[11] T. Kuroda. "Optimization and control of VDD and VTH for low-power, high-speed CMOS design." *2002 IEEE/ACM* International Conference *on Computer*-Aided Design, pp. 28–34, 2002.

[12] T. Kuroda, T. Fujita, S. Mita, T. Nagamatsu, S. Yoshioka, K. Suzuki, F. Sano, M. Norishima, M. Murota, M. Kako, M. Kinugawa, M. Kakumu, and T. Sakurai. (1996, Nov.). "A 0.9V 150MHz 10mW 4mm2 2-D discrete cosine transform core processor with variable-threshold-voltage scheme." *IEEE Journal of Solid-State Circuits. 31*(11), pp. 1770–1779.

[13] M. Takahashi, M. Hamada, T. Nishikawa, H. Arakida, T. Fujita, F. Hatori, S. Mita, K. Suzuki, A. Chiba, T. Terazawa, F. Sano, Y. Watanabe, K. Usami, M. Igarashi, T. Ishikawa, M. Kanazawa, T. Kuroda, and T. Furuyama. (1998, Nov.). "A 60-mW MPEG4 video codec using clustered voltage scaling with variable supply-voltage scheme." *IEEE Journal of Solid-State Circuits. 33*(11), pp. 1772–1780.

[14] H. Fuketa, T. Yasufuku, S. Iida, M. Takamiya, M. Nomura, H. Shinohara, and T. Sakurai. "Device-circuit interactions in extremely low voltage CMOS designs (invited)." *2011 IEEE International Electron Devices Meeting*, pp. 559–562, Dec. 2011.

[15] V. De. "Energy efficient computing in nanoscale CMOS: Challenges and opportunities (plenary)," *2014 IEEE Asian Solid-State Circuits Conference*, pp. 5–8, Nov. 2014.

[16] T. Ishigaki, R. Tsuchiya, Y. Morita, N. Sugii, and S. Kimura. (2010). "Ultralow-power LSI technology with silicon on thin buried oxide (SOTB) CMOSFET," in *Solid State Circuits Technology*, J. W. Swart, Ed. Croatia: INTECH, 2010, ch. 7, pp. 145–156.

[17] R. O. Topaloglu and H.-S. P. Wong, Eds. (2015). *Beyond-CMOS Technologies for Next Generation Computer Design*. Cham, Switzerland: Springer.

[18] K. Fant. (2005). *Logically Determined Design: Clockless System Design with NULL Convention Logic*. Hoboken: Wiley-Interscience.

[19] T. Maruyama, M. Hamada, and T. Kuorda. (2018, Aug.). "Comparative performance analysis of dual-rail domino logic and CMOS logic under near-threshold operation," *IEEE International Midwest Symposium on Circuits and Systems*, pp. 25–28.

[20] J. D. Rowley. (2018, Mar.). "Venture funding into AI and machine learning levels off as tech matures." Retrieved on Jun. 12, 2019. Available: news.crunchbase.com/news/venture-funding-ai-machine-learning-levels-off-tech-matures/.

[21] M. Tyson. (2018, Jun.). "Intel 10nm density is 2.7× improved over its 14nm node." Retrieved on Jun. 12, 2019. Available: https://hexus.net/tech/news/cpu/119699-intel-10nm-density-27x-improved-14nm-node

[22] J. Hruska. (2018, Jun.). "As chip design costs skyrocket, 3nm process node is in jeopardy." Retrieved on Jun. 12, 2019. Available: www.extremetech.com/computing/272096-3nm-process-node

[23] D. Manners. (2014, Apr.). "EUV cost is $14bn and counting." Retrieved on Jun. 12, 2019. Available: www.electronicsweekly.com/news/business/finance/euv-cost-14bn-counting-2014-04/

[24] S. Moore. (2018, Jan.). "EUV lithography finally ready for chip manufacturing." Retrieved on Jun. 12, 2019. Available: https://spectrum.ieee.org/semiconductors/nanotechnology/euv-lithography-finally-ready-for-chip-manufacturing

[25] H. Aochi, R. Katsumata, and Y. Fukuzumi. (2011). "BiCS flash memory for realization of ultrahigh-density nonvolatile storage devices." In Japanese. *Toshiba Review. 66*(9), pp. 16–19.

[26] (2005, Oct.). "LSIは平面から立体へ チップを貫く伝送路で実現 [Shift from 2D to 3D LSI integration realized using through-chip interconnects]." In Japanese. *Nikkei Electronics,* (2005/10/10) pp. 82–91.

[27] Y. Yano, T. Sugiyama, S. Ishihara, Y. Fukui, H. Juso, K. Miyata, Y. Sota, and K. Fujita. "Three-dimensional very thin stacked packaging technology for SiP." *52nd Electronic Components and Technology Conference,* pp. 1329–1334, May 2002.

[28] K. Matsudera and K. Kawasaki. (2016). "World's first 16-die stacked NAND flash memory package fabricated using TSV technology." *Toshiba Review. 71*(6), pp. 20–23.

[29] W. Worwag and T. Dory. "Copper via plating in three dimensional interconnects." *2007 Electronic Components and Technology Conference,* pp. 842–846, May 2007.

[30] D. James. (2014). "3D ICs in the real world." *25th Annual SEMI Advanced Semiconductor Manufacturing Conference,* pp. 113–119.

[31] T. Ezaki, K. Kondo, H. Ozaki, N. Sasaki, H. Yonernura, M. Kitano, S. Tanaka, and T. Hirayarna. "A 160Gb/s interface design configuration for multichip LSI." *2004 IEEE International Solid-State Circuits Conference*, pp. 140–141, Feb. 2004.

[32] J. Burns, L. McIlrath, C. Keast, C. Lewis, A. Loomis, K. Warner, and P. Wyatt. "Three-dimensional integrated circuits for low-power, high-bandwidth systems on a chip." *2001 IEEE International Solid-State Circuits Conference*, pp. 268–269, Feb. 2001.

[33] K. Kanda, D. D. Antono, K. Ishida, H. Kawaguchi, T. Kuroda, and T. Sakurai. "1.27Gb/s/pin 3mW/pin wireless superconnect (WSC) interface scheme." *2003 IEEE International Solid-State Circuits Conference*, pp. 186–187, Feb. 2003.

[34] D. Mizoguchi, Y. B. Yusof, N. Miura, T. Sakura, and T. Kuroda. "A 1.2Gb/s/pin wireless superconnect based on inductive inter-chip signaling (IIS)." *2004 IEEE International Solid-State Circuits Conference*, pp. 142–143, Feb. 2004.

[35] K. Kumagai, C. Yang, H. Izumino, N. Narita, K. Shinjo, S. Iwashita, Y. Nakaoka, T. Kawamura, H. Komabashiri, T. Minato, A. Ambo, T. Suzuki, Z. Liu, Y. Song, S. Goto, T. Ikenaga, Y. Mabuchi, and K. Yoshida. "System-in-silicon architecture and its application to H.264/AVC motion estimation for 1080HDTV." *2006 IEEE International Solid-State Circuits Conference*, pp. 430–431, Feb. 2006.

[36] D. Hopkins, A. Chow, R. Bosnyak, B. Coates, J. Ebergen, S. Fairbanks, J. Gainsley, R. Ho, J. Lexau, F. Liu, T. Ono, J. Schauer, I. Sutherland, and R. Drost. "Circuit techniques to enable 430Gb/s/mm2 proximity communication." *2007 IEEE International Solid-State Circuits Conference*, pp. 368–369, Feb. 2007.

[37] Q. Gu, Z. Xu, J. Ko, and M.-C. F. Chang. "Two 10Gb/s/pin low-power interconnect methods for 3D ICs." *2007 IEEE International Solid-State Circuits Conference*, pp. 448–449, Feb. 2007.

[38] N. Miura, D. Mizoguchi, M. Inoue, H. Tsuji, T. Sakurai, and T. Kuroda. "A 195Gb/s 1.2W 3D-stacked inductive inter-chip wireless superconnect with transmit power control scheme." *2005 IEEE International Solid-State Circuits Conference*, pp. 264–265, Feb. 2005.

[39] N. Miura, D. Mizoguchi, M. Inoue, K. Niitsu, Y. Nakagawa, M. Tago, M. Fukaishi, T. Sakurai, and T. Kuroda. "A 1Tb/s 3W inductive-coupling transceiver for inter-chip clock and data link." *2006 IEEE International Solid-State Circuits Conference*, pp. 424–425, Feb. 2006.

[40] N. Miura, H. Ishikuro, T. Sakurai, and T. Kuroda. "A 0.14pJ/b inductive-coupling inter-chip data transceiver with digitally-controlled precise pulse shaping." *2007 IEEE International Solid-State Circuits Conference*, pp. 358–359, Feb. 2007.

[41] N. Miura, Y. Kohama, Y. Sugimori, H. Ishikuro, T. Sakurai, and T. Kuroda. "An 11Gb/s inductive-coupling link with burst transmission." *2008 IEEE International Solid-State Circuits Conference*, pp. 298–299, Feb. 2008.

[42] S. Denda. (2015.). 半導体の高次元化技術 *[Enabling Technology for Higher Dimensional Semiconductors], in Japanese.* Japan: Tokyo Denki University Press.

[43] J. Kim. "The future of graphic and mobile memory for new applications." *2016 IEEE Hot Chips 28 Symposium,* Aug. 2016.

[44] A. P. Karmarkar, X. Xu, and V. Moroz. "Performance and reliability analysis of 3D-integration structures employing Through Silicon Via (TSV)." *IEEE International Reliability Physics Symposium,* pp. 682–687, Apr. 2009.

[45] S. K. Samal, D. Nayak, M. Ichihashi, S. Banna, and S. K. Lim. "Monolithic 3D IC vs. TSV-based 3D IC in 14nm FinFET technology." *IEEE SOI-3D-Subthreshold Microelectronics Technology Unified Conference,* pp. 1–2, 2016.

[46] I. Bolsens. (2011). "2.5D ICs: Just a stepping stone or a long term alternative to 3D." Retrieved Aug. 8, 2019. Available: www.xilinx.com/publications/about/3-D_Architectures.pdf.

[47] L. Madden, E. Wu, N. Kim, B. Banijamali, K. Abugharbieh, S. Ramalingam, and X. Wu. "Advancing high performance heterogeneous integration through die stacking." *2012 European Solid-State Circuit Conference,* pp. 18–24, Sep. 2012.

[48] W. C. Chen, C. Hu, K. C. Ting, V. Wei, T. H. Yu, S. Y. Huang, V.C.Y. Chang, C. T. Wang, S. Y. Hou, C. H. Wu, and D. Yu. "Wafer level integration of an advanced logic-memory system through 2nd generation CoWoS® technology." *2017 Symposium on VLSI Technology*, pp. T54–T55, Jun. 2017.

[49] N. Miura and T. Kuroda. (2008). 3次元実装のための低電力・広帯域誘導結合通信 [Low-power and wideband inductive-coupling communication for 3D integration]. In Japanese. *エレクトロニクス実装学会誌 Journal of The Japan Institute of Electronics Packaging. 11*(3), pp 174–181. Available: www.jstage.jst.go.jp/article/jiep1998/11/3/11_3_174/_pdf/-char/ja.

[50] H. Ishikuro and T. Kuroda. (2010, Oct.). "Wireless proximity interfaces with a pulse-based inductive coupling technique." *IEEE Communications Magazine.* 48(10), pp. 192–199.

[51] T. Kuroda. "System integration in a package for cloud and edge." *2017 IEEE Electron Devices Technology and Manufacturing Conference*, pp. 42–43, Feb. 2017.

[52] (2015). International Technology Roadmap for Semiconductors 2.0, 2015 Edition, Heterogeneous Integration. Available: www.dropbox.com/sh/3jfh5fq634b5yqu/AADYT8V2Nj5bX6C5q764kUg4a?dl=0&preview=2_2015+ITRS+2.0+Herogeneous+Integration.pdf.

[53] Y. Take, N. Miura, and T. Kuroda. "A 30 Gb/s/Link 2.2 Tb/s/mm2 inductively-coupled injection-locking CDR for high-speed DRAM interface." *2010 IEEE Asian Solid-State Circuits Conference*, pp. 81–84, Nov. 2010.

[54] N. Miura, T. Shidei, Y. Yuxiang, S. Kawai, K. Takatsu, Y. Kiyota, Y. Asano, T. Kuroda. "A 0.7V 20fJ/bit inductive-coupling data link with dual-coil transmission scheme." *2010 Symposium on VLSI Circuits*, pp. 201–202, Jun. 2010.

[55] M. Saito, N. Miura, and T. Kuroda. "A 2Gb/s 1.8pJ/b/chip inductive-coupling through-chip bus for 128-Die NAND-Flash memory stacking." *2010 IEEE International Solid-State Circuits Conference*, pp. 440–441, Feb. 2010.

[56] M. Saito, Y. Sugimori, Y. Kohama, Y. Yoshida, N. Miura, H. Ishikuro, T. Sakurai, and T. Kuroda. (2010, Jan.). "2 Gb/s 15 pJ/b/chip inductive-coupling programmable bus for NAND Flash memory stacking." *IEEE Journal of Solid-State Circuits. 45*(1), pp. 134–141.

[57] K. Niitsu, Y. Shimazaki, Y. Sugimori, Y. Kohama, K. Kasuga, I. Nonomura, M. Saen, S. Komatsu, K. Osada, N. Irie, T. Hattori, A. Hasegawa, and T. Kuroda "An inductive-coupling link for 3D integration of a 90nm CMOS processor and a 65nm CMOS SRAM." *2009 IEEE International Solid-State Circuits Conference*, pp. 480–481, Feb. 2009.

[58] K. Osada, M. Saen, Y. Okuma, K. Niitsu, Y. Shimazaki, Y. Sugimori, Y. Kohama, K. Kasuga, I. Nonomura, N. Irie, T. Hattori, A. Hasegawa, and T. Kuroda. "3D system integration of processor and multi-stacked SRAMs by using inductive-coupling links." *2009 Symposium on VLSI Circuits*, pp. 256–257, Jun. 2009.

[59] A. Kosuge. "伝送線路結合器を用いた高信頼非接触インタフェース [High reliability contactless interface using transmission line coupler]," in Japanese, Ph.D. dissertation, Department of Electronics and Electrical Engineering, Keio University, Yokohama, Kanagawa Prefecture, Japan, 2016.

[60] M. Meghelli, S. Rylov, J. Bulzacchelli, W. Rhee, A. Rylyakov, H. Ainspan, B. Parker, M. Beakes, A. Chung, T. Beukema, P. Pepeljugoski, L. Shan, Y. Kwark, S. Gowda, and D. Friedman. "A 10Gb/s 5-Tap-DFE/4-Tap-FFE transceiver in 90nm CMOS." *2006 IEEE International Solid-State Circuits Conference*, pp. 213–214, Feb. 2006.

[61] S. Shekhar, J. E. Jaussi, F. O'Mahony, M. Mansuri, and B. Casper. "Design considerations for low-power receiver front-end in high-speed data links." *2013 IEEE Custom Integrated Circuits Conference*, pp. 1–8, Sep. 2013.

[62] T. Musah, J. E. Jaussi, G. Balamurugan, S. Hyvonen, T.-C. Hsueh, G. Keskin, S. Shekhar, J. Kennedy, S. Sen, R. Inti, M. Mansuri, M. Leddige, B. Horine, C. Roberts, R. Mooney, and B. Casper. (2014, Dec.). "A 4-32 Gb/s bidirectional link with 3-Tap FFE/6-Tap DFE and collaborative CDR in 22 nm CMOS." *IEEE Journal of Solid-State Circuits. 49*(12), pp. 3079–3090.

[63] A. Kosuge, J. Kadomoto, and T. Kuroda. (2016, Jun.). "A 6 Gb/s 6 pJ/b 5 mm-distance non-contact interface for modular smartphones using two-fold transmission line coupler and high EMC tolerant pulse transceiver." *IEEE Journal of Solid-State Circuits. 51*(6), pp. 1446–1456.

[64] A. Kosuge, J. Hashiba, T. Kawajiri, S. Hasegawa, T. Shidei, H. Ishikuro, T. Kuroda, and K. Takeuchi. "Inductively-powered wireless solid-state drive (SSD) system with merged error correction of high-speed non-contact data links and NAND flash memory." *2015 Symposium on VLSI Circuits*, pp. c218–c219, Jun. 2015.

[65] A. Kosuge, S. Ishizuka, M. Taguchi, H. Ishikuro, and T. Kuroda. (2015, Aug.). "Analysis and design of an 8.5-Gb/s/link multi-drop bus using energy-equipartitioned transmission line couplers." *IEEE Transactions on Circuits and Systems. 62*(8), pp. 2122–2131.

[66] N. Miura, M. Saito, M. Taguchi, and T. Kuroda. (2013, Feb.). "A 6nW inductive-coupling wake-up transceiver for reducing standby power of non-contact memory card by 500×." *2013 IEEE International Solid-State Circuits Conference*, pp. 214–215.

[67] W. Mizuhara, T. Shidei, A. Kosuge, T. Takeya, N. Miura, M. Taguchi, H. Ishikuro, and T. Kuroda. "A 0.15mm-thick non-contact connector for MIPI using vertical directional coupler." *2013 IEEE International Solid-State Circuits Conference*, pp. 200–201, Feb. 2013.

[68] A. Kosuge, S. Ishizuka, J. Kadomoto, and T. Kuroda. "A 6Gb/s 6pJ/b 5mm-distance non-contact interface for modular smartphones using two-fold transmission-line coupler and EMC-qualified pulse transceiver." *2015 IEEE International Solid-State Circuits Conference*, pp. 176–177, Feb. 2015.

[69] A. Kosuge, S. Ishizuka, L. Liu, A. Okada, M. Taguchi, H. Ishikuro, and T. Kuroda. "An electromagnetic clip connector for in-vehicle LAN to reduce wire harness weight by 30%." *2014 IEEE International Solid-State Circuits Conference*, pp. 496–497, Feb. 2014.

[70] A. Kosuge, S. Ishizuka, M. Abe, S. Ichikawa, and T. Kuroda. "A 6.5Gb/s shared bus using electromagnetic connectors for downsizing and lightening satellite processor system by 60%." *2015 IEEE International Solid-State Circuits Conference*, pp. 434–435, Feb. 2015.

2 ThruChip Interface
A Wireless Chip Interface

2.1 TCI Fundamentals

As discussed in the previous chapter, TCI is an inductive near-field coupling interconnect technology offering a low-cost alternative to TSV for 3D IC integration and for delivering high interface bandwidth at low power. In this chapter, we first drill down into the design and electrical characteristics of the interface to lay the foundation, then go into discussion of architectural and circuit implementation trade-offs to achieve low power, high area efficiency, high per-channel data rate, and high interface bandwidth in various applications.

2.1.1 Basic Structure and Operating Principle

Figure 2.1 illustrates the basic structure of a TCI interconnect. While the TX and RX chips are shown to be stacked facing up, there is no constraint on whether the two chips are stacked facing the same direction or not. The transmitter circuit drives an inductor coil laid out at the top of the transmitting chip, which is magnetically coupled to an inductor coil at the top of the receiving chip, which in turn drives the receiver circuit. As we learned in our introductory physics class, whenever there is a change in current flowing through an inductor, a voltage is induced in a neighboring inductor through the resultant change in magnetic field linkage between the two.

Quantitatively, the received voltage V_R is related to the change in transmitter current I_T by the following equation:

$$V_R = M\frac{\mathrm{d}I_T}{\mathrm{d}t}, \tag{2.1}$$

where M is the mutual inductance between the two inductor coils. M in turn is related to the inductances of the transmitter and receiver coils, L_T and L_R respectively, by the following equation:

$$M = k\sqrt{L_T L_R}, \tag{2.2}$$

where k is the coupling coefficient determined by how much of the magnetic field from one coil is linked to the other.

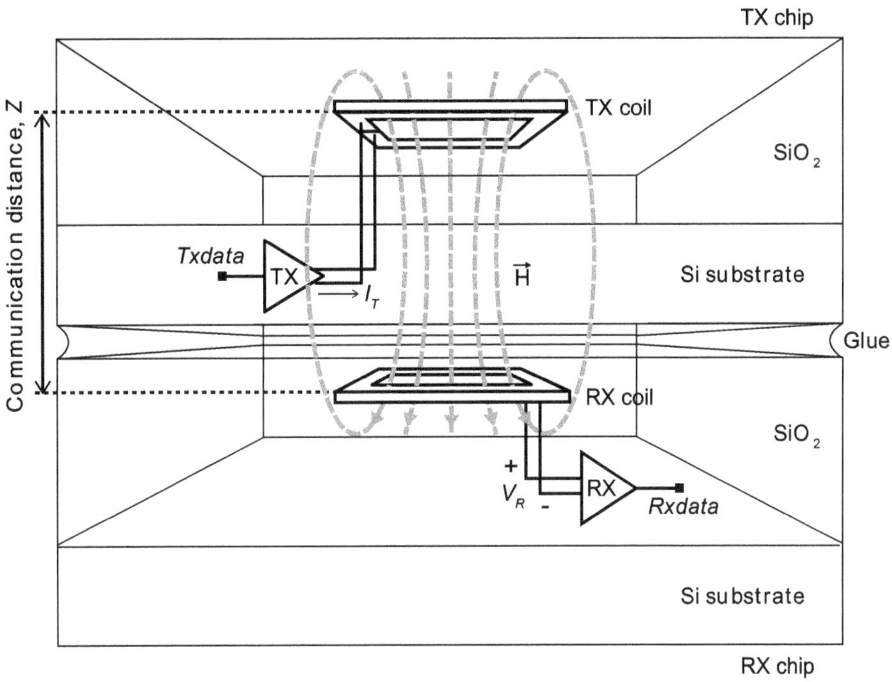

Figure 2.1 Basic structure of a TCI interconnect. Reproduced by permission from Muhannad S. Bakir, James D. Meindl, Integrated Interconnect Technologies for 3D Nanoelectronic Systems, Norwood: Artech House, Inc., 2008. © 2008 of publication by Artech House, Inc

By using *Txdata* to modulate the transmitter current I_T, information can be transmitted in the form of variation in the voltage V_R induced in the receiver coil, which can then be extracted by the receiver to generate *Rxdata*, thereby achieving wireless communication.

2.1.2 Transceiver Design

The TCI transceiver design is determined by how *Txdata* is used to modulate I_T. As previously explained, one merit of near-field coupling is that the transmitted energy is concentrated in a small space. For far-field communication such as in Long-Term Evolution (LTE) or Wi-Fi, to compensate for the high loss due to the spreading of transmitted energy, information is transmitted by modulating amplitude, phase, or frequency of a carrier signal to maintain a high signal-to-noise ratio (SNR). By contrast, since TCI uses near-field coupling over a short communication distance with small transmission loss, sufficient SNR can be achieved by using a simple pulse modulation scheme, thereby eliminating the need for oscillators and filters required for carrier modulation, resulting in a low-power, compact transceiver design.

Table 2.1 lists a few design options for the TCI transmitter and the trade-off in speed, power, and area of each option. The large switching noise of the single-ended

Table 2.1. TCI transmitter design options.

Option	Name	Schematic	Pros	Cons
1	CMOS H-bridge	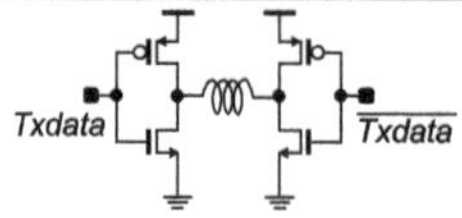	• Faster • Lower switching noise	• Larger power
2	NMOS push-pull	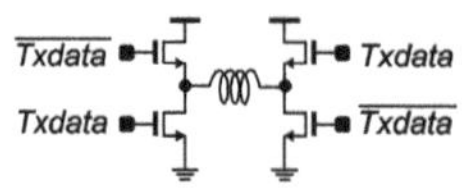	• Lower power • Lower switching noise • NMOS only ⇒ less variations	• Additional V_{DD}
3	Pulse	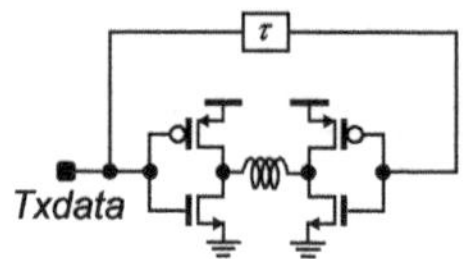	• Lower power	• Slower • Larger coil
4	Single-ended	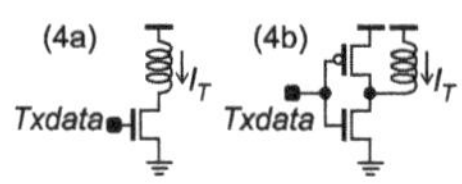	• Lower device count	• Larger switching noise • (4a): Ringing • Larger coil
5	Center-tapped coil and NMOS pseudodifferential driver	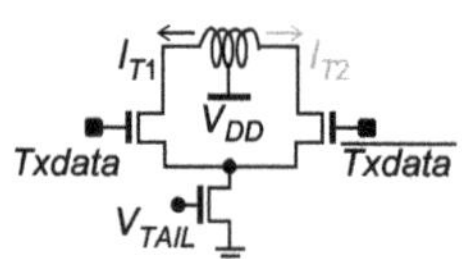	• Lower switching noise • NMOS only ⇒ less variations • V_{DD} < H-bridge (but > NMOS)	• Larger peak current • Very large coil

transmitter and large coil diameter required by the center-tapped coil transmitter are difficult challenges to overcome, making those options impractical. The most commonly used design is the CMOS H-bridge, which is explained in further detail next. NMOS push-pull and pulse transmitters are discussed in Section 2.2 as lower-power alternatives.

2.1.2.1 Standard Transceiver Design

Figure 2.2 depicts the standard TCI transceiver design and its simulated time domain waveforms. The standard transmitter is a CMOS H-bridge which consists of an H-bridge with two CMOS inverters and acts as a voltage-to-current converter. It converts digital data *Txdata* from a differential voltage signal to a current I_T to drive the transmitter coil. When *Txdata* is high, the n-channel metal–oxide–semiconductor (NMOS) on the left and the p-channel metal–oxide–semiconductor (PMOS) on the right turn on to cause a positive current $I_T(t) = +I_P$ to flow through the coil. Conversely, when *Txdata* is low, the NMOS on the right and PMOS on the left turn on to generate a current of the same magnitude flowing in the reverse direction. Hence $I_T(t) = -I_P$. As a result, apart from the polarity reversal during a transition, there is a constant current of magnitude I_P flowing through the transmitter coil irrespective of the state of *Txdata*.

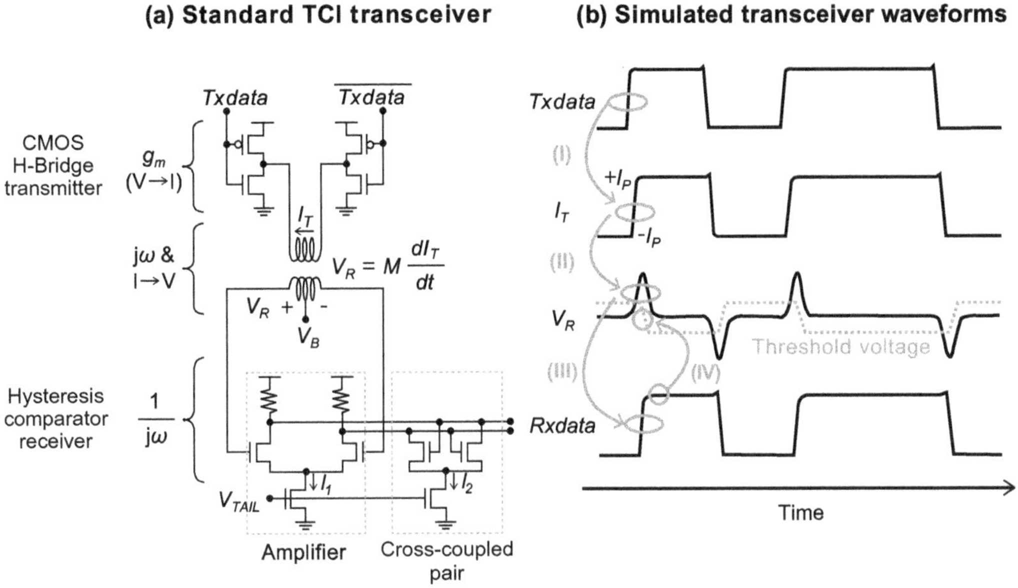

Figure 2.2 Standard transceiver design and simulated waveforms.

The coupled coils, acting as a current-to-voltage converter, cause a voltage V_R to be induced in the receiver coil through first-order differentiation of $I_T(t)$ per equation (2.1).

The receiver circuit performs first-order integration of voltage pulses induced in the receiver coil. The standard receiver is a two-stage hysteresis comparator that consists of a differential amplifier with a resistive load followed by a latch formed from a cross-coupled NMOS pair. The first stage amplifies the low-amplitude voltage pulse to drive the second-stage latch. The cross-coupled pair in the latch then performs an integration to generate *Rxdata*. Due to feedback from the cross-coupled pair, threshold voltage of the inverters in the amplifier is shifted low when *Rxdata* is high, and high when *Rxdata* is low, creating hysteresis to help the comparator hold the value of *Rxdata* against noise disturbances (the dotted line in Figure 2.2b).

Data transmission is therefore achieved through the following sequence of events (Figure 2.2b):

(I) When *Txdata* switches from low to high, the transmitter current changes from $-I_P$ to $+I_P$.

(II) The coupled coils cause a positive voltage pulse to be induced in the receiver coil.

(III) The receiver circuit integrates the positive voltage pulse to drive *Rxdata* to switch high.

(IV) The high *Rxdata* causes the receiver threshold voltage to shift low, which in turn causes the comparator to hold *Rxdata* high against negative noise disturbances until *Txdata* switches from high to low.

2.1.2.2 Transceiver Characteristics

Figure 2.3 depicts the time and frequency domain characteristics of the transmitter current I_T and receiver voltage V_R of the standard TCI transceiver when the coupled coils are ideal inductors with mutual inductance M in between. Since the magnitude of the transmitter current is I_P except during transitions, the transmitter power P_{TX} can be approximated by

$$P_{TX} = V_{DD} I_P \tag{2.3}$$

The receiver voltage pulse can be approximated as a Gaussian pulse in time domain, resulting in a Gaussian frequency spectrum. Specifically,

$$V_R(t) = V_P \mathrm{e}^{-\frac{4t^2}{\tau^2}} \tag{2.4}$$

$$V_R(f) = \frac{\sqrt{\pi}\tau V_P}{2} \mathrm{e}^{-\left(\frac{\pi\tau f}{2}\right)^2}, \tag{2.5}$$

where τ is the pulse width of $V_R(t)$ such that

$$V_R\left(\pm\frac{\tau}{2}\right) = \frac{V_P}{\mathrm{e}}. \tag{2.6}$$

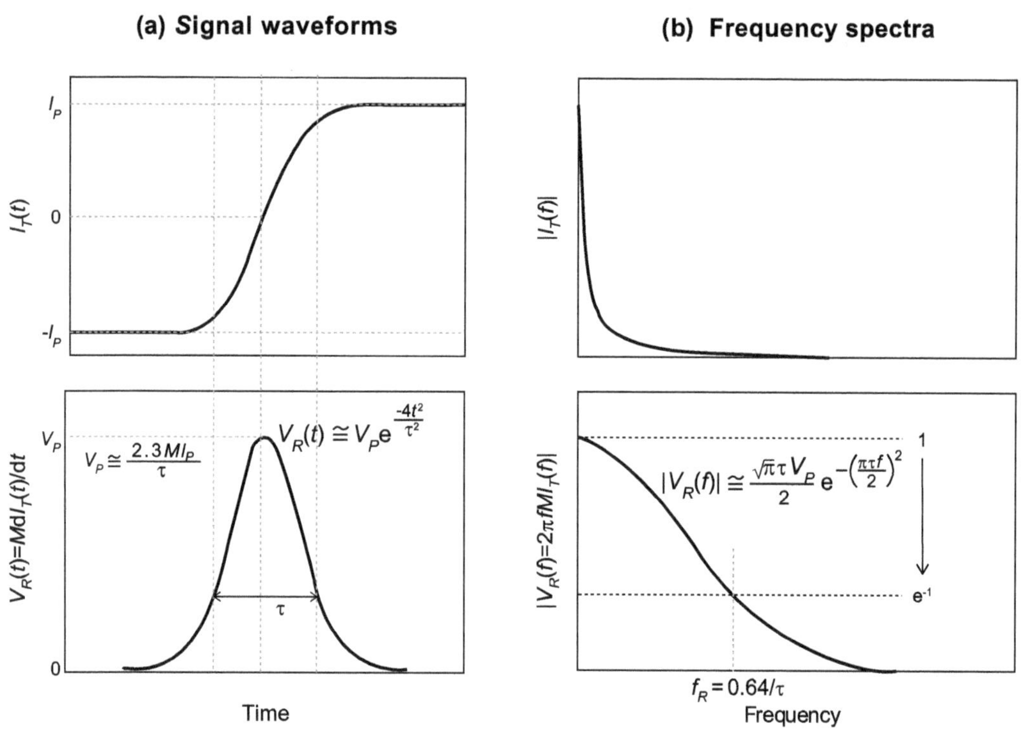

Figure 2.3 Standard transceiver signal time and frequency domain characteristics.

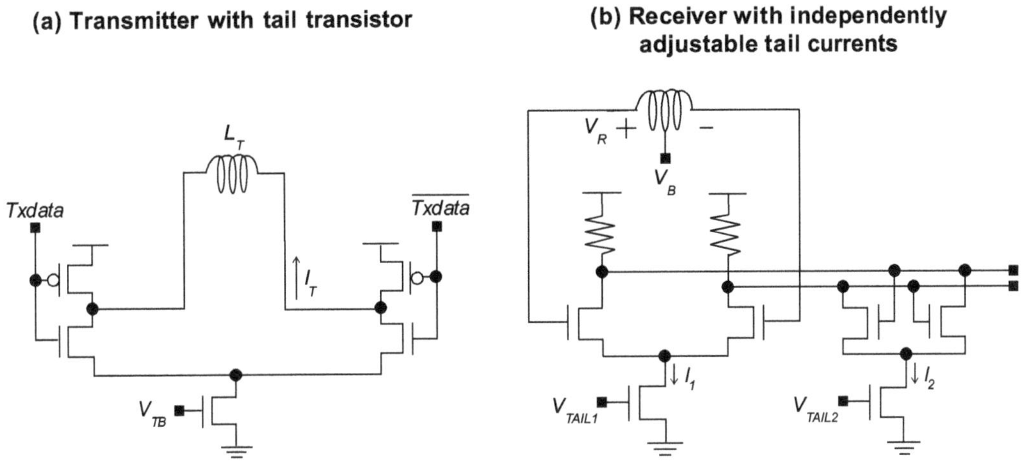

Figure 2.4 Transceiver design with performance tuning circuits.

As a result, the peak voltage V_P is approximately given by

$$V_p = \frac{2.3MI_p}{\tau}. \tag{2.7}$$

The pulse width is in turn inversely proportional to how fast I_T changes. In other words,

$$\tau \propto \frac{1}{\mathrm{d}I_T/\mathrm{d}t}. \tag{2.8}$$

TCI performance tuning can thus be achieved by adjusting the mutual inductance M of the coupled coils, peak current of the transmitter I_P, and the slew rate of the transmitter current $\mathrm{d}I_T/\mathrm{d}t$. By adding a tail transistor to the standard transmitter design as shown in Figure 2.4a, I_P can easily be adjusted by varying the bias voltage V_{TB} applied to its gate.

In the frequency domain, $|V_R(f)|$ decreases with frequency and falls to 1/e of its peak value at a frequency of

$$f_R = \frac{2}{\pi\tau} = \frac{0.64}{\tau}. \tag{2.9}$$

Therefore, most of the frequency content of V_R is at $f \leq f_R$. Consequently, f_R provides a measure of the bandwidth of V_R, and the channel cutoff frequency should be designed to exceed f_R to minimize signal distortion.

On the receiver side, the key performance characteristics are its hysteresis width or threshold voltage V_{TH} and response time $\tau_{TH,min}$. Threshold voltage defines the minimum amplitude of the received pulse required to switch the receiver, while response time is the minimum width of the pulse required at $V_R(t) = \pm V_{TH}$ (Figure 2.5a). For a given receiver design, its response time can be shortened to support a faster data rate by increasing the ratio of V_P/V_{TH} (Figure 2.5b). However,

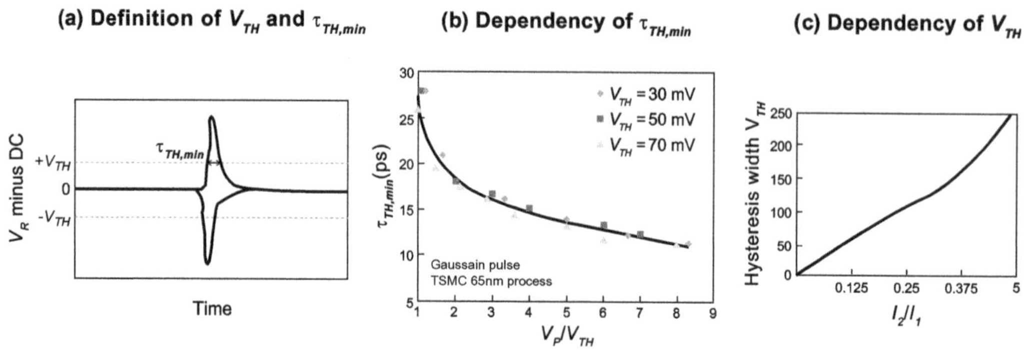

Figure 2.5 Receiver characteristics and their dependencies.

this means either a larger V_P resulting in higher power, or a smaller V_{TH} increasing susceptibility to noise. It thus represents a trade-off of power or noise immunity for speed.

The threshold voltage V_{TH} can be adjusted by varying the ratio of the two tail currents I_2/I_1 (Figure 2.5c). Using the enhanced receiver design in Figure 2.4b, bias voltages to the two tail transistors can be independently varied to adjust I_2/I_1 to optimize V_{TH}.

2.1.3 Coil Design

2.1.3.1 Coil Layout Options

Table 2.2 enumerates several layout options for the TCI coil and the merit of each option. While the words "coil" and "solenoid" are used to describe the layout, they are generally implemented as square loops to create the largest aperture for magnetic flux linkage to maximize mutual inductance for a given area. The one-layer spiral winds within one layer to create multiple turns (each turn being one complete loop) without using any vias, but it prevents other signals from being routed through. The XY spiral winds in two layers, using one exclusively for *x*- and the other exclusively for *y*-direction routing. In other words, it switches layers every time it bends. While this takes up an additional layer, it is generally preferred over the one-layer spiral since it allows other signals to be routed through, as illustrated in Figure 2.6, thus minimizing routing channel penalty.

In contrast to spiral coils, solenoid coils repeat their looping vertically instead of horizontally (one turn per layer). This, however, is not practical since it requires too many layers. More practical layouts combine it with a spiral layout. The double-spiral is a combination of a one-layer spiral and solenoid – it winds inward in one layer, crosses layers, and winds outward in another layer. On the other hand, the XY stack is a combination of an XY spiral and a solenoid – it winds inward in two layers for separate *x*- and *y*-direction routing, crosses layers, and winds outward in another pair of layers for separate *x*- and *y*-direction routing. Both combinations double the number

Table 2.2. TCI coil layout options.

Type	Name	# of Layers	Description	Image	Merit
Spiral	One-layer	1	Coil spirals inward within one layer	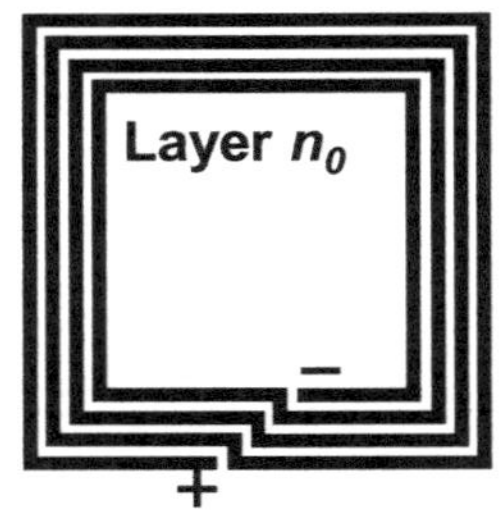	• One layer only • No via
	XY	2	Coil spirals inward using one layer for all *x*-, another for all *y*-direction routing		• Possible routing of other signals through coil to reduce routing channel penalty
Solenoid		≥2	Coil winds across multiple layers, with one turn (loop) per layer	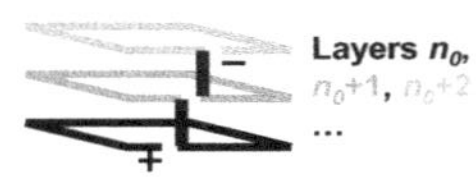	• Smallest area

Table 2.2. *(cont.)*

Type	Name	# of Layers	Description	Image	Merit
Combo	Double-spiral	2	Coil spirals inward in one layer, then spirals outward in the same direction in a second layer	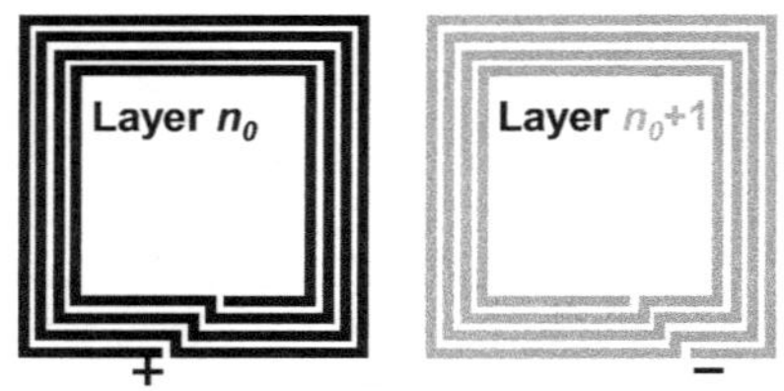	• Twice as many turns for the same area and aperture
	XY stack	4	Coil spirals inward using XY routing in two layers, then spirals outward in the same direction using XY routing in two other layers	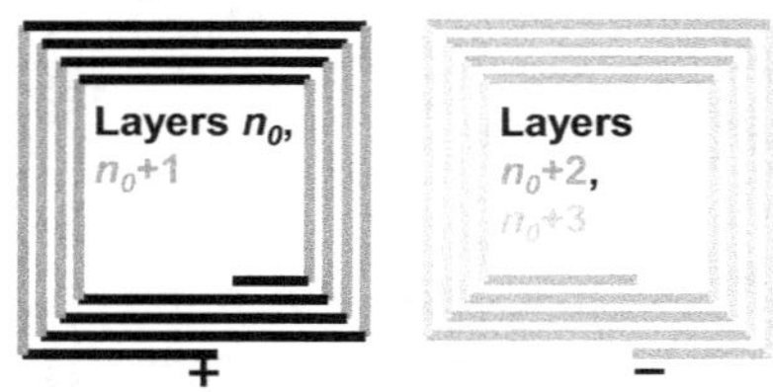	• Twice as many turns for the same area and aperture • Routing of other signals through coil possible

of turns for the same area and coil aperture compared to their spiral counterparts, at the expense of doubling the required number of routing layers.

2.1.3.2 Coil Design Parameters and Electrical Characteristics

Figure 2.7 depicts a generic layout of an array of coupled coils, and together with Table 2.3 defines the key physical design parameters. Each coil consists of a square spiral with n turns in each of m metal layers, where one turn corresponds to one

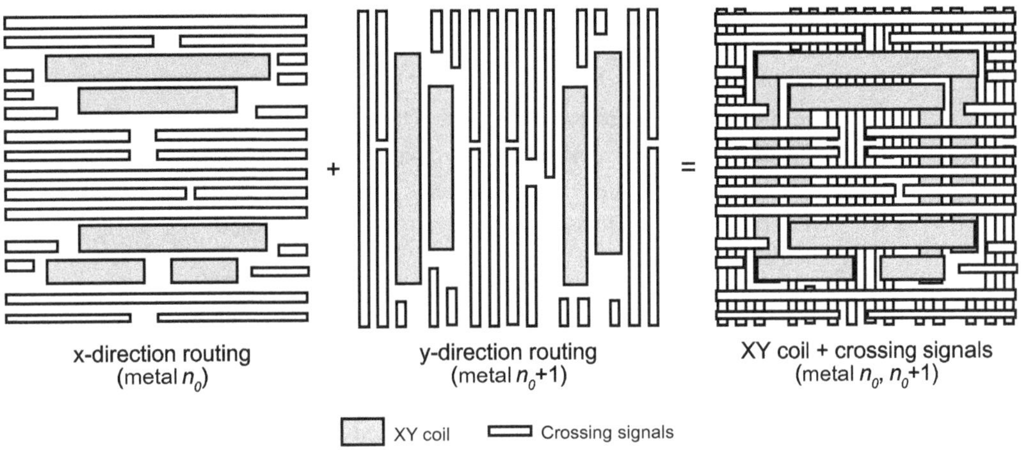

Figure 2.6 Layout of XY spiral coil and crossing signals. © 2009 IEEE. Reprinted, with permission, from [1]

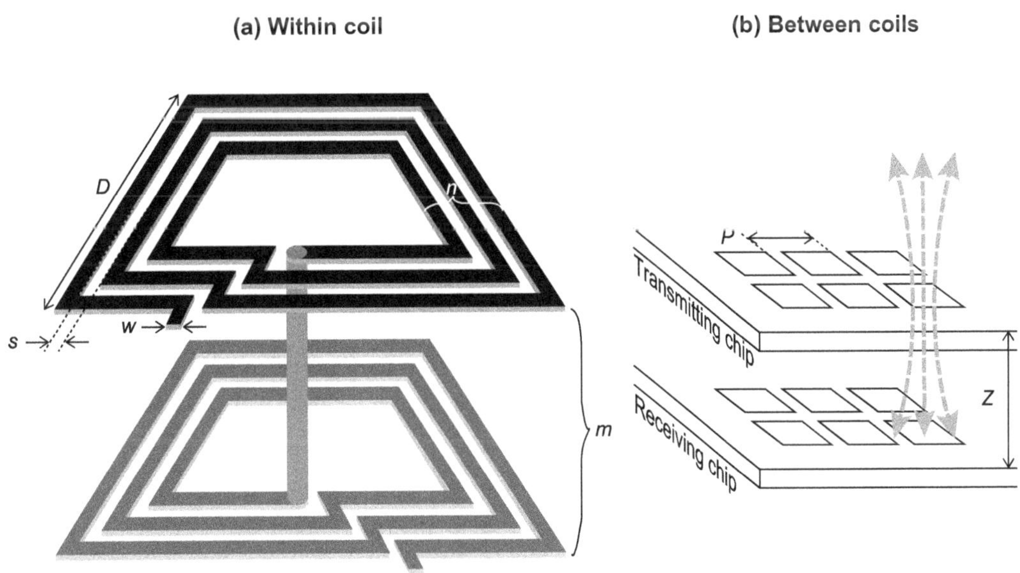

Figure 2.7 Definition of coil physical design parameters.

Table 2.3. Definition of coil physical design parameters.

Parameter		Symbol
Within coil	Wire width	w
	Wire spacing	s
	Diameter	D
	# of turns/loops per layer	n
	# of layers	m
Between coils	Communication distance	Z
	Array pitch	P

complete loop. The two coils are separated by a distance of Z, which defines the communication distance. In an array implementation, the coil layout is repeated with a pitch of P. Note that although the coil layout is generally a square loop, for simplicity, the outermost length of its side is referred to as its diameter D, as if the coil were circular.

Because the input impedance of the first-stage amplifier in the receiver is large, the current in the receiver coil is negligible, and hence any voltage induced back in the transmitter coil from the receiver can be ignored. However, parasitic (unintended) resistance and capacitance of the metal lines have significant effect on the electrical characteristics of the coils. Therefore, the coupled coils are modeled with an RLC network as shown in Figure 2.8a instead of two ideal inductors. Consequently, the transimpedance of the channel relating V_R to I_T is the product of three frequency responses: (1) second-order low-pass filter (LPF) response of the transmitter coil; (2) frequency response of the mutual inductance M; and (3) LPF of the receiver coil. In other words,

$$\frac{V_R}{I_T}(\omega) = \frac{j\omega M}{\left[(1-\omega^2 L_T C_T) + j\omega R_T C_T\right]\left[((1-\omega^2 L_R C_R) + j\omega R_R C_R)\right]}. \tag{2.10}$$

The LPF of a coil, the frequency response of mutual inductance M, and the transimpedance of the coupled coils are plotted in Figure 2.8b. The LPF response peaks at its LC self-resonance frequency f_{SR} and falls off rapidly afterward, where

$$f_{SR} = \frac{1}{2\pi\sqrt{LC}}. \tag{2.11}$$

As a result, the channel must be designed such that f_{SR} is significantly larger than f_R, the bandwidth of V_R, to minimize signal distortion. In practice, the channel is designed to achieve

$$f_{SR} \geq 1.5 f_R. \tag{2.12}$$

Equation (2.12) ensures that most of the frequency content of V_R is at a frequency lower than the resonance frequency of the channel. Nevertheless, V_R may still have some, albeit small, amount of energy around f_{SR}, which is amplified by channel resonance. This will manifest itself as high-frequency ringing in the transmitted signal.

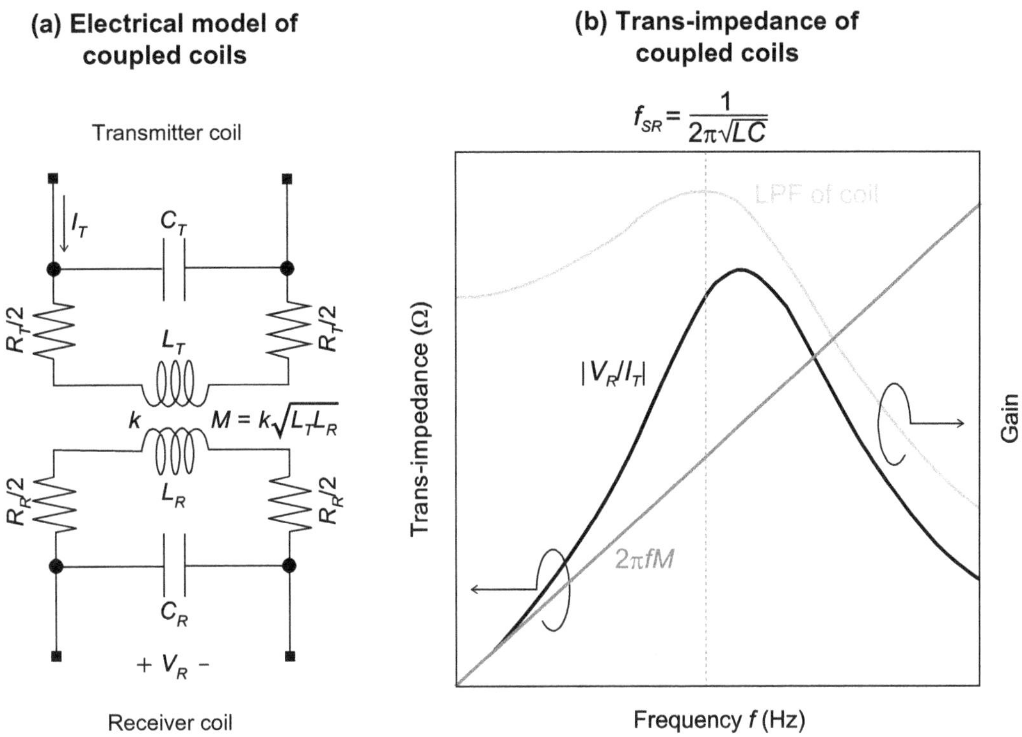

Figure 2.8 Electrical model and transimpedance of the coupled coils.

The amplitude of the ringing is determined by the sharpness of the resonance peak in the frequency response of the coil, which is quantified by its Q-factor, where

$$Q = \frac{\sqrt{L/C}}{R}. \tag{2.13}$$

A large Q indicates sharp resonance, resulting in more ringing (Figure 2.9b). The coils must therefore be designed to have a small Q to suppress ringing, which requires a large R. On the other hand, a large R results in large resistive loss, reducing the amplitude of the transmitted signal (Figure 2.9a). It is therefore necessary to choose an optimal Q to balance between large loss and large ringing. Typically, Q is chosen to be (as shown in Figure 2.9c)

$$1 \leq Q \leq 2.5. \tag{2.14}$$

On the transmitter side, since the on-resistance of a PMOS and an NMOS are on the current path, they provide additional damping to suppress ringing. Meanwhile, the large current in the transmitter coil requires a low resistance to minimize IR drop. As a result, the transmitter coil can be designed with a larger Q than (2.14) to reduce IR drop without causing excessive ringing. It is therefore common to implement the transmitter coil with a wide trace and the receiver coil a narrow trace.

Table 2.4. Conversion of physical to electrical design parameters.

Electrical parameter	Dependency on Physical Parameters		Remark
Coil inductance	$L = \frac{L_0 D^{4/3} n^{5/3}}{(w+s)^{1/3}}$	(2.15)	L_0 = per unit length inductance of metal trace
Coil capacitance	$C = \left(C_G w - \frac{C_c}{s}\right) Dn$	(2.16)	C_G = per unit area capacitance between GND and metal trace C_C = per unit area capacitance between traces
Coil resistance	$R = \frac{4R_0 Dn}{w}$	(2.17)	R_0 = sheet resistance of metal trace
Coil pair coupling coefficient	$k = \begin{cases} -\left(\frac{Z}{D}\right)^3 + 0.205 \; for \frac{D}{Z} < 2 \\ 0.065\frac{D}{Z} - 0.05 \; for 2 \le \frac{D}{Z} \le 3 \end{cases}$	(2.18)	k is reduced by ~ 20% if there is silicon substrate and power/signal mesh between the coil pair

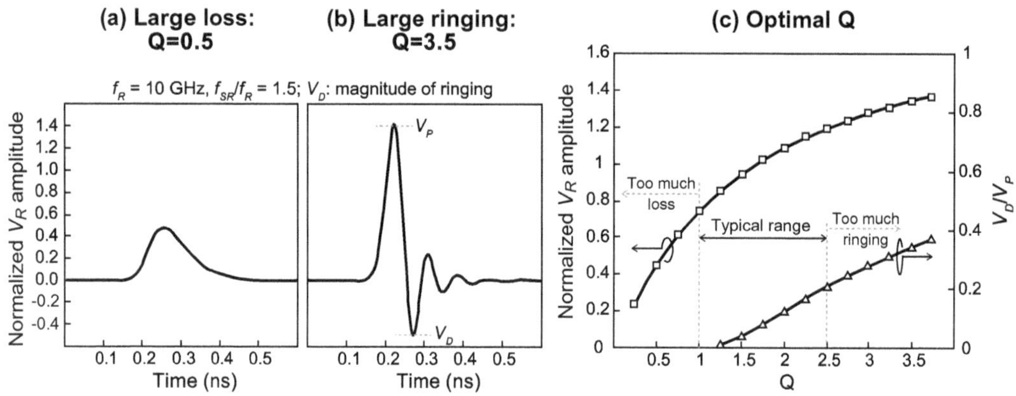

Figure 2.9 Optimization of the Q-factor of coil self-resonance.

Table 2.4 lists approximate closed-form equations for computing the electrical model parameters of spiral (one-layer and XY) coils from its physical design parameters. While they are not accurate enough for design verification, they are useful for creating the initial coil design, and for providing insight into how electrical characteristics can be tuned by adjusting the coil physical design parameters. The latter is especially helpful in understanding trade-offs during design optimization.

The most important coil design parameter is its diameter D, since every electrical parameter depends on it. In particular, the coupling coefficient k, which determines how strongly the coils are coupled and hence the communication efficiency of the channel, depends only on D and Z, and no other design parameters. The communication distance Z is defined by the thickness of the stacked chips, and therefore cannot be varied arbitrarily. Consequently, D is the most important knob for design optimization.

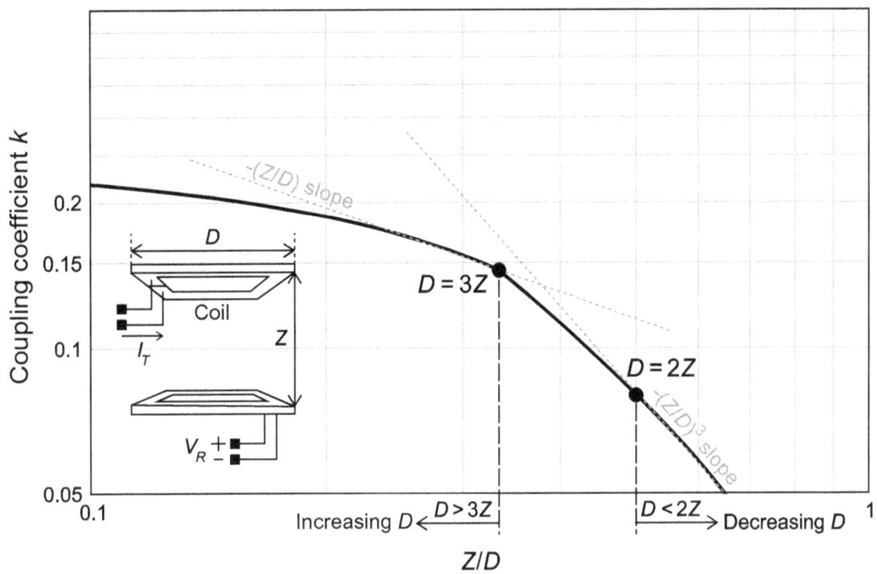

Figure 2.10 Coupling coefficient k as a function of coil diameter D.

Figure 2.10 depicts the dependency of coupling coefficient k on coil diameter D. Two important boundaries are $D = 3Z$ and $D = 2Z$. In general, a larger D is desirable to achieve better coupling, but that is at the expense of larger silicon area and hence lower area efficiency. When $D > 3Z$, k increases less than linearly as D is increased. Therefore, increasing D beyond $3Z$ results in a diminishing return on k while lowering area efficiency. When $D < 2Z$, k falls off as a function of $(Z/D)^3$ as D decreases, rapidly making the channel ineffective. As a result, D is generally selected to be between $2Z$ and $3Z$ to achieve the best balance between area and coupling efficiencies. Nevertheless, $D < 2Z$ is sometimes used to achieve aggressive area efficiency, in which case compensating measures such as higher transmitter power or higher receiver sensitivity may be necessary.

2.1.4 Crosstalk between Channels

To maximize interface bandwidth in a given area, parallel channels should be placed at the minimum pitch possible, while keeping crosstalk within acceptable levels. To understand how crosstalk varies with pitch, crosstalk from the aggressor array in Figure 2.11c on a transmitter chip to a victim coil on a receiver chip is computed. Figure 2.11a plots the crosstalk from one aggressor as a function of lateral offset Y from the victim. It reveals that crosstalk falls off rapidly as a function of $1/Y^3$. As the array size grows, the number of aggressors grows only in a quadratic fashion. Therefore, crosstalk contribution from the additional aggressors diminishes. Consequently, as illustrated in Figure 2.11b, which plots the aggregate crosstalk from

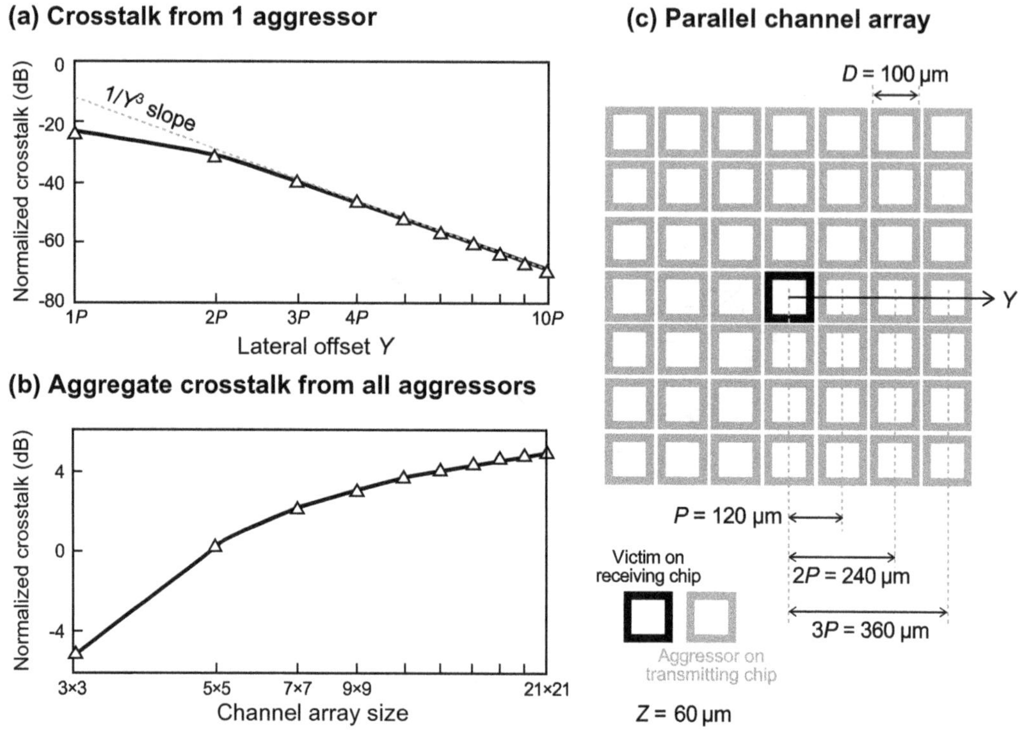

Figure 2.11 Computed crosstalk as a function of horizontal offset and array size. © 2007 IEEE. Reprinted, with permission, from [2]

all aggressors as a function of transmitter coil array size, aggregate crosstalk gradually levels off as array size increases. This is the merit of operating in the near-field region, as previously discussed in Section 1.4.1. As a result, TCI can be used to implement a highly parallel chip interface with hundreds and thousands of channels while keeping aggregate crosstalk bounded.

Figure 2.12 plots aggregate crosstalk as a function of normalized pitch (*P/D*) for a one- and two-dimensional array respectively. To suppress aggregate crosstalk to between –15 and –25 dB, the recommended pitch is $P_{Line} = 1{\sim}2D$ for a one-dimensional line array, and $P_{Array} = 2{\sim}3D$ for a two-dimensional area array.

A straightforward approach to reducing crosstalk is to limit the number of simultaneous switching aggressors. A systematic solution is phase-division multiplexing (PDM), where switching of adjacent channels are multiplexed in time. In a two-phase PDM (2PDM) scheme, the parallel channels are divided into two groups, with the first group consisting of channels in a checkerboard pattern, and the second group the complementary channels. By switching the first group on the rising edge, at phase $\emptyset$, and the second group the falling edge of the clock, at phase $\bar{\emptyset}$, the number of simultaneous switching aggressors is reduced by half (Figure 2.13b). Figure 2.13a plots the normalized aggregate crosstalk as a function of pitch for two different values

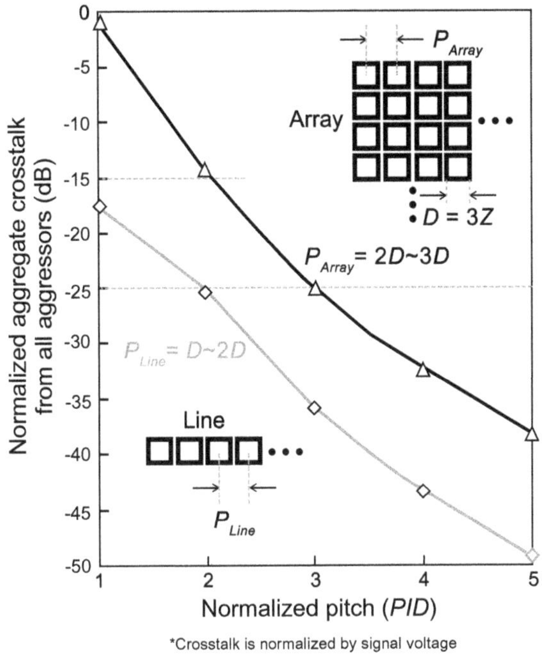

Figure 2.12 Aggregate crosstalk vs. pitch of a one- and two-dimensional array.

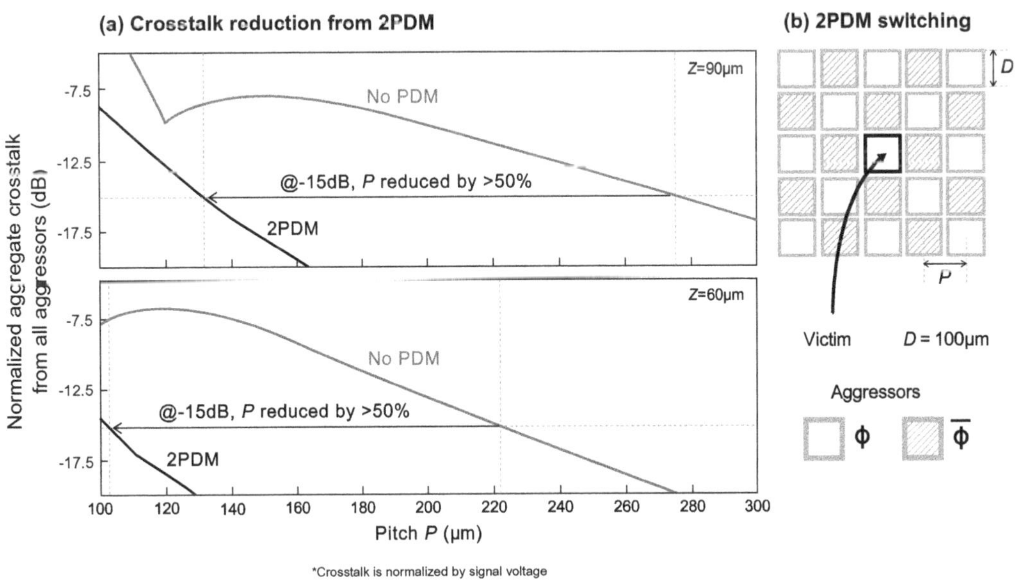

Figure 2.13 Aggregate crosstalk as a function of pitch and PDM. © 2007 IEEE. Reprinted, with permission, from [2]

of Z, evaluating the improvement of such a 2PDM scheme. In both cases, for the same aggregate crosstalk of −15 dB, the pitch is more than halved when 2PDM is applied.

With 2PDM, it takes twice the time for all the channels to switch. The interface bandwidth is therefore halved. However, since the pitch can be halved, in an area array the number of channels per unit area is quadrupled. Therefore, the net result is a twofold increase in bandwidth per unit area, or a doubling of area efficiency.

2.1.5 Electromagnetic Compatibility

Since TCI is a wireless technology, it is necessary to ensure that there are no electromagnetic compatibility (EMC) issues, whether it is electromagnetic interference (EMI) from TCI, or electromagnetic susceptibility (EMS) of TCI.

A coil placed concentric to a TCI coil with a vertical offset is expected to be most susceptible to magnetic interference from the TCI coil. Figure 2.14 shows the computed noise voltage induced by the changing magnetic field of a TCI coil on a concentric coil of the same size underneath, as a function of the vertical offset Z_{EMI}. The noise voltage can be reduced to less than 1 mV when the separation is more than three times the coil diameter ($Z_{EMI} \geq 180$ μm). Furthermore, the plot reveals that the noise voltage falls off quickly with separation as a function of $1/Z_{EMI}^3$. As a result, beyond a separation that is on the order of 1 mm the noise voltage becomes less than the noise floor of a low-noise amplifier (LNA). Therefore, EMI from TCI outside the chip package can be safely ignored. Incidentally, this means that eavesdropping from outside the package is difficult.

Another structure susceptible to interference from a TCI coil is a signal line running in parallel to a coil edge. Figure 2.15a shows the computed mutual inductance

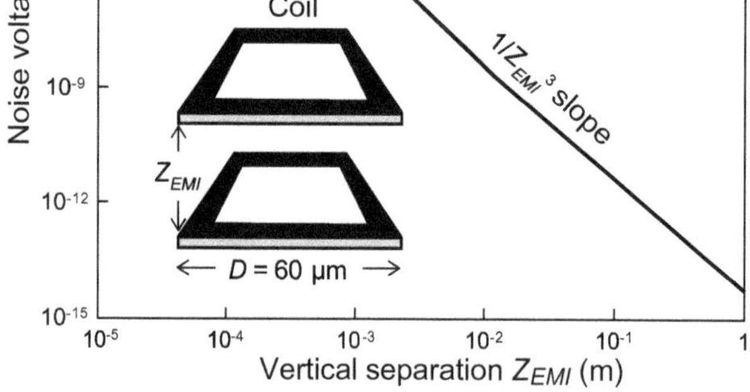

Figure 2.14 Noise induced by the TCI coil in another coil. © 2007 Noriyuki Miura [3]

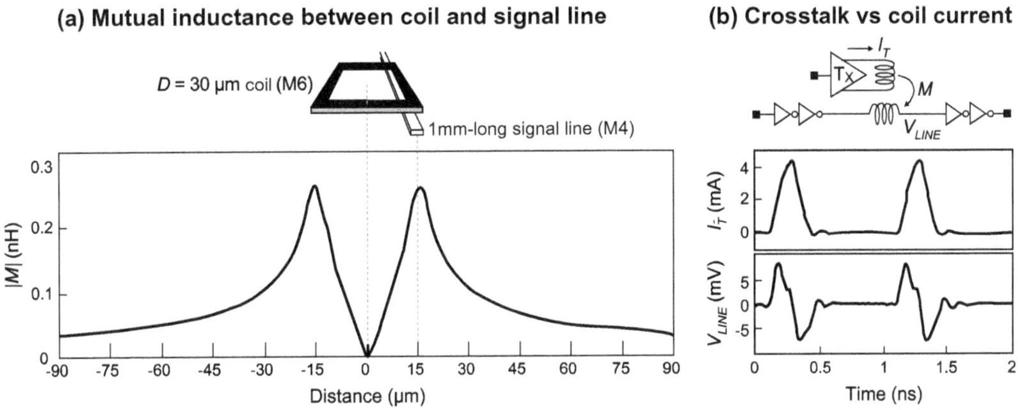

Figure 2.15 Noise induced by TCI coil in a signal line. Reproduced by permission from Muhannad S. Bakir and James D. Meindl, Integrated Interconnect Technologies for 3D Nanoelectronic Systems, Norwood, MA: Artech House, Inc., 2008. © 2008 of publication by Artech House, Inc

between a TCI coil and a signal line in a neighboring layer, as a function of horizontal offset from the center of the coil. The mutual inductance reaches its maximum as the signal line moves into the shadow of a coil edge ($Y = \pm 15$ μm). Figure 2.15b shows the noise voltage induced by the changing magnetic field of the TCI coil on the signal line when the mutual inductance peaks. While this ± 10 mV noise can be ignored by a digital signal with amplitude on the order of 1 V, it can interfere with small amplitude analog signals. Therefore, sensitive analog circuits should be placed some distance away. The noise voltage can be reduced to less than 1 mV when the distance is more than twice the coil diameter.

While more mass production–level validation is necessary, results from [4] suggest that TCI interference of SRAM circuits is negligible. Preliminary testing also showed that TCI coils placed over DRAM memory cells cause no appreciable degradation to DRAM operation. Therefore, while sensitive analog circuits should be placed at a horizontal offset of more than two times the coil diameter, digital circuits and potentially memory circuits can be placed anywhere with no keep-out from TCI coils.

In evaluating electromagnetic susceptibility of TCI, the maximum acceptable magnetic field strength defined in the Radio Radiation Protection Guidelines from Japan's Ministry of Internal Affairs and Communication, B_{EMI}, is used to define the aggressor signal [5]. Figure 2.16 shows the computed noise voltage induced by B_{EMI} penetrating through a TCI coil as a function of frequency. Even at its peak, the induced voltage is only on the order of 0.1 mV and can thus be safely ignored.

2.1.6 Closing Thoughts

TCI achieves near-field, wireless communication through magnetic coupling between the transmitter and receiver coils. The simple operating principle and short

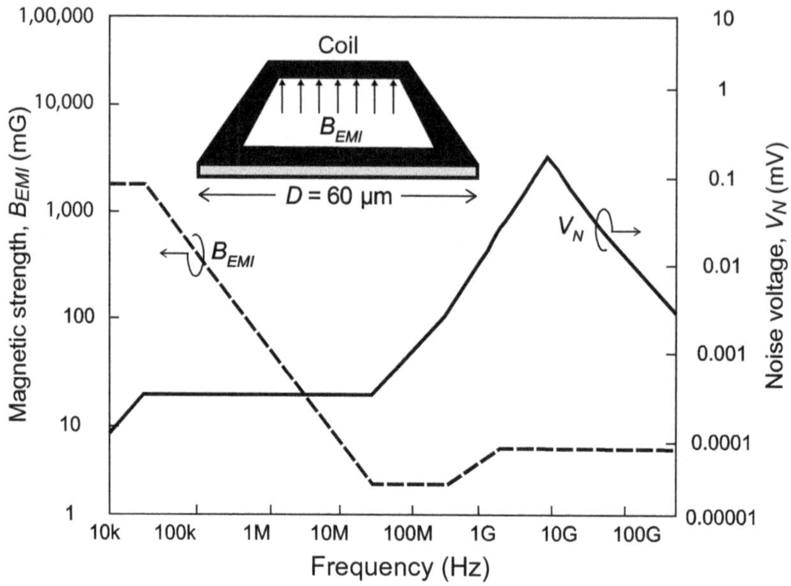

Figure 2.16 Noise induced by external magnetic field in TCI coil. © 2007 Noriyuki Miura [3]

communication distance allow the use of simple pulse modulation scheme and simple I/O design in terms of both transceiver and coil design, resulting in a compact, low-power, and high-speed interface. Furthermore, approximate closed-form equations can be used to translate physical coil layout parameters into electrical characteristics, making design trade-offs intuitive.

Since TCI coils are implemented in IC interconnect layers, whose RLC characteristics do not vary significantly from process to process, coil designs are fairly portable and vary mostly as a function of communication distance, which decreases as chip thinning technology advances.

Because of the rapid falloff in energy with distance in near-field coupling, aggregate crosstalk from simultaneous switching aggressors in a coil array levels off as the array size grows. Suppression of crosstalk from adjacent aggressors can be achieved through PDM of transmission. For instance, two-phase PDM enables a net twofold increase in area efficiency defined as interface bandwidth per unit area. Furthermore, with four-phase PDM, as discussed in Section 2.2.3.1, 1,024 channels each at 1 Gb/s can be packed into a 1 mm^2 area delivering total interface bandwidth of 1 Tb/s while consuming 3 pJ/b, at a BER of 10^{-13} [6].

While sensitive analog circuits should be placed at more than a horizontal offset of two times the diameter from a TCI coil to minimize interference, digital and memory circuits can potentially be placed underneath TCI I/Os with no keep-out to reduce area overhead. Furthermore, signal leakage through the package is minimal, alleviating concern of EMI and eavesdropping outside the package.

Nevertheless, two technical hurdles remain for wide adoption of TCI. First, chip thinning technology needs to continue to advance to provide performance scalability to TCI. TCI performance is a strong function of communication distance Z, which is dictated by chip thickness. A thinner chip results in a smaller Z, which in turn leads to a smaller coil D for the same coupling coefficient. A smaller D improves not only area in a quadratic relationship, but also RLC (Table 2.4). Smaller LC in turn increases the channel cutoff frequency to support a higher data rate.

While chip thickness has not garnered much attention in the last several decades, with the emergence of 3D integration, chip thinning technology development is expected to accelerate going forward.

The second hurdle to wide adoption of TCI is power delivery. Power delivery vertically through a chip can be achieved with highly doped silicon via (HDSV), to be introduced in Section 2.5. However, it is still under research and not yet ready for mass production. In the interim, power can be delivered by TSVs in high-performance systems and by bond wires in lower-cost systems. While such a hybrid solution does not take full advantage of the merits of TCI, the use of TCI for data still offers significant advantage in cost, reliability, and packaging simplicity.

2.2 TCI Design Trade-offs and Optimization

Tables 2.1 and 2.2 enumerate a few possible variations of the transmitter and coil design. In addition, the coil layout design parameters, in particular coil diameter D and number of turns n, provide additional dimensions of freedom to optimize the TCI channel design. At the interface level, switching of parallel channels can be spread out in time using PDM to reduce aggregate crosstalk to enable smaller pitch P for better area efficiency. In this section, we will discuss key transceiver and coil design optimization ideas and trade-offs for power and area efficiency using these parameters and techniques. We will also present a few design examples for illustration.

2.2.1 Design Solution Space

Table 2.5 recapitulates the key interdependencies between TCI channel electrical and physical design parameters. While some of the equations are approximate, they provide insight into the trade-offs when the TCI channel design is optimized to meet different application requirements.

Since dependency on communication distance Z exists only in the coupling coefficient k, the selection of the optimal Z is straightforward – choose the smallest possible Z based on the smallest manufacturable chip thickness to maximize k. Adjustment of other design parameters results in trade-offs. Hence selection of the optimal value depends on the design objective, whether it is lower power, higher speed, or higher density and interface bandwidth.

A key TCI electrical parameter that determines channel robustness is the voltage V_R induced in the receiver coil. Since $V_R \propto M\frac{dI_T}{dt}$ and $M \propto k\sqrt{L_T L_R}$, channel design optimization is driven by optimization of $k\sqrt{L_T L_R}\frac{dI_T}{dt}$. Meanwhile, $f_R \propto \frac{1}{\tau}$ and $\tau \propto \frac{1}{dI_T/dt}$ imply that $f_R \propto \frac{dI_T}{dt}$. Given that $f_{SR} \propto \frac{1}{\sqrt{LC}}$, to maintain a certain ratio of

Table 2.5. Dependency of TCI electrical On physical design parameters.

Electrical parameter	Dependency	Equation
Coil inductance	$L = \frac{L_0 D^{4/3} n^{5/3}}{(w+s)^{1/3}}$	(2.15)
Coil capacitance	$C = \left(C_G w - \frac{C_c}{s}\right) Dn$	(2.16)
Coil resistance	$R = \frac{4R_0 Dn}{w}$	(2.17)
Coil pair coupling coefficient	$k = \begin{cases} -\left(\frac{Z}{D}\right)^3 + 0.205\ for \frac{D}{Z} < 2 \\ 0.065\frac{D}{Z} - 0.05\ for 2 \leq \frac{D}{Z} \leq 3 \end{cases}$	(2.18)
Coupling mutual inductance	$M = k\sqrt{L_T L_R}$	(2.2)
Receiver voltage	$V_R = M\frac{\mathrm{d}I_T}{\mathrm{d}t}$	(2.1)
Peak induced voltage	$V_p = \frac{2.3MI_p}{\tau}$	(2.7)
Transmitter power	$P_{TX} = V_{DD} I_P$	(2.3)
Pulse width of receiver voltage	$\tau \propto \frac{1}{\mathrm{d}I_T/\mathrm{d}t}$	(2.8)
Bandwidth of receiver voltage	$f_R = \frac{0.64}{\tau}$	(2.9)
Channel resonance frequency	$f_{SR} = \frac{1}{2\pi\sqrt{LC}}$	(2.11)
Resonance Q factor	$Q = \frac{\sqrt{L/C}}{R}$	(2.13)

channel cutoff frequency to signal bandwidth to minimize signal distortion, $\frac{f_{SR}}{f_R} \propto \frac{1}{\sqrt{LC}\mathrm{d}I_T/\mathrm{d}t}$ must remain fairly constant as the design is optimized. In other words, any increase in $\sqrt{L_T L_R}$ must be accompanied by a decrease in $\frac{\mathrm{d}I_T}{\mathrm{d}t}$ and vice versa. As a result, $\sqrt{L_T L_R}\frac{\mathrm{d}I_T}{\mathrm{d}t}$ cannot be arbitrarily varied without affecting signal quality. Therefore, a more effective way to optimize V_R is to vary k by tuning the coil diameter D. Nevertheless, since D alters power and area efficiencies in opposite ways, tuning of D results in a trade-off between the two. The optimal value of D is therefore determined by the target balance between power and area efficiencies. The trade-off is quantified in Table 2.6. If D is increased (scaled by a factor of $\beta > 1$) while other layout design parameters remain unchanged and V_R is kept constant, transmitter power decreases in a larger than quadratic fashion (scaled by $1/\beta^{2.33}$), while area increases in a quadratic fashion (scaled by β^2). In other words, when D is varied while V_R stays constant, transmitter power is traded off against area in a smaller than 1:1 fashion. Note that varying D changes f_{SR} and Q as well. The associated changes in signal quality must be

Table 2.6. Trade-off between power and area efficiencies as a function of D.

If D is scaled by a factor of		$\boldsymbol{\beta}$
Resultant scaling	L	$\beta^{1.33}$
	C	β
	R	β
	k ($2 \leq D/Z \leq 3$)	β
	M	$\beta^{2.33}$
	I_P	$1/\beta^{2.33}$
	V_R	**1**
	f_{SR}	$1/\beta^{1.17}$
	Q	$1/\beta^{0.83}$
	P_{TX}	$\boldsymbol{1/\beta^{2.33}}$
	Area	$\boldsymbol{\beta^2}$

evaluated and compensated for if necessary. For instance, a decrease in f_{SR} can be compensated for by slowing down the transmitter slew rate to reduce f_R proportionally.

Another important coil design parameter is n, the number of turns in each layer, since RLC all vary linearly or more with n. In fact, effect of changes in D on RLC can be effectively compensated for by similar changes in n in the opposite direction since RLC depend on both equally or almost equally.

Our discussion so far has assumed the use of spiral coils. If combo coils are used, the total effective number of turns for the same area and coil aperture can be increased at the expense of additional routing layers. While more research is warranted for better understanding of the effect of increasing the number of routing layers, for first-order estimation, the increase in coil R and L is as follows:

$$R \propto m_{eff} \tag{2.19}$$

$$L \propto {m_{eff}}^{1.8}, \tag{2.20}$$

where m_{eff} is the effective number of layers and is equal to m (number of layers) for a double spiral and $m/2$ for an XY stack. C is relatively independent of m_{eff}.

Table 2.7 compares the simulated electrical parameters of three receiver coil designs, namely an XY coil with four (XY-4) and eight turns (XY-8) respectively, and an XY stack from the stacking of two four-turn XY coils for effectively a total of eight turns (XY-Stack-4). The transmitter coil design is the same in all cases.

With the XY coil layout, doubling n from 4 to 8 approximately doubles R and triples L, while causing a moderate increase in k, resulting in a roughly doubling of M without significant change in C. This is however achieved with a trade-off of an approximately 43% lower f_{SR} and a moderately smaller Q. If the eight turns are instead implemented by stacking two four-turn XY coils as in XY-Stack-4, there is a further 16% increase in k and hence M, without additional trade-off in f_{SR} and Q. The downside is the requirement of two additional metal layers.

The electrical parameters in Table 2.7 are extracted from each of the layouts using a 3D field solver. Table 2.8 compares the increase in L and R of XY-8 and XY-Stack-4 relative

Table 2.7. Electrical parameters comparison of three coil designs.

Name	m	n	Image	L(nH)	C(fF)	$R(\Omega)$	k	M(nH)	F_{SR}(GHz)	Q
XY-4	2	4	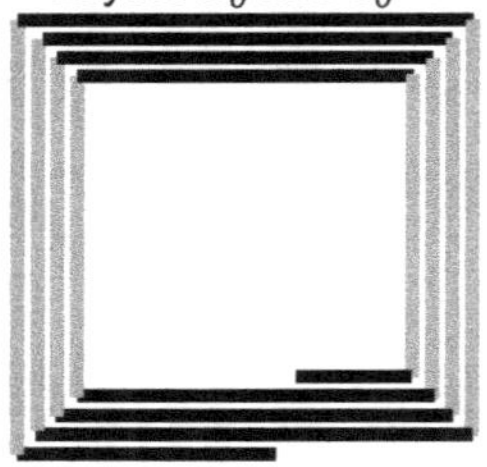	2.12	7.52	215	0.138	0.214	39.9	2.5
XY-8	2	8	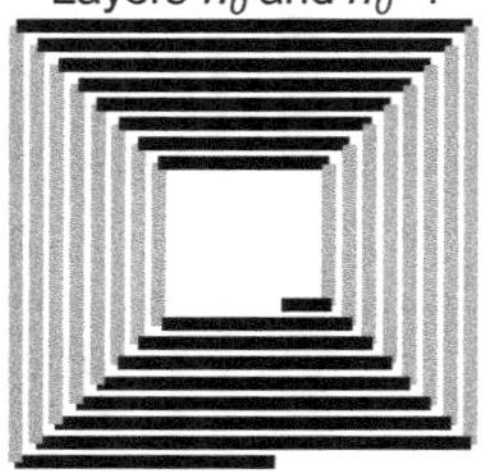	6.49	7.56	414	0.148	0.403	22.7	2.2
XY-Stack-4	4	4	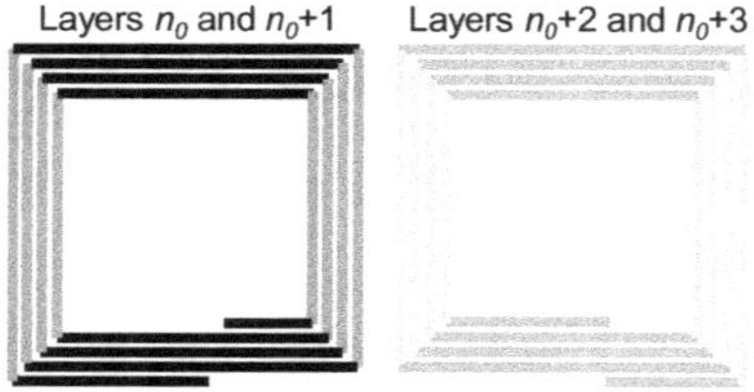	6.46	7.90	432	0.172	0.467	22.3	2.1

Table 2.8. Approximate dependency formulae vs. simulation.

			Normalized L			Normalized R		
Name	m	n	**Simulated**	**Estimated** $\propto n^{5/3}$, $\propto (m/2)^{1.8}$	**|Error|**	**Simulated**	**Estimated** $\propto n$, $\propto m/2$	**|Error|**
XY-4	2	4	1	1	–	1	1	–
XY-8	2	8	3.06	3.18	3.9%	1.93	2	3.6%
XY-Stack-4	4	4	3.05	3.48	14.1%	2.01	2	0.5%

Table 2.9. Design knobs for improving TCI power efficiency.

Knob	To improve power efficiency	Trade-off	Remark
Transmitter design	Lower V_{DD}	Additional supply voltage	Transmitter accounts for 70–80% of total I/O power ⇒ optimizing transmitter design effectively reduces I/O power
	Turn off I_T after switching	Lower data rate	
SerDes	Serialize data for higher data rate per channel	High-speed clock design + SerDes circuit area and power	H-bridge transmitter power is mostly independent of data rate ⇒ smaller pJ/b at higher date rate
Coil size	Increase D	Larger area	See Table 2.6
Sleep mode	Put inactive channels to sleep	Wake-up latency	Standard transceiver consumes power even when idle ⇒ sleep mode effectively reduces power
Timing calibration	Reduce timing offsets due to PVT and channel skew to reduce required timing margin, enabling a narrower pulse	Overhead in calibration time and circuit area	Narrower pulse enables higher data rate for the same power

to XY-4 estimated using the dependency formulae in Table 2.5, (2.19) and (2.20), with those from the field solver simulation. This confirms that the dependency formulae are sufficiently accurate for first pass estimation and for gaining insight into trade-offs.

2.2.2 Design for Power Efficiency

Table 2.9 enumerates key design knobs for improving power efficiency of TCI implementation as well as their trade-offs. Optimization of transmitter design, adoption of SerDes, and sleep mode are discussed in further detail next.

2.2.2.1 Low V_{DD} Transmitter Design: NMOS Push-Pull Transmitter

Given that the CMOS H-bridge transmitter power is proportional to V_{DD} (2.3), lowering V_{DD} results in proportionally lower transmitter power. However, as V_{DD} is lowered, the voltage between the metal–oxide–semiconductor (MOS) transistor drain

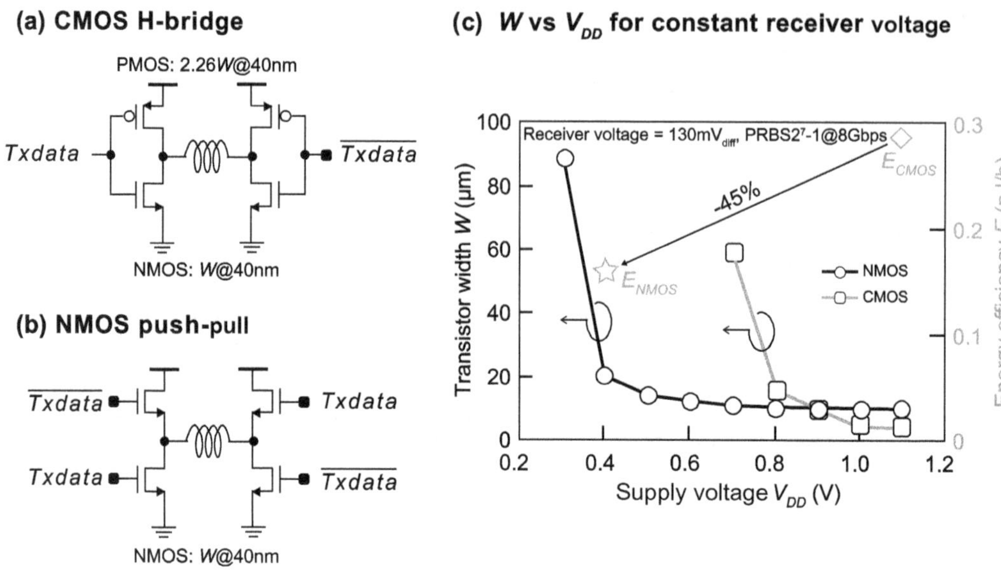

Figure 2.17 Comparison of NMOS push-pull to CMOS H-bridge transmitter. © 2018 Tomoki Miyata [7]

and source, V_{DS}, decreases, eventually pushing the transistor out of saturation into the linear operating region, resulting in a lower transconductance and hence lower on-current. Therefore, in order to reduce the transmitter power without degrading channel performance, the transistor size would need to be increased to restore the original magnitude of the transmitter current, resulting in larger circuit area and capacitive loading. Except for advanced ($\leq$7 nm) processes, the transition into linear region occurs at a much higher V_{DD} for PMOS than for NMOS. As a result, to achieve the smallest V_{DD} for the most power reduction, PMOS transistors should be removed from the transmitter design in order to enable a sufficiently low V_{DD}.

The NMOS push-pull transmitter shown in Figure 2.17b is proposed in [7] to enable low V_{DD}. To quantify the power reduction, its performance is compared with the standard CMOS H-bridge transmitter shown in Figure 2.17a. The PMOS transistor width of the CMOS transmitter is set to 2.26W, where W is the NMOS transistor width. The value of W required to maintain a differential receiver voltage of 130 mV as V_{DD} is lowered is simulated using parameters from the TSMC 40 nm process and PRBS2^7–1 data patterns. The results are shown in Figure 2.17c. For both transmitters, there is a sharp increase in W when V_{DD} is lowered to a certain value, corresponding to the transition from saturation to linear operating mode. However, the transition occurs at a much lower V_{DD} of 0.4 V for the NMOS transmitter than the 0.8 V for the CMOS transmitter. Furthermore, above the transition point, W increases more slowly for the NMOS transmitter as V_{DD} is lowered. Consequently, using the NMOS transmitter, V_{DD} can be lowered to 0.4 V without incurring a large penalty in the form of larger transistors. At this supply voltage, the NMOS transmitter achieves a 45% improvement in energy efficiency compared to the CMOS transmitter operating at the nominal V_{DD} of 1.1 V. The NMOS transmitter is therefore an effective means for power reduction. Since other digital circuits including buffers for the transmitter still operate at the nominal voltage, the trade-off is that an additional supply voltage is required.

2.2.2.2 Pulse Transmitter Design

As noted in Table 2.1, the drawback of the standard CMOS H-bridge transmitter is its higher power, due to the constant current flow regardless of the state of *Txdata*. If current flow is limited to only during data transitions, power consumption can be significantly reduced, especially at lower data rate. Figure 2.18b depicts a pulse transmitter proposed in [8] that draws current only when *Txdata* switches. Its signal waveforms and power consumption are compared against those of the standard transmitter shown in Figure 2.18a. The pulse transmitter differs from the standard transmitter in that $\overline{Txdata}$ is replaced by *Txdatad*, a delayed version of *Txdata*, as input to the right side of the H-bridge. Data transmission is achieved through the following sequence of events:

(I) When *Txdata* switches high, *Txdatad* remains low, resulting in transistors MNL and MPR being on and MPL and MNR being off. As a result, a positive current I_T is generated.

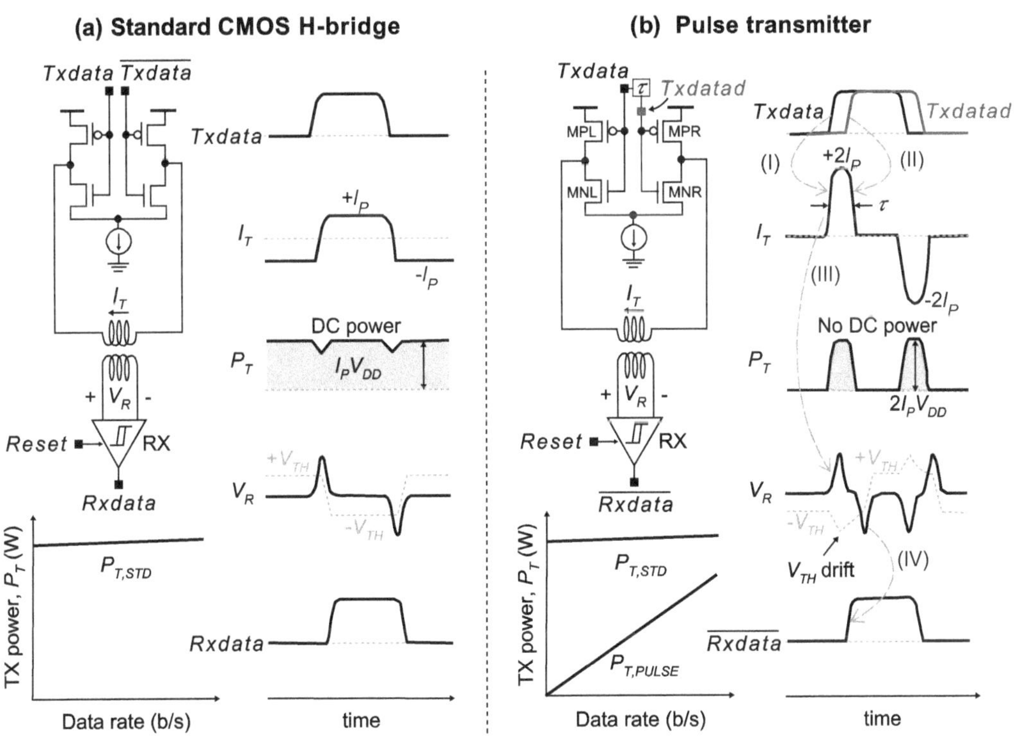

Figure 2.18 Comparison of pulse to CMOS H-bridge transmitter. Copyright 2011 The Japan Society of Applied Physics [8]

(II) After a delay of τ, *Txdatad* switches high, turning MPR off and MNR on. As a result, I_T turns off. Therefore, a rising edge in the transmitted data generates a positive I_T pulse with a pulse width equal to τ.

(III) With a rising edge followed by a falling edge, the I_T pulse in turn induces a positive followed by a negative pulse in V_R.

(IV) The trailing negative V_R pulse is integrated to create $\overline{Rxdata}$ to reproduce *Txdata*.

Note that since I_T switches from 0 to high or low, instead of from one polarity to another as in the CMOS H-bridge, the peak current of the pulse transmitter must be doubled to $2I_P$ to create the same ΔI.

Because I_T flows only during a data transition, there is no DC power. The transmitter power varies linearly with how frequently data switches as well as the I_T pulse width τ, as illustrated in the comparison in Figure 2.18. Therefore, the amount of power reduction achieved by the pulse transmitter depends on the data rate.

Figure 2.19a plots the channel BER and transmitter power as a function of I_T pulse width τ, at a data rate of 1.5 Gb/s for the test chip in [8]. It demonstrates how transmitter power scales linearly with pulse width. The minimum pulse width to achieve BER $< 10^{-12}$ is $\tau_{min} = 130$ ps at a transmitter power of $P_{TX} = 4.2$ mW. Increasing τ beyond 130 ps results in increasing the positive timing margin at the expense of linearly rising transmitter power.

Figure 2.19b plots and compares transmitter power between the standard CMOS H-bridge and pulse transmitter as a function of data rate, while keeping the pulse transmitter τ at 130 ps and BER $< 10^{-12}$. Compared to the CMOS H-bridge, the use of the pulse transmitter results in 1/60× the power at 0.1 Gb/s and 1/4× the power at 1.5 Gb/s.

By properly resetting the initial state of the standard hysteresis comparator receiver to ignore the leading pulse, the trailing V_R pulse can be sampled to generate $\overline{Rxdata}$.

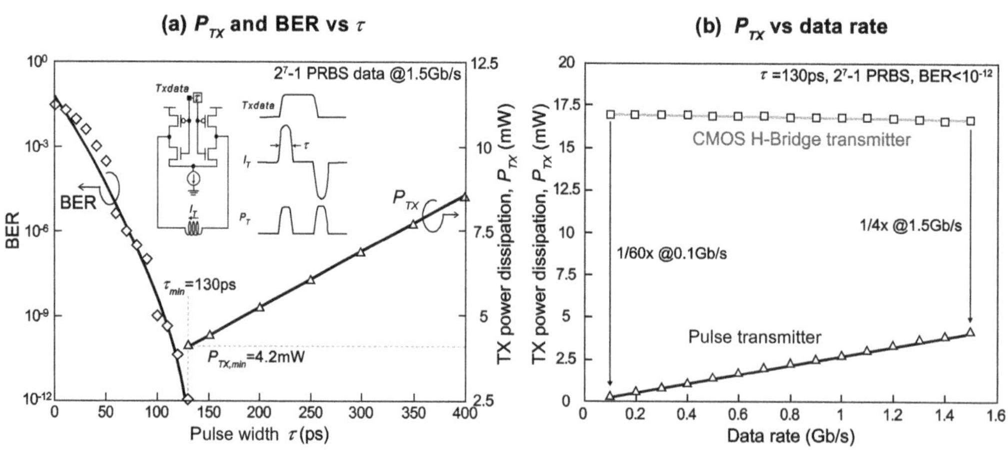

Figure 2.19 Power and BER as a function of pulse width and data rate. Copyright 2011 The Japan Society of Applied Physics [8]

However, since the leading pulse creates a drift in V_{TH}, sufficient time must be inserted before the trailing pulse arrives to allow V_{TH} to return to its nominal level to avoid degradation in receiver operating margin. The time interval between the leading and trailing pulses is determined by τ. Since a longer τ results in larger transmitter power, optimization of τ involves a trade-off between receiver margin and transmitter power.

This reveals the fundamental drawback of the pulse transmitter – a lower data rate. Since each I_T pulse results in two V_R pulses, the minimum I_T pulse width is twice the width of each V_R pulse. Meanwhile, robust receiver operation requires a time interval to be inserted within the double V_R pulse, further increasing the I_T pulse width and hence reducing the data rate. Conversely, for the same data rate, the V_R pulse for a pulse transmitter is much narrower than the standard transmitter, resulting in tight sampling timing margin. Therefore, a synchronous transceiver architecture may be required to support a higher data rate.

2.2.2.3 SerDes for High Per-Channel Data Rate

Because transmitter power accounts for most of the interface power of TCI, and it is relatively independent of data rate for the standard transmitter, maximizing the per-channel data rate results in minimum energy required to transmit each bit, resulting in the maximum power efficiency. Meanwhile, since TCI employs a simple digital transceiver with no ESD and hence low capacitive loading, with proper coil design to keep self-resonance frequency f_{SR} high, it can operate as fast as the process technology allows. For instance, an 11 Gb/s/channel can be achieved in a 180 nm process [9], and a 30 Gb/s/channel in a 65 nm process [10]. By adopting a SerDes architecture where multiple data streams are serialized for transmission by each channel at the highest possible data rate, TCI power efficiency can be significantly improved.

The trade-off is the addition of multiplexer and demultiplexer circuits for serialization and deserialization on the transmitter and receiver respectively, plus the generation and distribution of a high-speed clock used for the SerDes operation. This results in not only increased design complexity but also additional circuit area and power, which must be accounted for in evaluating the net benefit in power efficiency. Section 2.2.4.2 describes a scheme to minimize the added costs of clock generation and distribution.

2.2.2.4 Sleep Mode

Since both the standard transmitter and receiver draw constant current, enabling sleep mode when the transceiver is not in use enables power saving. If the transmitter includes a tail transistor as shown in Figure 2.4a, it can be put in sleep mode by simply turning off the tail transistor. Otherwise, a gate can be added to the input terminal to support sleep mode. The NAND and NOR gates in Figure 2.20a allow the enable signal to directly turn off both the PMOS and NMOS and create an open in the transmitter coil. Note that because any change in transmitter current results in an induced V_R pulse, the entrance into and exit from sleep mode where transmitter current is switched off and on respectively can cause false switching in the receiver.

Therefore, the receiver should be turned off first when entering sleep mode, and the transmitter turned on first when exiting the mode.

Because both the amplifier and cross-coupled pair in the standard receiver includes a tail transistor, it can be put into sleep mode by turning off the tail transistors. However, when the receiver exits sleep mode, its initial state is undefined. Therefore, it is necessary to reset the receiver upon sleep mode exit and before transmission begins, using the reset circuit in Figure 2.20b.

2.2.3 Design for Area Efficiency

Table 2.10 enumerates key design knobs for improving area efficiency of TCI implementation as well as their trade-offs. PDM and coil size are discussed in further detail next.

Table 2.10. Design knobs for improving TCI area efficiency.

Knob	To improve area efficiency	Trade-off	Remark
PDM	Reduce number of simultaneously switching transmitters and hence crosstalk to enable smaller coil array pitch	4-phase clock generation for 4PDM + additional V_{TH} control; reduced data rate	4PDM enables overlapping coil placement for an effective pitch that is smaller than D
SerDes	Serialize data to reduce number of parallel channels and hence coil array size	High-speed clock design + SerDes circuit area and power	
Coil size	Reduce D only	Larger power	
	Reduce D and increase n and/or m	Lower f_{SR}	Increase in n/m compensates for decrease in M due to smaller D

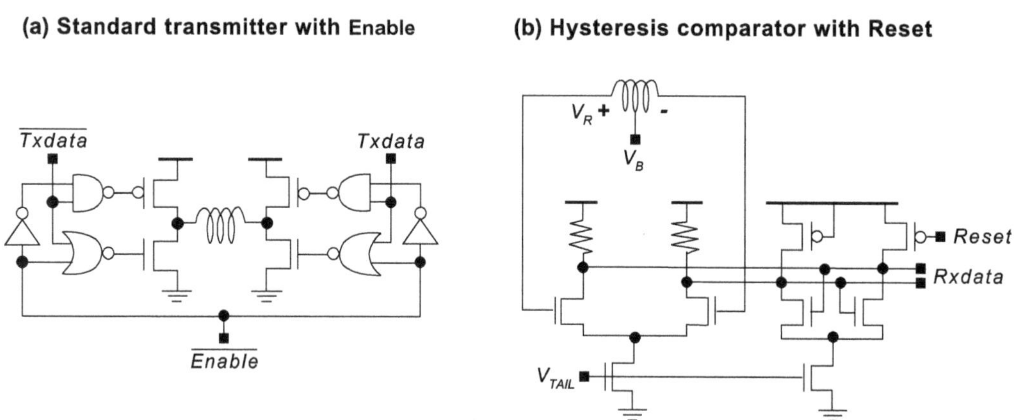

Figure 2.20 Transceiver circuits with support for sleep mode.

2.2.3.1 PDM: Phase Division Multiplexing

Section 2.1.4 introduces phase division multiplexing (PDM) as a means to reduce crosstalk by dividing the simultaneously switching of parallel channels into different phases, thus reducing crosstalk and enabling a smaller pitch for the coil array. In particular, 2PDM (two-phase PDM) enables the pitch to be halved, resulting in a net twofold increase in area efficiency. Since the generally recommended pitch for a two-dimensional coil array is $P_{Array} = 2 \sim 3D$, 2PDM enables the pitch to be reduced down to D, resulting in minimal airgap between coils. By the same reasoning, 4PDM (four-phase switching) should enable a pitch smaller than D for additional improvement in area efficiency. This enables two two-dimensional coil arrays to be placed overlapping with each other.

Figure 2.21 depicts the 4PDM implementation proposed in [11]. Two two-dimensional XY coil arrays with the same D and P are laid out in the same two metal layers (M3 for x-, M2 for y-direction routing), with an offset of $P/2$ in both directions. Each array is divided into two complementary groups of coils with checkerboard placement. The total of four groups are switched in four different phases 90 degrees ($\pi/2$) apart respectively for a fourfold reduction in the number of simultaneously

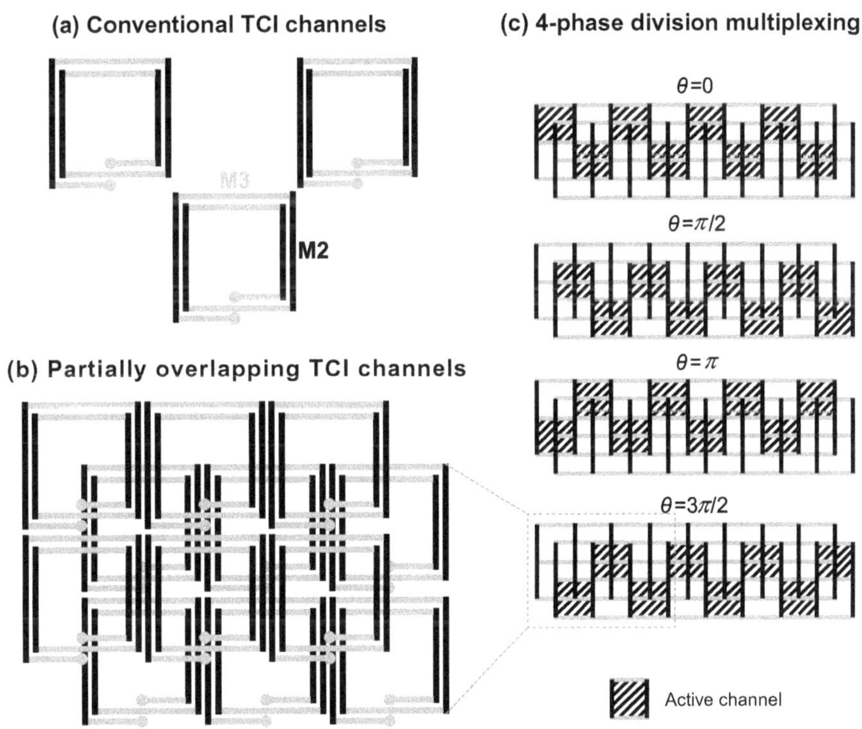

Figure 2.21 4PDM of TCI channels in overlapping arrays. Copyright © 2014 IEEE. Reprinted, with permission, from [11]

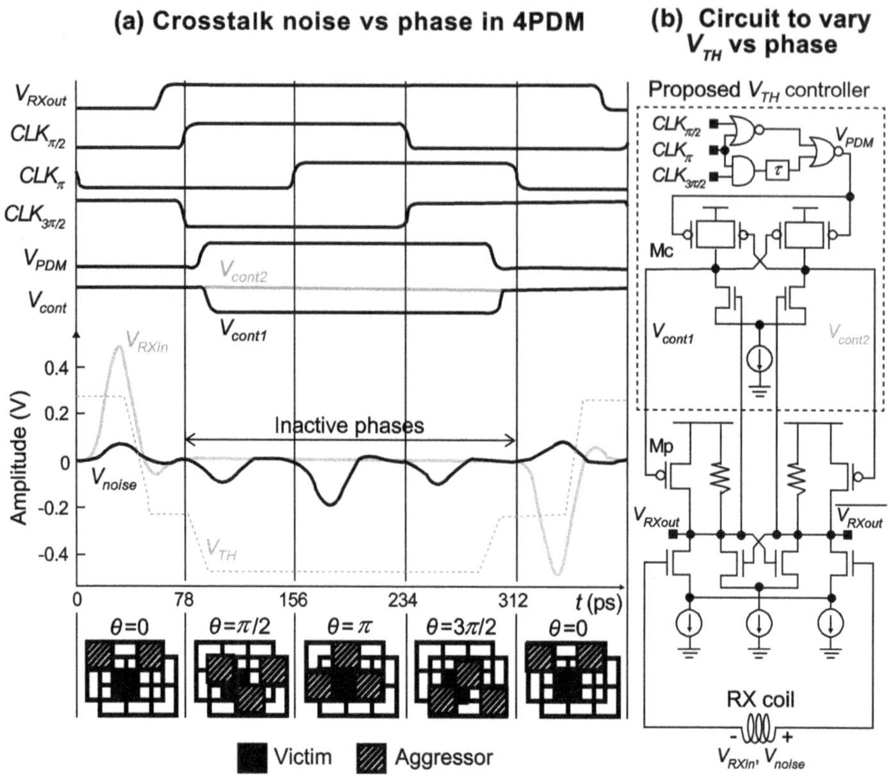

Figure 2.22 4PDM crosstalk noise and V_{TH} adjustment vs. phase. Copyright © 2014 IEEE. Reprinted, with permission, from [11]

switching aggressors. The resultant crosstalk reduction is sufficient to enable an 8 Gb/s data rate despite the high coil density.

Each receiver is subject to crosstalk noise even outside its active phase of communication (Figure 2.22a). To improve receiver noise immunity during inactive phases, a controller circuit can be added to vary V_{TH} of the receiver over time such that it is moved farther away from the polarity of the bit being held. Figure 2.22b illustrates the controller circuit for a receiver that communicates during phase 0. The clock signals for phases $\pi/2$, π, and $3\pi/2$, namely $CLK_{\pi/2}$, CLK_{π}, and $CLK_{3\pi/2}$, are used as input to the logic at the top of the circuit to activate the control signal V_{PDM} during inactive phases. V_{PDM} in turn pulls V_{TH} low for better noise immunity given that V_{RXout} is being held high in the illustration.

2.2.3.2 Coil Size

Table 2.11 reveals how various TCI channel electrical parameters change as coil diameter D is reduced (scaled by $1/\beta$, where $\beta > 1$) to reduce area. The key assumption is that V_R remains constant so there is no reduction in voltage and timing margins. Under this assumption, there are two scenarios. In the first scenario, M is allowed to

Table 2.11. Coil size vs. power and self-resonance frequency.

		Reduced M	**Constant M**
If	D scaled by a factor of	$1/\beta$	$1/\beta$
	n scaled by a factor of	1	$\beta^{1.4}$
Then resultant scaling of	L	$1/\beta^{1.33}$	β
	C	$1/\beta$	$\beta^{0.4}$
	R	$1/\beta$	$\beta^{0.4}$
	k	$1/\beta$	$1/\beta$
	M	$1/\beta^{2.33}$	1
	I_P	$\beta^{2.33}$	1
	V_P	1	1
	f_{SR}	$\beta^{1.17}$	$\mathbf{1/\beta^{0.7}}$
	Q	$\beta^{0.83}$	$\mathbf{1/\beta^{0.1}}$
	P_{TX}	$\boldsymbol{\beta^{2.33}}$	**1**
	Area	$1/\beta^2$	$1/\beta^2$

decrease and I_P is increased proportionally to restore V_R. In the second scenario, other coil design parameters are adjusted to keep M constant. In particular, the right column of Table 2.11 shows the case when n is increased to restore M. Increasing m instead of n will have a similar effect on M since the dependency of L on m is similar to n, but the effect on f_{SR} and Q will be different since the dependency of C on m is different.

Under the first scenario (reduced M), when D is reduced by a factor of β, area is reduced by a factor of β^2, while transmitter power is increased by a factor of $\beta^{2.33}$. Therefore, the trade-off of power for area is more than 1:1. The increase in f_{SR} is an advantage, while the increase in Q may require compensation by increasing R if it exceeds the acceptable range.

Under the second scenario (constant M), when D is reduced by a factor of β, n needs to be increased by a factor of $\beta^{1.4}$ to restore M. As a result, transmitter power remains unchanged, while f_{SR} is reduced by a factor of $\beta^{0.7}$. The trade-off of f_{SR} for area is significantly less than 1:1. Nevertheless, the impact on signal quality at the target data rate must be evaluated to determine if it is acceptable. If not, other design changes must be made such as reducing w to reduce C, thereby increasing f_{SR}. This will affect L and R as well, so rebalancing all these parameters is necessary. The comparatively smaller decrease in Q should be evaluated as well.

It is also possible to combine the two scenarios to mitigate increase in transmitter power and decrease in f_{SR}.

2.2.4 Implementation Examples

This section uses a few implementation examples to illustrate how TCI design can be adapted to meet different interface requirements to enable adoption in a variety of applications.

2.2.4.1 Design for Low Power

In HDTV applications, 1080 H.264 video decoding requires a memory bandwidth of up to 20 Gb/s [12], while the entire decoder chip consumes about 100 mW of power [13]. To keep the memory interface power low, for example, at 10% of total chip power, the energy consumption needs to be within 10 mW/20 Gb/s = 0.5 pJ/b. Using a pulse transmitter to reduce power, the TCI transceiver in [14] implemented in a 180 nm CMOS process achieved an energy consumption of 2.8 pJ/b. That is still more than five times the amount required for the HDTV example. Therefore, for such high-bandwidth, low-power applications, it is necessary to drastically improve the energy efficiency of TCI.

The energy consumption of the TCI channel in [14] can be broken down into transmitter and receiver energy consumption of 2.2 pJ/b and 0.6 pJ/b respectively. Therefore, transmitter energy consumption dominates. Energy consumption in the pulse transmitter is mainly in the H-bridge, which generates the transmitter current I_T. If I_T is approximated as a triangular pulse with peak I_P and width τ, the energy per bit, E_{TX}, which is proportional to the total amount of charge delivered by V_{DD} or the area under the current pulse, can be approximately expressed as

$$E_{TX} = \frac{V_{DD} I_P \tau}{2} \tag{2.21}$$

as shown in Figure 2.23. Because the induced voltage in the receiver is $V_R = M\frac{\mathrm{d}I_T}{\mathrm{d}t}$, its peak value V_P depends only on the slew rate of I_T, which is $S_P = \frac{\mathrm{d}I_T}{\mathrm{d}t}$, and not I_P nor τ.

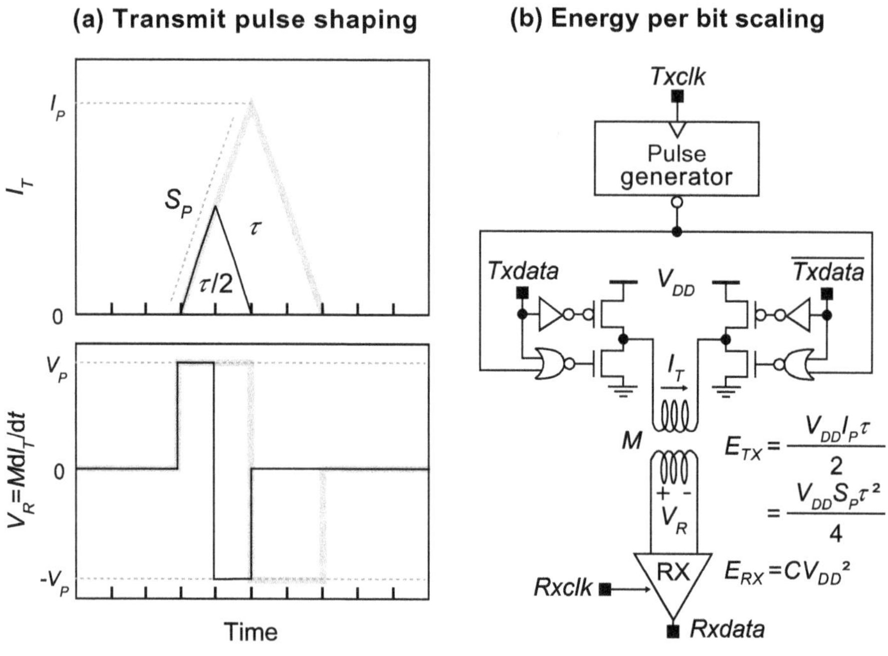

Figure 2.23 Dependence of receiver voltage on transmitter current pulse shape. (b) Copyright © 2008 IEEE. Reprinted, with permission, from [15]

Therefore, E_{TX} can be reduced while keeping V_P and hence channel voltage margin unchanged by shortening the pulse without changing the slew rate. Expressing E_{TX} in terms of S_P,

$$E_{TX} = \frac{V_{DD} S_p \tau^2}{4}. \tag{2.22}$$

Energy per bit is therefore a quadratic function of the transmitter current pulse width, making the latter an effective knob for improving energy efficiency. However, since reducing transmitter current pulse width also proportionally reduces receiver voltage pulse width, robust timing design against process, voltage, and temperature (PVT) variations becomes critical to ensure the lower receiver timing margin does not degrade channel BER.

Figure 2.24 depicts a pulse width control circuit as well as a sampling timing control circuit based on the same common design, that have been developed to optimize transmitter current pulse shape for low power. The pulse width control circuit shapes the current pulse by adjusting its width, slew rate, and amplitude. The sampling timing control circuit enables calibration to compensate for sampling timing shift due to PVT variations, thereby minimizing BER degradation due to reduced pulse width.

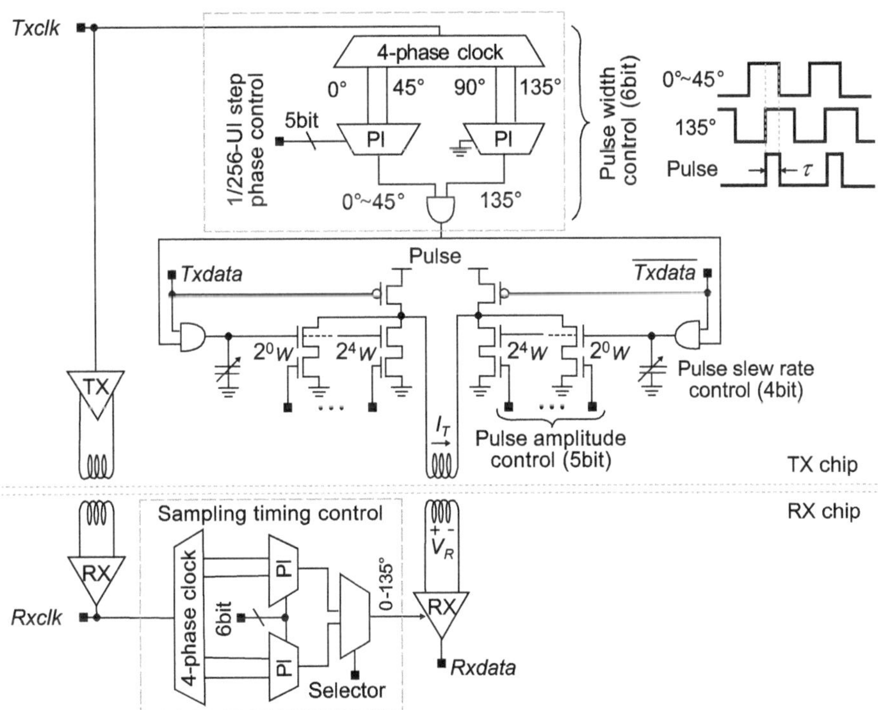

Figure 2.24 Pulse shaping and sampling timing control circuits for low power. Copyright © 2007 IEEE. Reprinted, with permission, from [16]

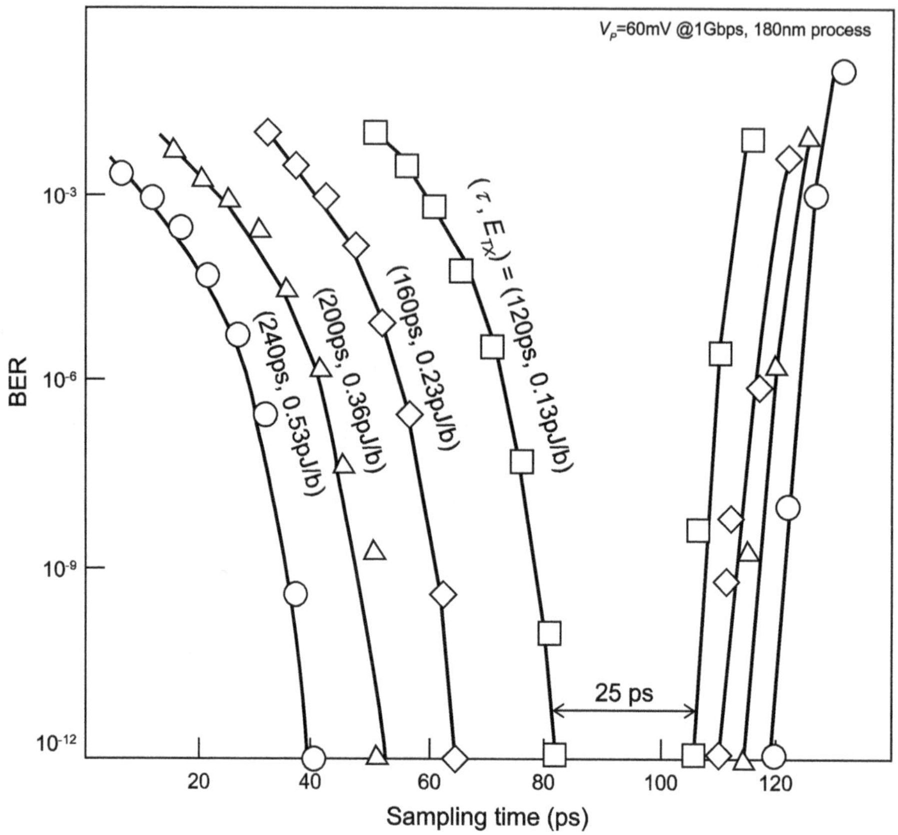

Figure 2.25 BER and timing margin vs. transmitter pulse width. Copyright © 2007 IEEE. Reprinted, with permission, from [16]

The design was implemented in a 180 nm process in [16] to evaluate its energy efficiency. Figure 2.25 plots the measured BER and timing margin as a function of transmitter pulse width and the resultant transmitter power. It confirms that E_{TX} is a quadratic function of τ. With this transceiver design, E_{TX} can be reduced to 0.13 pJ/b with a timing margin of 25 ps at BER $<10^{-12}$ when τ is reduced to 120 ps, representing a $>17\times$ improvement over the previous design implemented in the same process node. Total link energy consumption is 0.33 pJ/b, making it possible to achieve the high energy efficiency required for HDTV applications.

2.2.4.2 Design for High Data Rate

Because TCI is a contactless interface, ESD protection circuits can be removed, resulting in not only smaller area but also lower loading capacitance. For comparison, a 30 μm diameter microbump channel including ESD has a loading capacitance of about 50 fF [17], while a capacitively or inductively coupled channel can have a loading capacitance of less than 10 fF due to the elimination of ESD [18]. TCI is therefore suited for high-speed operation. In fact, as long as the coil is designed to

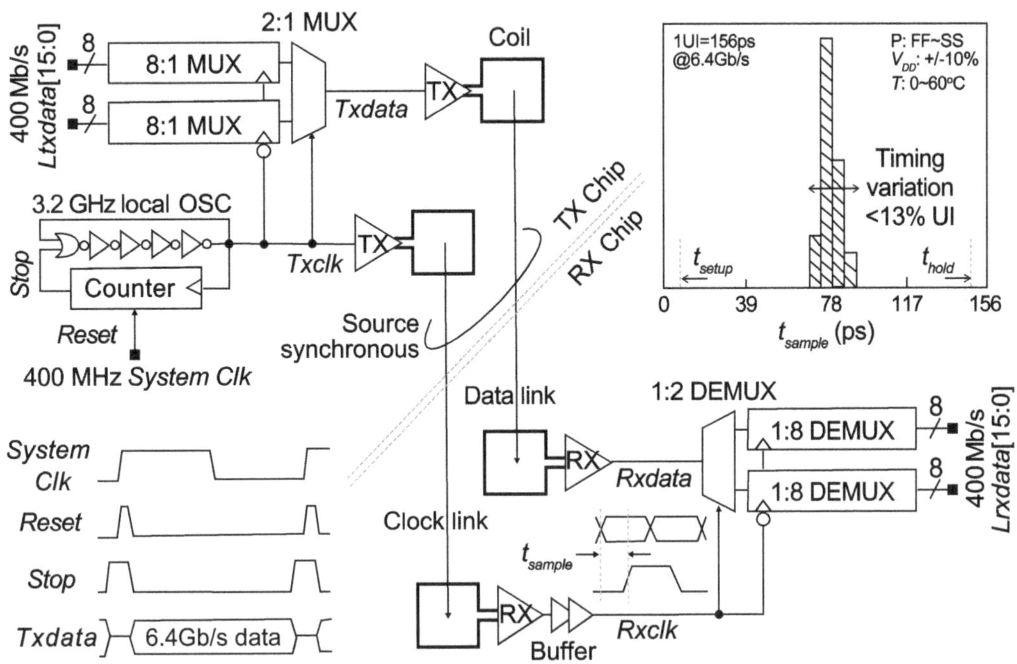

Figure 2.26 Transceiver block diagram for burst transmission. Copyright © 2008 IEEE. Reprinted, with permission, from [19]

keep its self-resonance frequency f_{SR} high, TCI can operate as fast as the transistors allow.

An I/O design based on the standard transceiver design that achieves a maximum 11 Gb/s data rate in a 180 nm CMOS process is reported in [9]. The asynchronous scheme allows complicated timing control circuits to be eliminated, enabling the high data rate. Nevertheless, since the asynchronous design has lower noise immunity, the coil size needs to be large to boost SNR, resulting in a larger area of implementation. By adopting SerDes for burst transmission to take advantage of the high data rate, the number of channels and hence the total interface area can be minimized.

Figure 2.26 depicts the transceiver block diagram for burst transmission used in [9]. Two groups of 8-bit, 400 Mb/s data streams are multiplexed to create a 6.4 Gb/s *Txdata* stream. A 3.2 GHz clock, *Txclk*, used to provide timing for the multiplexers, is generated using a local ring oscillator (OSC). This avoids the power and area penalty of a PLL. A counter counts the number of clock pulses and stops the OSC when enough pulses have been generated for multiplexing. *Txclk* is transmitted together with *Txdata* using two parallel TCI channels. The received clock *Rxclk* is then used to demultiplex the received data *Rxdata*. The source-synchronous nature of the transmission enables common mode rejection of OSC jitter as well as sampling timing tracking over PVT.

Table 2.12 summarizes the test chip results from [9], comparing the use of SerDes high-speed burst transmission using one data and one clock channel against transmission using parallel 400 Mb/s channels. In a 180 nm process, burst transmission results

Table 2.12. Area efficiency of burst transmission. Copyright © 2009, IEEE. Reprinted, with permission, from [9].

	Parallel TCI		Serial TCI	
CMOS process	180 nm		180 nm	90 nm
Coil diameter, D	90 μm		120 μm	120 μm
Multiplexing ratio	N/A		16:1	38:1
# of data links	16	38	1	1
Per-link data rate (Gb/s)	0.4	0.4	6.4	15.2
Aggregate BW (Gb/s)	6.4	15.2	6.4	15.2
Area (mm^2)	0.3	0.72	0.1*	0.08*
Normalized area	1	1	1/3 (= 0.1/0.3)	1/9 (= 0.08/0.72)

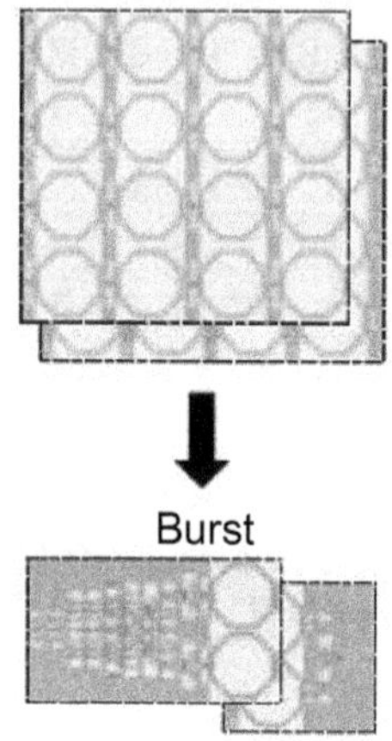

* Note: Including SerDes clock link.

in a 3× reduction in area for the same aggregate bandwidth even with a larger coil diameter. With scaling to a 90 nm process, the burst transmission data rate can be further increased to 15.2 Gb/s, enabling serialization of 38 parallel links, resulting in an even larger reduction of 9× in area.

Table 2.12 also illustrates an important advantage of TCI over conventional wired high-speed interfaces – it takes full advantage of performance improvement through process scaling. Through scaling from a 180 to 90 nm process, the TCI channel data rate improves 2.375× from 6.4 to 15.2 Gb/s. This is close to being the same as the 2.5× improvement in transistor transition frequency f_T. Furthermore, the circuit area is reduced due to process scaling, resulting in an area efficiency improvement of 3× from 180 to 90 nm, even though the data rate improvement is only 2.375×.

2.2.4.3 Design for High-Density and High-Interface Bandwidth

The Wide I/O (WIO) standard was previously proposed as a low-power, high-bandwidth DRAM interface solution. It achieves low power by keeping the channel short and data rate low. At the same time, it achieves high total interface bandwidth by utilizing TSV to implement hundreds or more of parallel channels. TCI can be used to achieve even higher bandwidth with better area and power efficiency, as demonstrated in [11]. And since it does not require TSV, it can be implemented at lower costs. It utilizes SerDes and 4PDM to simultaneously achieve high per-channel data rate and high channel density.

Figure 2.27 depicts the TCI channel configuration that is compatible with the JEDEC WIO2 standard, as well as an overlapping coil placement enabled by 4PDM and the V_{TH} control circuit in Figure 2.22. There are four groups of 20 channels, each including 11 data channels as well as clock and control signals.

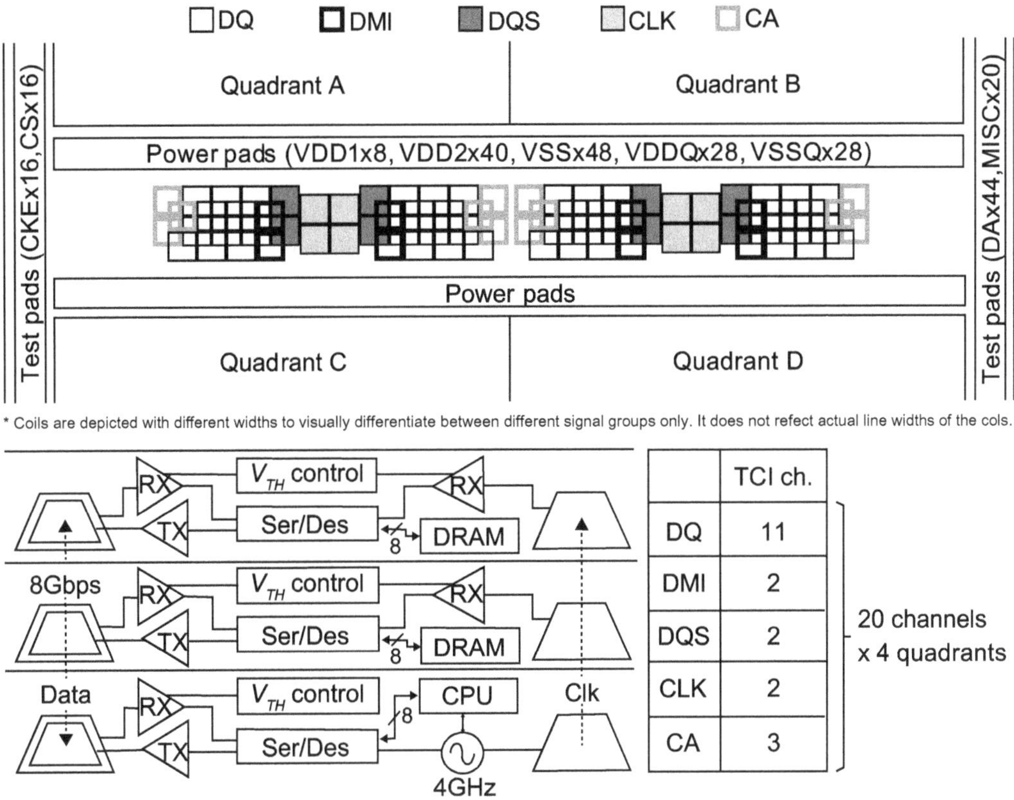

Figure 2.27 WIO2-compatilbe TCI channel configuration and overlapping coil placement. © 2014 IEEE. Reprinted, with permission, from [11]

Table 2.13. Comparison of TCI to TSV for WIO2 implementation. Copyright © 2014 IEEE. Reprinted, with permission, from [11].

WIO2 implementation	TSV	4PDM SerDes TCI	Improvement
Aggregate BW (Gb/s)	102.4	352.0	3.4×
Total # of data links	512	44	–
Data rate/link (Gb/s)	0.2	8	–
IO layout area (mm^2)	2.70	2.08	–23%
Area efficiency (mm^2/Gb/s)	0.0264	0.0059	–78%
Power efficiency (pJ/b)	0.78	0.71	–9%

Table 2.13 compares the TCI implementation to the TSV implementation reported in [20]. The TCI implementation delivers more than three times the interface bandwidth in a 23% smaller area, resulting in an almost 80% improvement in area efficiency, while being 9% more power efficient.

2.2.5 Closing Thoughts

Because of its simple operating principle – vary the transmitter current in some way to create induced voltage pulses in the receiver through magnetically coupled coils – TCI can be implemented with a variety of transmitter and coil designs. For instance, a coil can be created by winding it all within one layer, or in two layers using one for x-, the other for y-direction routing, and repeated across layers. Furthermore, the simple transceiver circuits allow them to run as fast as the transistors allow, without prohibitively high power. Their simplicity also allows the transceiver circuits to be easily turned on and off without much latency. Therefore, there are multiple degrees of freedom in design optimization – in transmitter architecture, coil routing layer utilization, data rate, number of parallel channels, sleep mode, as well as physical layout parameters including coil diameter D, pitch P, and number of turns n. The minimum pitch can be reduced by using phase division multiplexing (PDM) to reduce the number of simultaneously switching aggressors to suppress crosstalk, together with the use of simple controller circuits to dynamically vary the receiver V_{TH} to improve its noise immunity. Meanwhile, the simple transceiver and coil design enable the use of closed-form equations to estimate interdependence of physical and electrical parameters, providing designers with insight to facilitate design trade-offs. As a result, designers are empowered to make design trade-offs to optimize TCI design based on their specific objective, be it high power efficiency, high area efficiency, or high interface bandwidth, within their design constraints such as number of routing layers and available routing channels, transistor speed, availability of low V_{DD} supply, as well as high-speed and multiphase clock.

2.3 TCI 3D/2.9D/2.5D/2D IC Integration

In this section, we will use an example application to elaborate on some of the challenges and solutions in applying TCI to 3D integration of more than two IC chips. Furthermore, we will introduce TCI as a lower-cost and higher-reliability alternative to TSV in 2.5D integration due to its low area overhead and contactless property. In fact, TCI enables additional IC integration configurations including 2.9D (almost 3D) and potentially 2D integration, where two chips placed side by side on a package substrate can communicate with each other directly and wirelessly without going through an interposer. The key ideas of each configuration will be explained.

2.3.1 TCI-Enhanced 3D IC Integration: Processor + SRAM

One of the bottlenecks in a high-performance computing system is in processor access to large memory capacity. While the use of TCI enables large bandwidth communication between a processor and a memory chip, large memory capacity necessitates the stacking of multiple memory chips, which presents additional challenges. First, power supply and control signals still require the use of wirebonding. (Wireless power

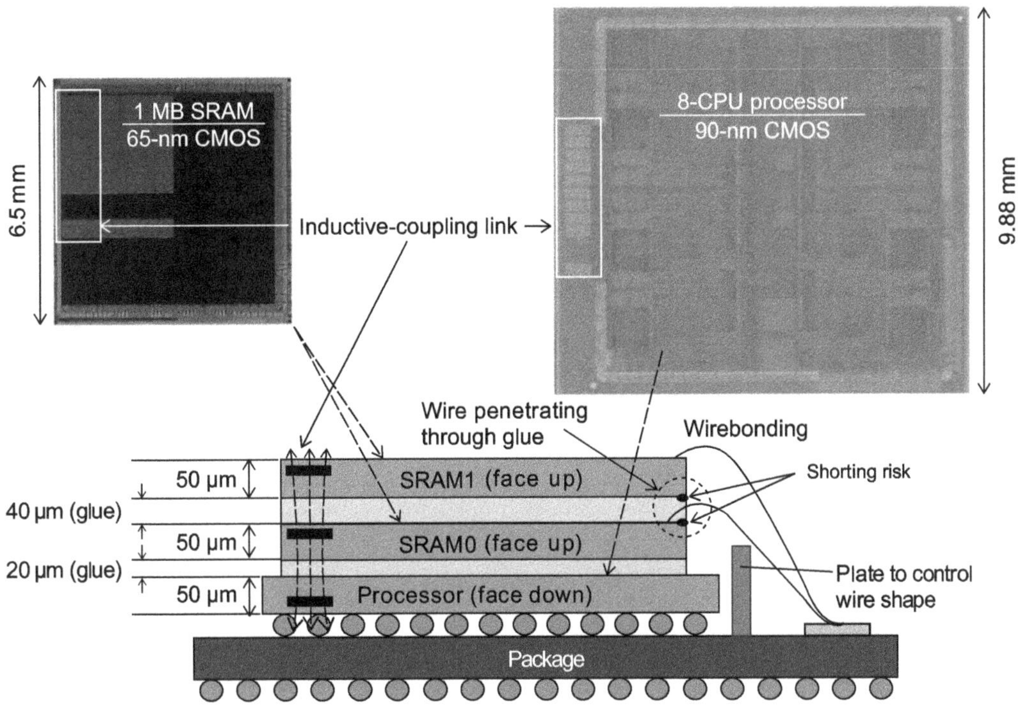

Figure 2.28 TCI test system with two SRAM chips stacked on top of a processor. Copyright 2009 The Japan Society of Applied Physics [21]

delivery and an advanced wired power delivery solution under development will be discussed in Sections 2.4 and 2.5.) If spacers are used between adjacent memory chips to create vertical clearance for wirebonding, overall height of the stack as well as TCI communication distance would increase, resulting in a large coil diameter and hence large I/O area. Second, during communication between the processor and the target memory chip, current changes in the transmitter coil induce voltages not only in the intended receiver coil but also coils on unselected memory chips that act to weaken the transmitted signal. The degradation multiplies as the number of memory chips in the stack increases.

A solution to each of these two challenges using a test system with two SRAM chips stacked on top of a processor chip (see Figure 2.28) is proposed and confirmed in [21]. All three chips are back grinded to a thickness of 50 µm. The processor chip is mounted face down on a flip-chip package substrate. The two SRAM chips are stacked and mounted face up on the back of the processor chip. Additional SRAM chips can be mounted face up in a similar fashion. The bottom SRAM chip, SRAM0, is attached to the processor chip using glue that is 20 µm thick. However, the thickness of the glue bonding the upper SRAM chip, SRAM1, to SRAM0 is increased to 40 µm to create clearance for wirebonding. Wirebonding is achieved through a special technique where the bond wire penetrates a specially formulated glue such as the

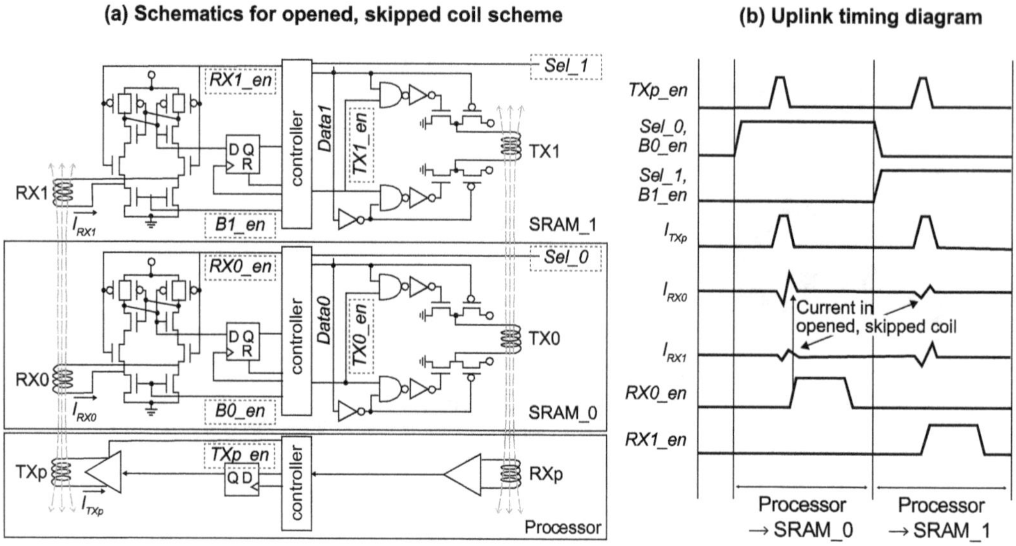

Figure 2.29 Circuit and timing diagrams of opened, skipped coil scheme. Copyright 2009 The Japan Society of Applied Physics [21]

Hitachi Chemical HS series glue. Furthermore, plates are placed on the package substrate around the processor chip to control the wire loop shape to avoid shorting through contact with the backside of SRAM1 and the edge of SRAM0. A coil diameter $D = 240$ μm is chosen for this design, which is 2× and 1.14× the communication distance between the processor and SRAM0 and SRAM1 respectively.

When the processor is communicating with SRAM1, the coil on SRAM0 is skipped over. Nevertheless, a current is still induced in the skipped coil, which acts to reduce the strength of the penetrating magnetic field, resulting in a degradation of the transmitted signal. Since the degradation is due to unintended current flow in the skipped coil, it can be suppressed by creating an "open" in the coil by effectively inserting a pass gate in the current loop controlled by an enable signal. Such an opened, skipped coil scheme is implemented in the circuit solution shown in Figure 2.29a. The timing diagram for uplink – processor to SRAM – communication is depicted in Figure 2.29b. The receiving SRAM is designated by a chip select signal (*Sel_0/1*) communicated through bond wires. The receiver coil on the unselected SRAM is opened by disabling both the bias and receiver-enable signals *B0/1_en* and *RX0/1_en*. Similarly, the transmitter coil on the unselected SRAM is opened by disabling the transmitter-enable signal *TX0/1*. Uplink communication from the processor to SRAM0 is started by activating *Sel_0* and *TXp_en* of the processor transmitter. The resulting current flow in the transmitter coil, I_{TXp}, induces a current in both the receiver coils on SRAM0 and SRAM1, I_{RX0} and I_{RX1}. Even with an opened coil, there is some minimal I_{Rx1} due to parasitic capacitance. Data transfer is completed by activating *RX0_en* during the sampling window to capture the induced I_{RX0} pulse.

Table 2.14. Simulated improvement of opening a skipped coil. Copyright 2009 The Japan Society of Applied Physics [21].

Link		Processor to SRAM0	Processor to SRAM1	SRAM0 to processor	SRAM1 to processor
Communication distance (μm)		120	210	120	210
Normalized signal strength	No degradation	1.000	0.389	0.997	0.389
	With degradation due to skipped coil	0.964	0.347	0.964	0.350
	After opening skipped coil	0.997	0.366	0.994	0.365
Degradation due to skipped coil		3.6%	10.8%	3.3%	10.0%
Degradation due to opened, skipped coil		0.3%	5.9%	0.3%	6.2%

Table 2.14 shows the simulated degradation due to a skipped coil as well as the improvement achieved by opening the skipped coil. As intuition would suggest, degradation due to an intervening, skipped coil on SRAM0 during communication between the processor and SRAM1 is more severe at 10–11% than degradation due to SRAM1 when the processor communicates with SRAM0. This is also the weaker link due to the longer communication distance. When the skipped coil is opened, the degradation is reduced to about 6%.

Since a larger number of skipped coils results in a larger amount of degradation, improvement from applying the opened, skipped coil scheme also increases. Figure 2.30 shows how the percentage of improvement increases as the number of SRAM chips increases. For fair comparison, the worst communication distance is kept constant at $Z = 210$ μm. In other words, as more chips are stacked, the thickness of each chip is further reduced, such that the distance from the processor surface to the surface of the top SRAM remains constant. When the number of stacked SRAM chips increases from two to four, the improvement increases from 6 to 25%. The opened, skipped coil scheme is therefore effective in supporting a larger number of stacked SRAM chips to provide scalability in memory capacity.

Figure 2.31 and Table 2.15 summarize performance of the TCI-enhanced processor-SRAM stack. Each link direction consists of 16 data channels each running at 1.2 Gb/s, resulting in an aggregate bandwidth of 19.2 Gb/s. Despite the downlink from SRAM1 to the processor having 27% less timing margin than from SRAM0, all links are able to communicate successfully.

This implementation example illustrates an important nature of magnetic coupling – a changing magnetic field will induce a current in any closed conducting path, whether it is part of the intended receiver or not. This is known as eddy current and can degrade the performance of TCI. If there is any large, closed conducting structure in the vicinity of a TCI coil, its effect on TCI must be evaluated and accounted for to ensure that TCI performs as intended.

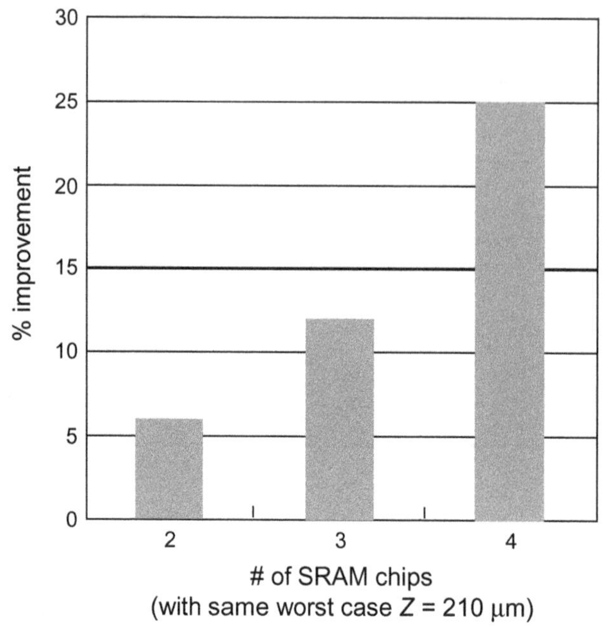

Figure 2.30 Improvement from opened-skipped-coil vs. number of SRAM chips. Copyright 2009 The Japan Society of Applied Physics [21]

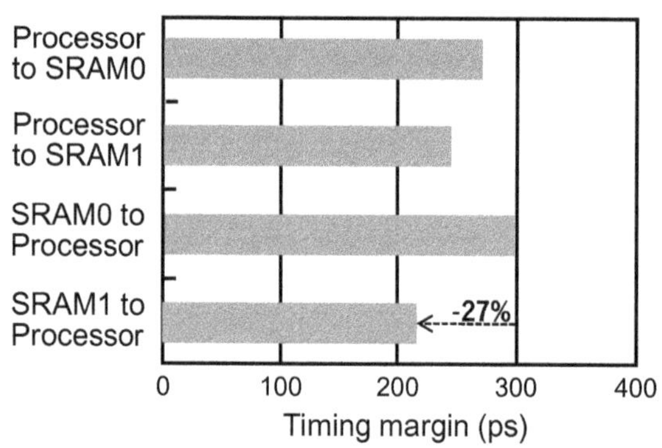

Figure 2.31 TCI processor-SRAM stack performance summary. Copyright 2009 The Japan Society of Applied Physics [21]

2.3.2 TCI-Enhanced 2.9D IC Integration

Since silicon is not an efficient thermal conductor, it is challenging to remove heat from deep in a 3D chip stack. This is especially a concern when a high-power chip is part of the stack, such as a high-performance processor. To circumvent the heat removal issue, a TCI-enhanced 2.9D (almost 3D) IC integration solution is proposed, where all but the high-power chip is directly stacked in a 3D configuration as shown in Figure 2.32. The 3D chip stack is sufficiently thinned to allow it to slide under the

Table 2.15. Processor + SRAM TCI channel performance summary. Copyright 2009 The Japan Society of Applied Physics [21].

Process technology	Process: 90 nm; SRAM: 65 nm
Die thickness	50 μm
Communication distance, *Z*	SRAM0: 120 μm; SRAM1: 210 μm
Coil diameter, *D*	Data: 240 μm; clock: 350 μm
Channel pitch	*x*: 243 μm, *y*: 320 μm
Energy efficiency	1 pJ/b
Data rate	1.2 Gb/s
Total bandwidth	19.2 Gb/s

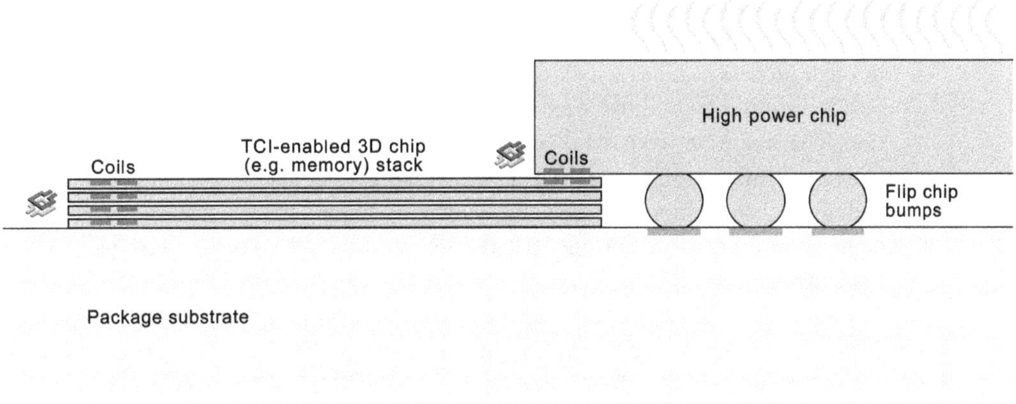

Figure 2.32 TCI-enhanced 2.9D integration.

high-power chip, which has been flip-chip attached to a package substrate. The high-power chip communicates with the chip stack through inductive coupling using TCI I/Os that have been placed over the overlapping area. In this 2.9D integration, the backside of the high-power chip remains exposed to facilitate heat removal.

The challenge in mass production of this configuration is in accommodating variations in the flip-chip bump height. If the high-power chip is connected to the chip stack with microbumps, manufacturing variations in the height of both the microbumps and flip-chip bumps would create undue mechanical stress, causing significant reliability issues. The problem is further exacerbated by thermal stress. The use of TCI eliminates such issues since it is a contactless interface. By designing the TCI coil to have a sufficiently large coupling efficiency for the worst case flip-chip bump height, the interface will function correctly across manufacturing variations.

2.3.3 TCI-Enhanced 2.5D IC Integration

TCI-enhanced 2.9D integration requires the 3D chip stack to be thin enough to fit under the high-power chip. Furthermore, the number of interconnections can be

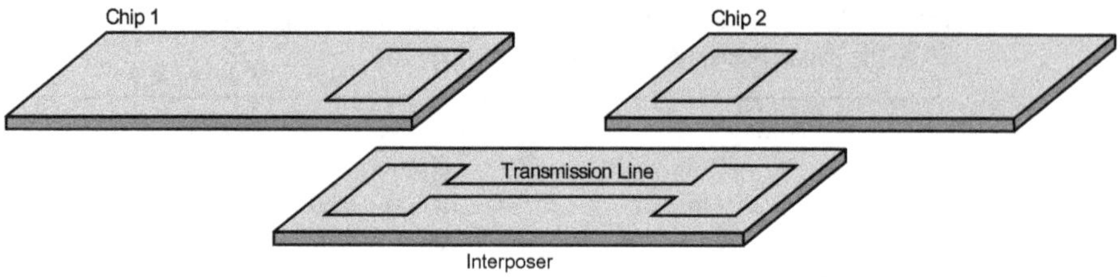

Figure 2.33 TCI interconnection in 2.5D integration.

limited. An alternate solution to circumvent these restrictions is 2.5D integration. By replacing TSV and microbumps in 2.5D integration, TCI offers the advantage of lower cost, lower power, higher performance, higher reliability, and performance scalability through process scaling, same as in 3D integration.

The insertion of an interposer alters the electrical characteristics of a TCI connection. The interposer consists of no active circuits, but two inductor coils connected with a transmission line for each channel, with one coil coupled to the transmitter coil, and the other the receiver coil, as shown in Figure 2.33. As a result, the TCI connection consists of not one, but two passes through a coupled coil pair, and a transmission line.

Figure 2.34 illustrates the electrical characteristics of the 2.5D TCI channel. Figure 2.34a depicts the equivalent circuit model of the TCI interconnect. As to be discussed later, unlike a TSV-based interposer, the TCI 2.5D interposer is very small, such that the interposer transmission line is short enough for a lumped model to be valid, which also means that impedance matching is not necessary. The channel frequency response can be divided into three regions as shown in Figure 2.34b. As frequency increases, the transimpedance of the channel, V_R/I_T, changes initially at a rate of –40dB/decade. As frequency increases beyond f_1, the slope of the response changes to –20dB/decade. This is because the induced voltage in the receiver, V_R, changes from being a second-order differential to a first-order differential of the transmitter current I_T. f_1 is defined by the following equation:

$$f_1 = \frac{R}{2\pi(L_1 + L_2)}. \tag{2.23}$$

As frequency continues to increase, the channel resonates when frequency reaches f_{SR}, where

$$f_{SR} = \frac{1}{2\pi\sqrt{(L_1 + L_2)C}}. \tag{2.24}$$

Figure 2.34c illustrates the time domain I_T and V_R waveforms. For proper operation, TCI must operate below f_{SR}. It may operate in first-order or second-order differential mode, depending on the data rate and the RLC characteristics of the interconnect.

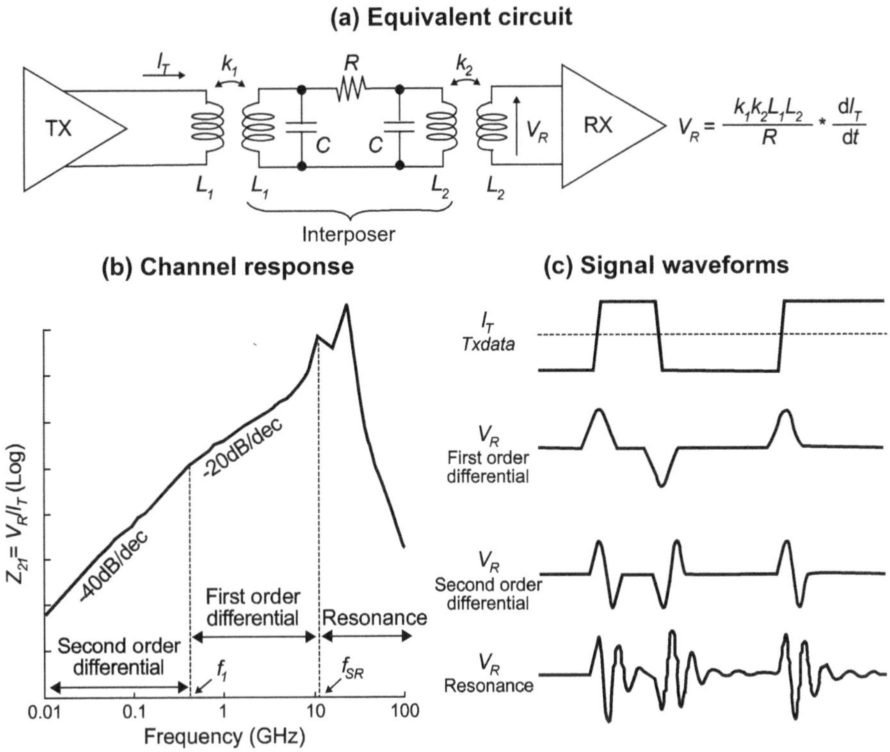

Figure 2.34 2.5D TCI interconnect model and simulated frequency and time domain responses. Copyright 2009 The Japan Society of Applied Physics [22]

Since theV_R waveform differs between the two modes, the transceiver must be designed according to the operating mode for accurate detection of the induced pulses.

Figure 2.35 depicts a dual-mode, edge-counting transceiver design proposed in [22] together with the time domain waveforms in both the first- and second-order differential operating modes. With this edge-counting transceiver, each transition in *Txdata* creates a short positive pulse in I_T, regardless of the direction of the transition. This in turn creates a double and triple pulse in V_R in first- and second-order differential mode respectively. A nonhysteresis comparator with offset threshold V_{TH} detects the negative pulse in both cases and toggles the data in the flip-flop to generate *Rxdata*. Since the transmitter consumes power only during a transition, this edge-counting scheme results in lower power. However, the short pulses require a higher channel cutoff frequency compared to the standard transceiver design that uses a hysteresis comparator in first-order differential mode. The dual-mode transceiver can operate as a standard transceiver in first-order differential mode, or an edge-counting transceiver in second-order differential mode, based on the value of the *MODE_SEL* bit.

Figure 2.36 illustrates a proposed implementation of TCI-enhanced 2.5D integration of SoC and memory chips from [23] and how it compares with a conventional TSV implementation in physical dimensions. The TSV implementation uses an

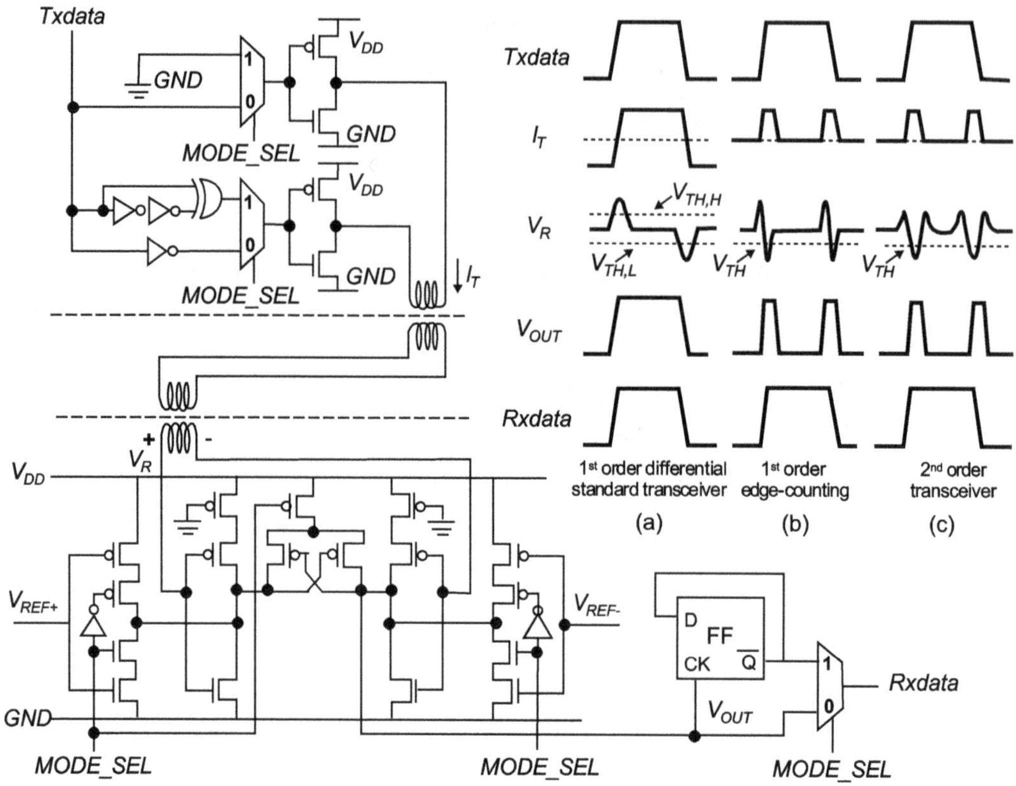

Figure 2.35 Dual-mode transceiver and time domain waveforms in each mode. Copyright 2009 The Japan Society of Applied Physics [22]

interposer with area that spans the sum of the areas of all chips, plus chip spacing and extra area to accommodate underfill bleeding. By contrast, the TCI implementation uses a small interposer that fits entirely underneath the overhangs of the interconnecting chips, which are flip-chip mounted on the package substrate. Using representative sizes of high-performance SoC and memory chips as an example – 21×36 mm^2 and 8×8 mm^2 respectively – an SoC and two memory chips would require a TSV interposer area of about 34×40 mm^2. By contrast, the TCI solution can be implemented with two pieces of interposer each with an area of 2.8×7.2 mm^2, resulting in a total area that is about 3% of the TSV interposer, representing significant cost saving. Furthermore, the TCI interposers do not add to the overall profile of the packaged system, which translates into $(0.22 - 0.08) = 0.14$ mm reduction in height.

Given that the TCI transceiver consists of digital CMOS circuits, and its coils are implemented using IC interconnect layers, its performance is expected to improve as process scales. If device size is scaled by a factor of $1/\alpha$, transfer speed and energy efficiency improve by factor of α. Furthermore, if communication distance is scaled by a factor of $1/\eta$, coil area can be reduced by a factor of $1/\eta^2$, and area efficiency improves by a factor of $\alpha\eta^2$.

Table 2.16. 2.5D TCI vs. advanced wired connection technologies. Copyright 2019 The Japan Society of Applied Physics [23].

	[24]	[25]	[26]	[27]	[28]	[23]	[23]
Year reported	2013	2016	2011	2013	2015	2019	2019
Area efficiency (Gb/s/mm^2)	250	267	167	94	500	317	1300
Energy efficiency (pJ/b)	0.26	0.94	2.6	0.10	0.30	1.2	0.22
Area ratio: interposer ÷ (SoC + memory)	N/A	N/A	1.54	1.54	1.54	0.045	–
Connection type	Bump	Bump	Bump/ TSV	Bump/ TSV	Bump/ TSV	Coil	Coil
Process (nm)	28	28	45	40	28	40	28

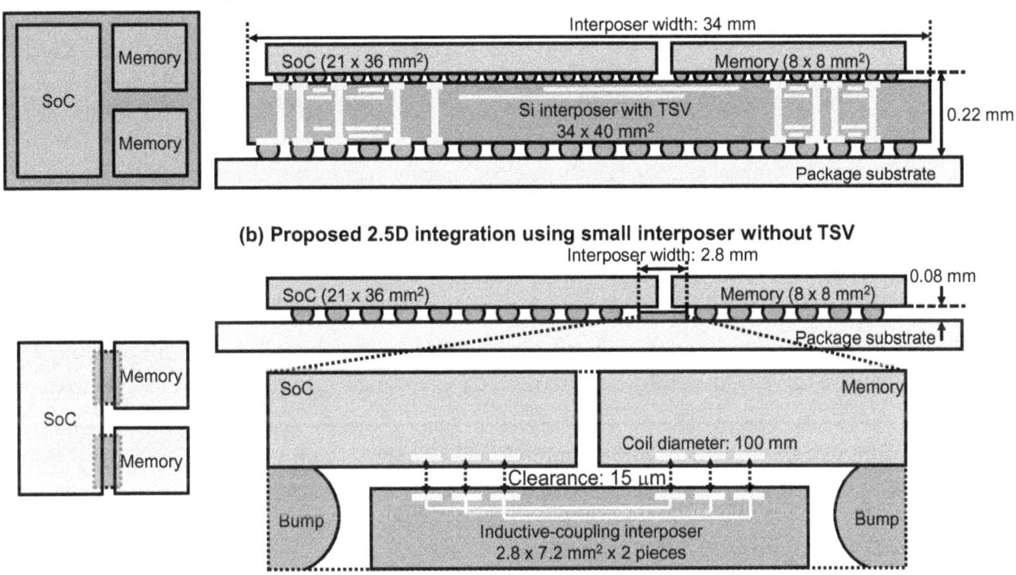

Figure 2.36 TCI-enhanced 2.5D integration of SoC and memory. (b) Copyright 2019 The Japan Society of Applied Physics [23]

A TCI-enhanced 2.5D integration test system using a transceiver chip fabricated in a 40 nm process is reported in [23]. The system employs the standard transceiver with an H-bridge transmitter and operates in the first-order differential mode at a data rate of 3.5 Gb/s. It achieves an area efficiency of 317 Gb/s/mm^2 and consumes 1.2 pJ/b of energy. This compares favorably with advanced wired interconnection technologies as shown in Table 2.16.

The last column in Table 2.16 shows the projected area efficiency and energy consumption of the 2.5D TCI channel in a 28 nm process based on simulation, which is better than performance improvement from the same process scaling achieved by

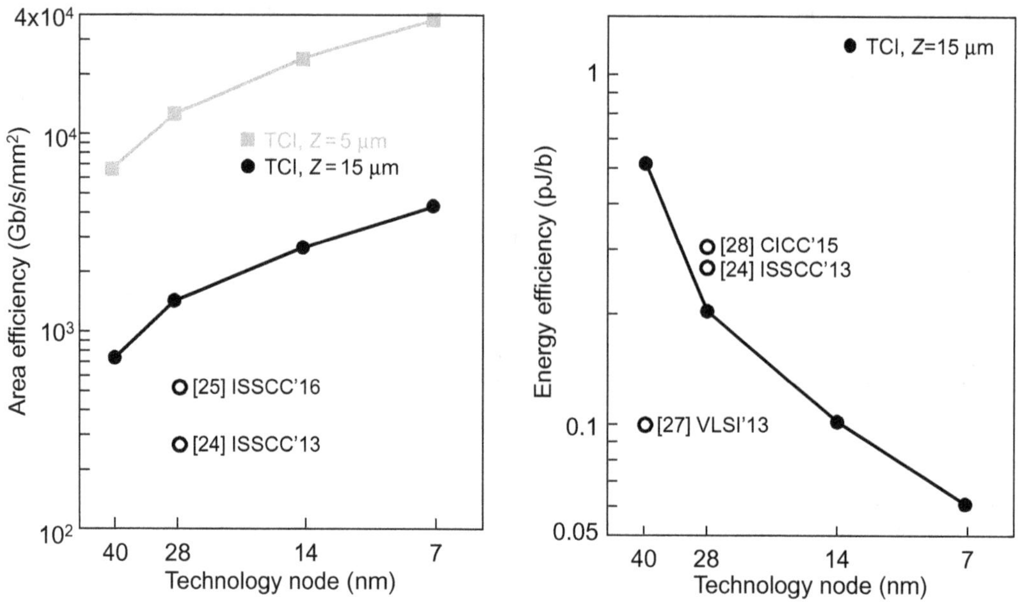

Figure 2.37 Simulated 2.5D TCI channel performance vs. process. Copyright 2020 The Japan Society of Applied Physics [29]

conventional technologies. These data plus additional simulated data points for 14 and 7 nm processes and a reduced communication distance $Z = 5$ μm are plotted in Figure 2.37, illustrating continuous TCI performance improvement as process continues to scale. Improvement from reduced communication distance can be achieved through improvement in die thinning and stacking technologies.

2.3.4 TCI-Enhanced 2D IC Integration

In the discussion thus far, the coupled TCI coils are placed face to face and concentrically to maximize the coupling coefficient. However, since the magnetic field emanating from the transmitter coil is omnidirectional, transmission is possible between coils arranged in other orientations. In particular, if the coils are placed side by side in a coplanar manner, it is possible to achieve direct communication between two chips mounted on the same package substrate, as illustrated in Figure 2.38. This would correspond to 2D IC integration being enhanced by TCI.

When the two coils are placed face-to-face and concentrically, coupling occurs on all four sides of the coils. By contrast, when the two coils are placed side by side, coupling occurs primarily between the facing sides, and hence the coupling strength is reduced to approximately ¼ for the same coupling distance. Recalling from Section 2.1 that $V_R = M\frac{dI_T}{dt}$ and $M = k\sqrt{L_T L_R}$, it can be seen that a smaller k resulting from the side-by-side orientation can be compensated by a larger dI_T/dt, L_T, and L_R. dI_T/dt can

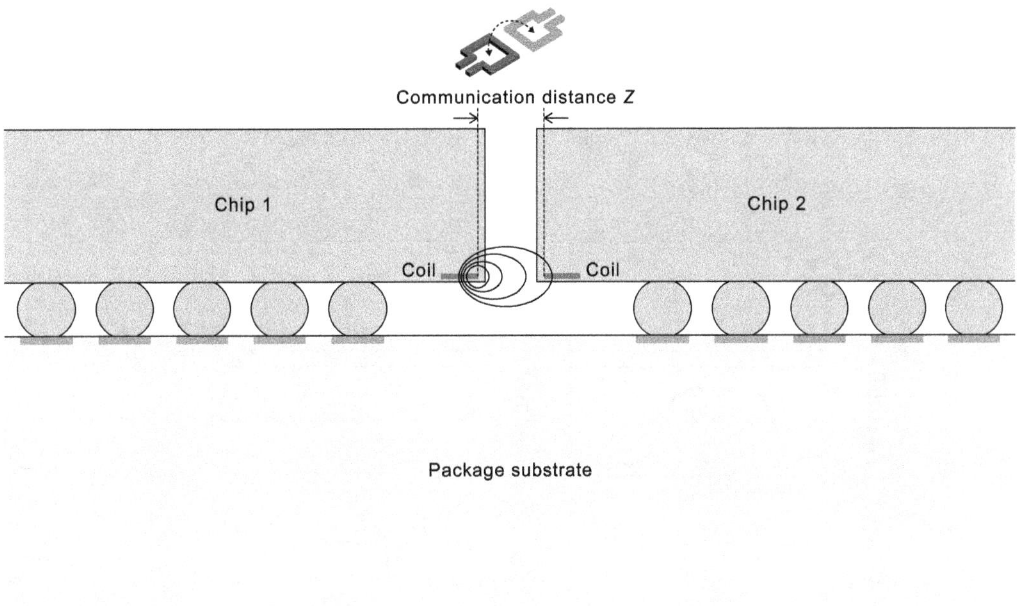

Figure 2.38 TCI-enhanced 2D IC integration.

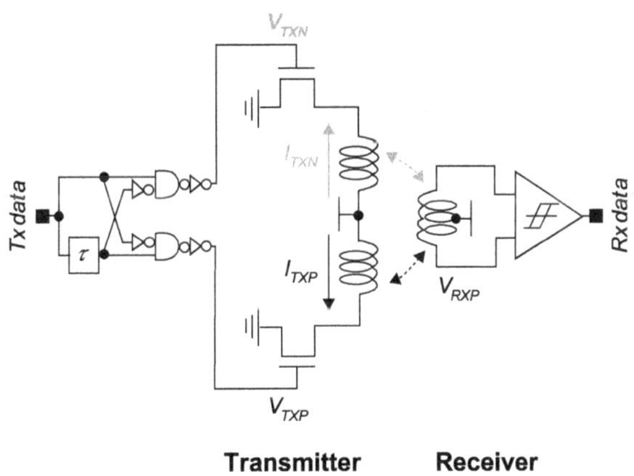

Figure 2.39 Rapid turn-off transceiver design for 2D integration. Copyright 2016 IEEE. Reprinted, with permission, from [30]

be increased by modifying the transceiver design, while L_T and L_R can be increased by modification of the physical layout of the coils.

Figure 2.39 depicts a rapid turn-off transceiver design proposed in [30] to achieve an increased dI_T/dt. The transmitter consists of two coils, each driven by an NMOS gate. Since the two NMOS gates are driven by complementary signals, they are turned on and off in a complementary manner. When an NMOS is being turned on, a counterelectromotive force is induced in the coil that lowers its drain voltage, putting

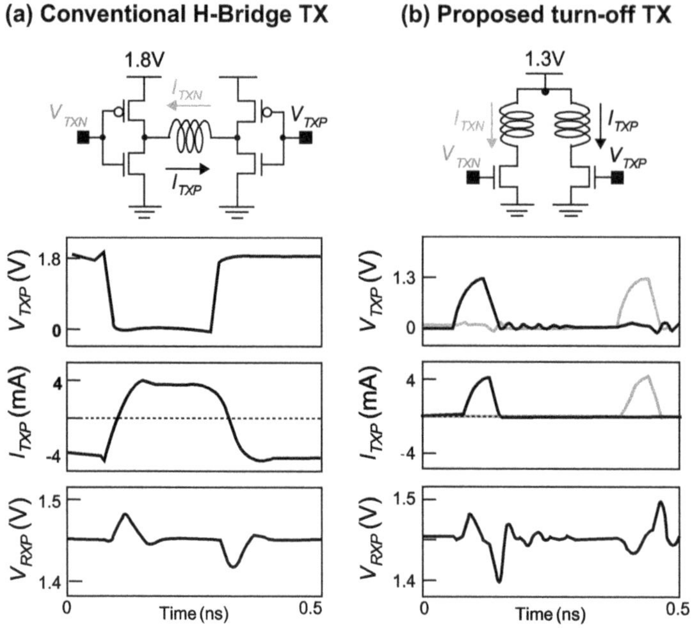

Figure 2.40 Comparison of rapid turn-off vs. H-bridge transmitter from simulation. Copyright 2016 IEEE. Reprinted, with permission, from [30]

the NMOS in its triode operating region. Conversely, when the NMOS is being turned off, the counterelectromotive force raises its drain voltage, putting it in the saturation region. As a result, the NMOS transconductance (βV_{DS}) is lower during turn-on and higher during turn-off, leading to rapid turn-off and hence larger $\mathrm{d}I_T/\mathrm{d}t$.

Figure 2.40 compares simulated transceiver waveforms between the rapid turn-off and conventional H-bridge transmitter, demonstrating a 1.5× faster $\mathrm{d}I_T/\mathrm{d}t$ achieved by the rapid turn-off design. Furthermore, the elimination of the PMOS from the H-bridge transmitter reduces the layout area by 2/3, allowing the entire transceiver to be placed within the coil area for a small coil diameter of $D = 40$ μm, even in a 180 nm CMOS process. In addition, the design allows the supply voltage to be lowered to 0.7 V_{DD} (voltage drop in the PMOS). Together with the replacement of edge transitions with pulses to avoid constant power consumption, total power consumption is reduced by 5/6. The coils are designed to have a low Q-factor (3–6) to suppress LC resonance when the NMOS is turned off.

An optimized coil layout to maximize coupling in 2D integration as shown in Figure 2.41 is also proposed in [30]. Since M is roughly proportional to $n^{1.7}$, where n is the number of turns, the design maximizes the number of turns both within an interconnect layer (spiral windings) and across interconnect layers (solenoid windings). Spiral windings, however, increase the communication distance. To minimize this increase, on the facing side that is used for coupling, the coil lines are made narrow. Since the other three sides of the coil have negligible effect on M, their lines

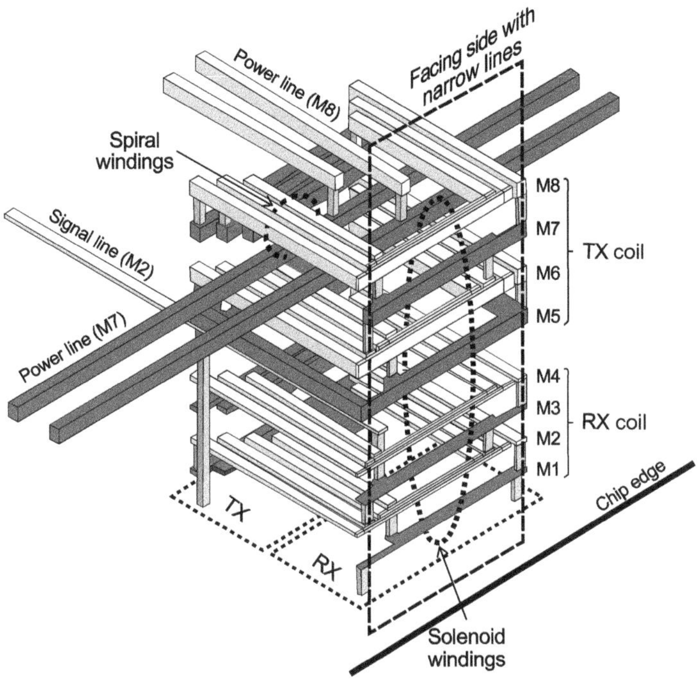

Figure 2.41 Optimized coil layout for 2D integration. Copyright 2016 IEEE. Reprinted, with permission, from [30]

are made wide to reduce the overall parasitic resistance of the coil. Furthermore, since there is no other wiring near the edge of the chip, all the wiring resources can be used for routing the facing side of the coil. Layout of the other three sides of the coil, however, must also accommodate layout of power and signal lines. Since the transmitter coil carries much larger current than the receiver coil, the former is routed on the thicker upper layers, while the latter is routed on the lower layers. For example, transmitter coil can be routed on layers M5–M8, while receiver coil is routed on layers M1–M4.

The weaker coupling between the coupled coils means that it is more sensitive to variations in communication distance Z (in this case, the distance between facing edges) and placement variations δy (lateral misalignment = relative shift along chip edge) and δz (variation in height). The use of laser ablation has made crack-free dicing possible, enabling metal wiring to be placed down to 1 μm from the chip edge [31]. Furthermore, bonding two chips with a minimum gap of 1 μm has been achieved in mass production [32]. As a result, the communication distance between side-by-side coupled coils can be as short as $Z = 3$ μm. On the other hand, when copper pillars are used to attach the chips to the package substrate, a height tolerance of ± 1.5 μm is achievable. Since the magnetic field from the facing side of the transmitter coil is circular, the effect on coupling due to a relative shift in height of $\delta z = 1.5$ μm between the facing coils is simply that of a slight increase in communication distance.

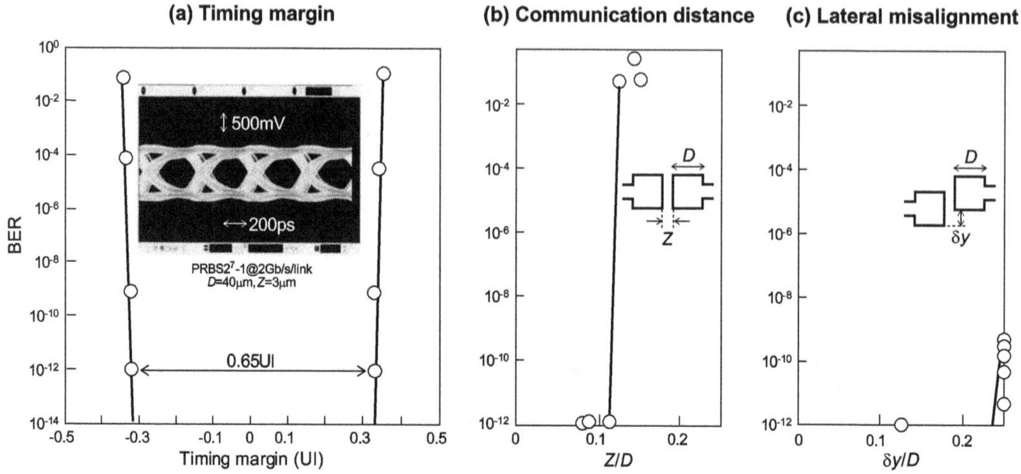

Figure 2.42 Measured BER as a function of coils placement. Copyright 2016 IEEE. Reprinted, with permission, from [30]

The sensitivity of the operation of side-by-side coupled coils to placement variations is investigated in [30] using a test chip fabricated in a 180 nm CMOS process and a TCI channel data rate of 2 Gb/s. The measured BER results are depicted in Figure 2.42 for a coil diameter of $D = 40$ μm. Timing margin at BER $= 10^{-12}$ is 0.65 UI for $Z = 3$ μm and $\delta y = 0$. Furthermore, communication is possible for $Z < 0.13D$ and $\delta y < 0.2D$. The test chip achieves an area efficiency of 1 Tb/s/mm^2 and energy consumption of 4 pJ/b.

In production, the dimensions and tolerances required for TCI-enhanced 2D integration, while achievable, are still difficult for mainstream application today. But when advances in mainstream manufacturing process finally enable the short communication distance and small tolerances required, a TCI-enhanced solution will become available to offer better performance at lower power and in a smaller footprint than conventional 2D integration.

2.3.5 Closing Thoughts

The contactless nature and omnidirectional property of magnetic coupling enable TCI to offer not only a lower cost, lower power, higher performance, and higher reliability solution than TSV in 3D and 2.5D integration, but also enhancement to IC integration in other configurations including 2.9D and 2D, thereby increasing design freedom to enable innovative system design to meet a variety of system requirements.

Figure 2.43 illustrates the enhancement from applying TCI to 3D integration of NAND and DRAM chips respectively. One order-of-magnitude or more improvement is achievable especially in chip stack height, energy, and area efficiencies.

Figure 2.44 depicts how TCI can be applied to build a DRAM memory system using the 2.9D integration concept described in Section 2.3.2 to enable a 100 GFLOPS/W-

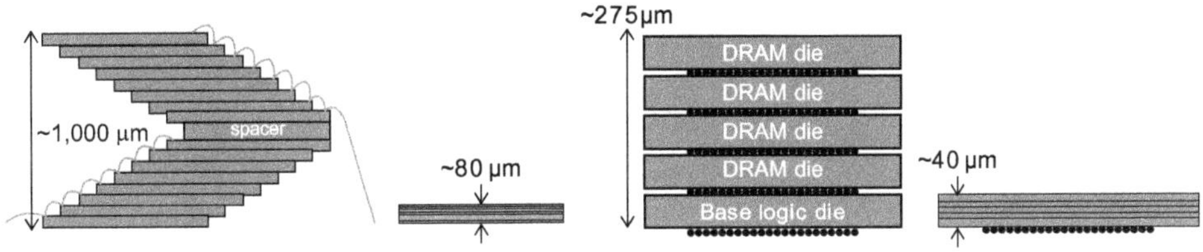

	NAND stacking		DRAM stacking	
	Wirebond	TCI, HDSV	TSV	TCI, HDSV
# of stacked die	16	16	5	5
Die pitch	50μ	5μ	55μ	8μ
Total height	~1000μ	~80μ	~275μ	~40μ
Die area	1x	~0.9x	1x	~0.9x
Data link	wirebond	TCI	TSV	TCI
Power delivery	wirebond	HDSV	TSV	HDSV
IO energy/bit	1x	< 1/400x	1x	< 1/10x
IO data rate/area			~ 200 Gb/s/mm²	~ 860 Gb/s/mm²

Figure 2.43 TCI-enhanced 3D integration of NAND and DRAM chips.

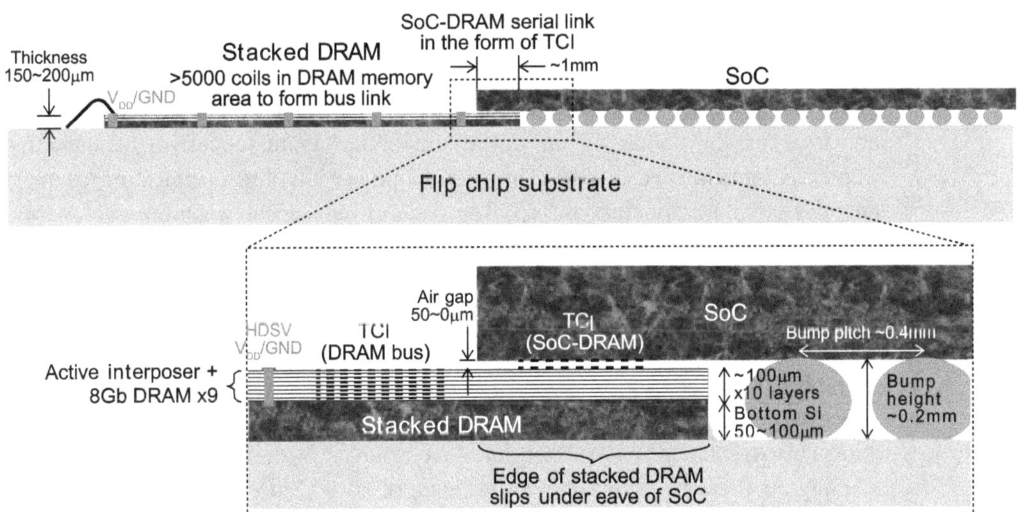

Figure 2.44 TCI applied to the development of a 100 GFLOPS/W computer in Japan [33].

class supercomputer. The concept was investigated in Japan from 2015 to 2019 as elaborated in [33]. It demonstrates the potential of TCI in enabling quantum leaps in both computational performance and power efficiency simultaneously.

The discussion in this section using different integration configurations as examples illustrates that there are various ways to optimize both the transceiver circuit and coil design to achieve different goals. Keeping the simple operating principle of TCI in mind, other transceiver design alternatives can be further explored to create additional applications of the technology.

2.4 Wireless Power Delivery

One of the benefits of using TCI as a chip interface is the elimination of physical connection between the communicating chips. The benefit can manifest itself in two ways. First, TCI eliminates reliability problems that result from stress in a physical connection such as thermal stress induced in microbump connections. Second, it avoids tear and wear resulting from physical contact, such as damages that probes inflict on contact pads of a chip during wafer testing. However, in addition to communication, a chip must draw power from the outside world as well. If physical connection is required for power delivery, it will partially negate the benefits of a contactless data interface. Furthermore, in the case where wirebonds are used for power delivery, a spacer may have to be inserted between two stacked chips, thereby increasing the communication distance and reducing the effectiveness of TCI coupling. A wireless power delivery technology is therefore a highly desirable complement to TCI.

2.4.1 Basic Structure of a Wireless Power Interface

Figure 2.45a depicts the basic structure of a wireless power interface based on inductive coupling, while Figure 2.45b shows the circuit model of the inductive coupler as implemented in [34]. The wireless power interface operates on the same inductive coupling principle as TCI. The main differences between the two and the rationale behind them are summarized in Table 2.17. An alternating pattern, in other words, a clock, is used instead of a data pattern to drive the transmitter coil to induce an alternating voltage in the receiver coil. The voltage is then passed through a rectifier to convert it into a DC voltage, which in turn drives the load R_{LOAD} with the help of a smoothing capacitor C_F.

With the use of a hysteresis comparator, TCI is able to sense a small received voltage to operate robustly. Therefore, the coupler is designed to have a small coupling coefficient k, typically in the range of 0.1–0.2, to achieve high area efficiency. In the case of the wireless power interface, however, a large k is needed to deliver sufficient power with high efficiency. For instance, the interface in [34] is designed with a coil diameter of $D = 700$ μm for a communication distance $Z = 15$ μm to achieve $k = 0.75$. This corresponds to a $D = 46.7\ Z$ compared to $D = 2\text{–}3\ Z$ recommended for TCI. Meanwhile, while the Q-factor of the TCI coupler is designed to be large enough to avoid excessive attenuation, yet small enough to avoid excessive ringing, it is designed to be large for a wireless power interface to maximize transmission efficiency. This in turn requires that resistance R of the coils to be small. Furthermore, the receiver coil resistance R_{L2} must be smaller than both the load and

Table 2.17. Key differences between wireless power interface and TCI.

	TCI	Wireless power	Rationale for difference
Driver pattern	Data pattern	Alternating pattern	To create an alternating receiver voltage
Receiver structure	Hysteresis comparator	Rectifier and capacitor	To convert receiver voltage from AC to DC
Coupling coefficient k	Small to achieve high area efficiency (typically 0.1–0.2)	Large	To deliver sufficient power
Q-factor	Not too small or large (typically 1–2.5)	Large	To maximize power transmission efficiency

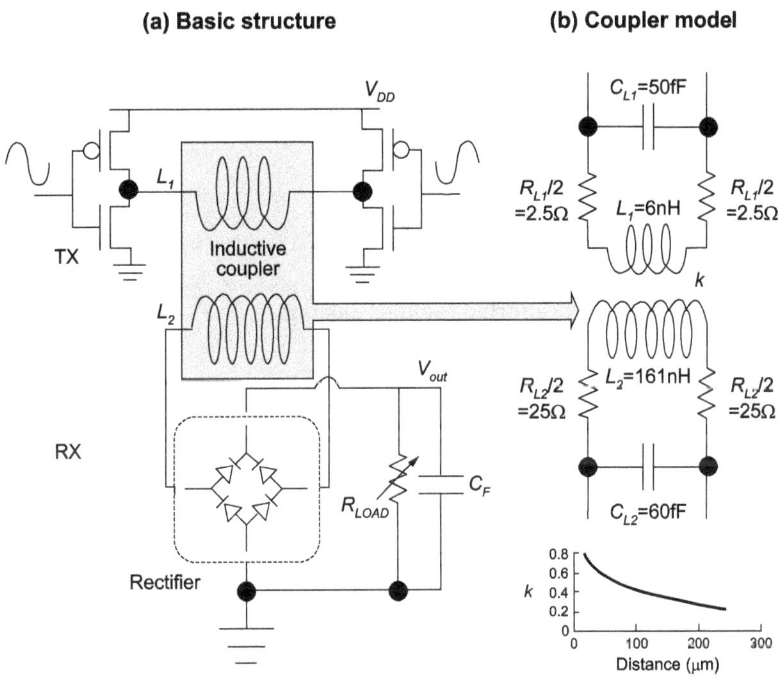

Figure 2.45 Basic structure of an inductively coupled wireless power interface. © 2007 IEEE. Reprinted, with permission, from [34]

resistance of the forward biased path through the rectifier to avoid excessive voltage drop in the output voltage V_{out}. As a result, while R of a TCI receiver coil is typically 100–300 Ω, it is designed to be 25 Ω in [34]. (As a reminder, R of the transmitter coil is smaller because the driver on-resistance is in series and hence adds to the resistance of the coil.) Nevertheless, making R small can reduce k. Hence, there may be a trade-off between Q, V_{out}, and k.

2.4.2 Optimization for Efficiency and Quality

2.4.2.1 Enhanced NCG Rectifier for Efficiency

Because of the trade-off between Q and k, there is a limit on how much they can be increased to improve power transmission efficiency. Furthermore, a large k requires a large D, degrading area efficiency. An alternate means to improve power transmission efficiency is optimizing structure of the rectifier. Figure 2.46a depicts two previously reported high-power rectifier designs from [35] and [36] respectively. The PMOS bridge (PB) rectifier consists of PMOS-based diodes only and is proposed for SiP application, while the NMOS cross-connected gate (NCG) rectifier has two NMOS transistors with cross-connected gates replacing the lower half of the PB rectifier and is used in biomedical implants. The PMOS-based diode has a main PMOS whose body effect is suppressed by two auxiliary PMOS transistors. The auxiliary transistors ensure that the N-well of the main transistor is connected to the higher potential between D+ and D–, which serves to eliminate substrate leakage and the risk of latch-up.

Figure 2.47 depicts the difference in terminal voltages of the lower-right M2 transistor between the PB and NCG rectifier designs when RFin1 > RFin2. In particular, the M2 transistor drive voltage, V_{GS2}, is as follows:

$$|V_{GS2,PB}| = |V_{DS2,PB}| \tag{2.25}$$

$$|V_{GS2,NCG}| = |V_{DS2,NCG}| + |V_{DS1,NCG}| + |V_{out,NCG}| \tag{2.26}$$

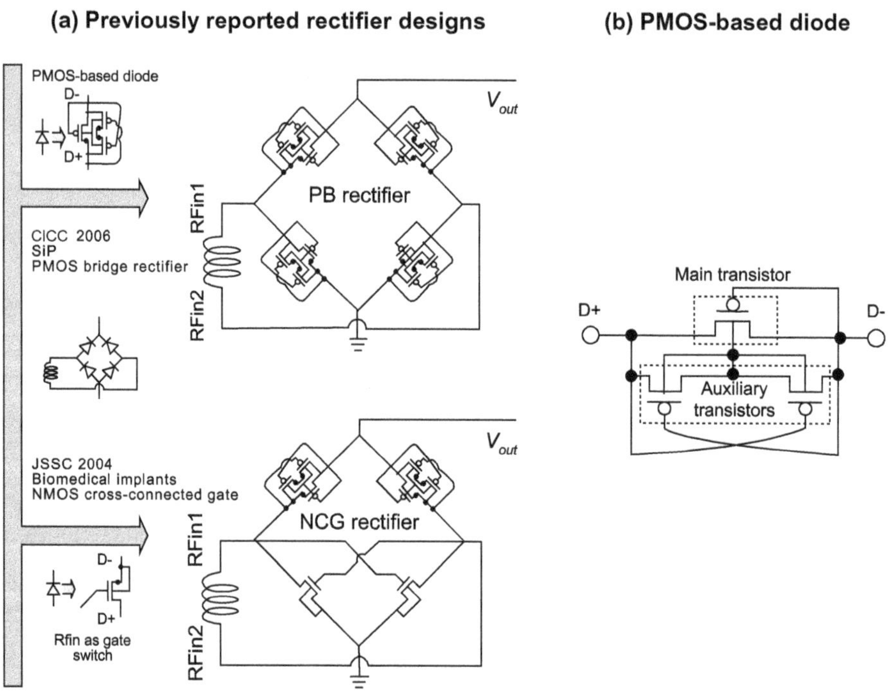

Figure 2.46 Previously reported PB and NCG rectifier designs. © 2007 IEEE. Reprinted, with permission, from [34]

Table 2.18. Comparison of PB and NCG rectifier designs.

	PB rectifier	**NCG rectifier**
Pro	• Smaller substrate leakage	• Smaller NMOS size • Lower turn-on input level • Lower on-resistance
Con	• Larger PMOS size • Higher turn-on input level • Higher on-resistance	• Higher switching loss especially when input voltage is high

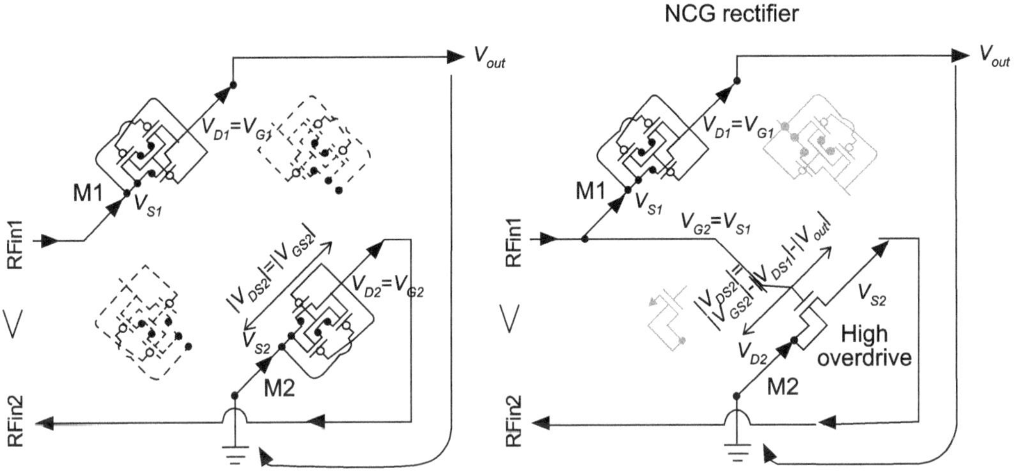

Figure 2.47 Overdrive voltage comparison between PB and NCG rectifier designs. © 2007 IEEE. Reprinted, with permission, from [34]

Therefore, the NCG rectifier has a much higher M2 overdrive voltage than PB. As a result, not only is the NCG rectifier smaller because NMOS is smaller than PMOS, but it also has lower turn-on input level and smaller on-resistance and hence lower voltage drop due to the larger overdrive. The NCG rectifier design generally suffers from higher switching loss, but it is not significant at our operating frequency of a few 100 MHz. The smaller voltage drop is, however, important given the low CMOS operating voltage, especially as process scales. The pros and cons of the two designs are summarized in Table 2.18.

The NCG rectifier design can be further enhanced by reducing the voltage drop V_{on} across the upper PMOS-based diodes. As is, the voltage drop is as follows:

$$|V_{on}| = |V_{SD}| = |V_{th}| + \sqrt{\frac{2I_D}{\mu C_{OX}(W_g/L_g)}}, \tag{2.27}$$

where V_{th} is the threshold voltage, I_D the drain current, μ the charge-carrier effective mobility, W_g the gate width, L_g the gate length, and C_{OX} the gate oxide capacitance per

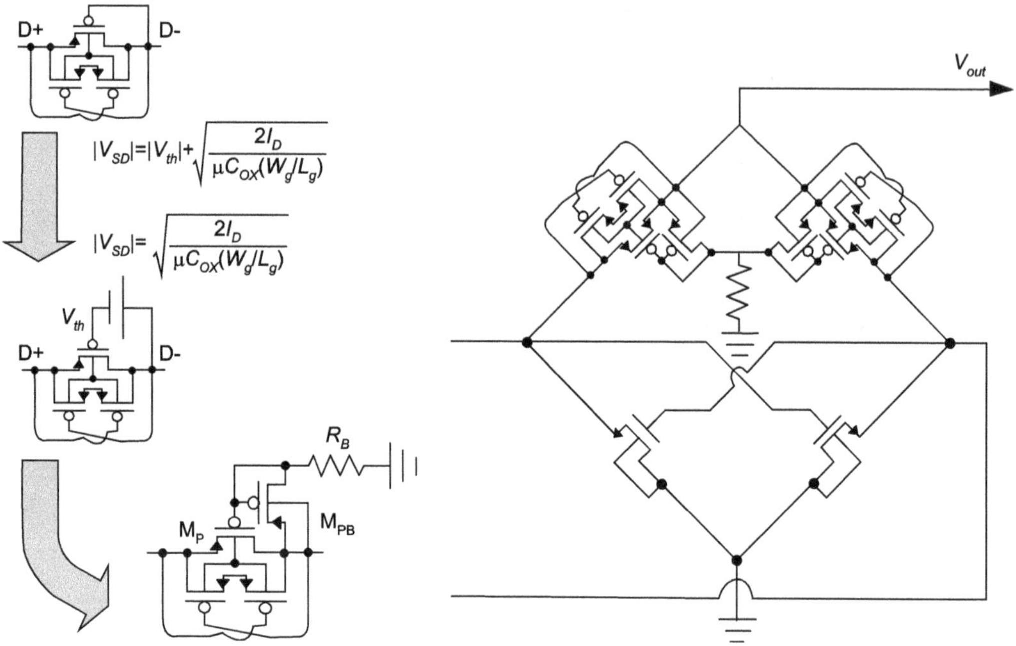

Figure 2.48 Enhanced NCG rectifier design. © 2007 IEEE. Reprinted, with permission, from [34]

unit area of the main transistor. As shown in [37], the threshold voltage can be canceled out by applying a bias voltage equal to V_{th} between the drain and the gate, resulting in

$$|V_{on}| = \sqrt{\frac{2I_D}{\mu C_{OX}(W_g/L_g)}}. \tag{2.28}$$

This drain-gate bias can be generated using an additional PMOS M_{PB} connected as shown in Figure 2.48. This bias serves to not only lower the voltage drop across the diode, but also reduce the turn-on time of the PMOS, resulting in enhanced high-frequency performance of the rectifier.

Figure 2.49 shows the measurement results from a test chip of the enhanced NCG rectifier design reported in [34]. The efficiency of power conversion from the input to the output of the rectifier, P_{out}/P_{in}, using a differential input with 50 Ω resistance and a load of 140 Ω, decreases with frequency but stays above 15% up to 350 MHz. The output resistance goes up as the input voltage decreases. It is as low as 20 Ω for an input of 1.8 V, and increases to close to 40 Ω for an input of 1 V. Both the frequency range and output resistance are adequate for the rectifier to be usable for our applications.

Figure 2.50 depicts the test system used in [34] to evaluate the basic wireless power interface structure combined with the enhanced NCG rectifier design. The output of

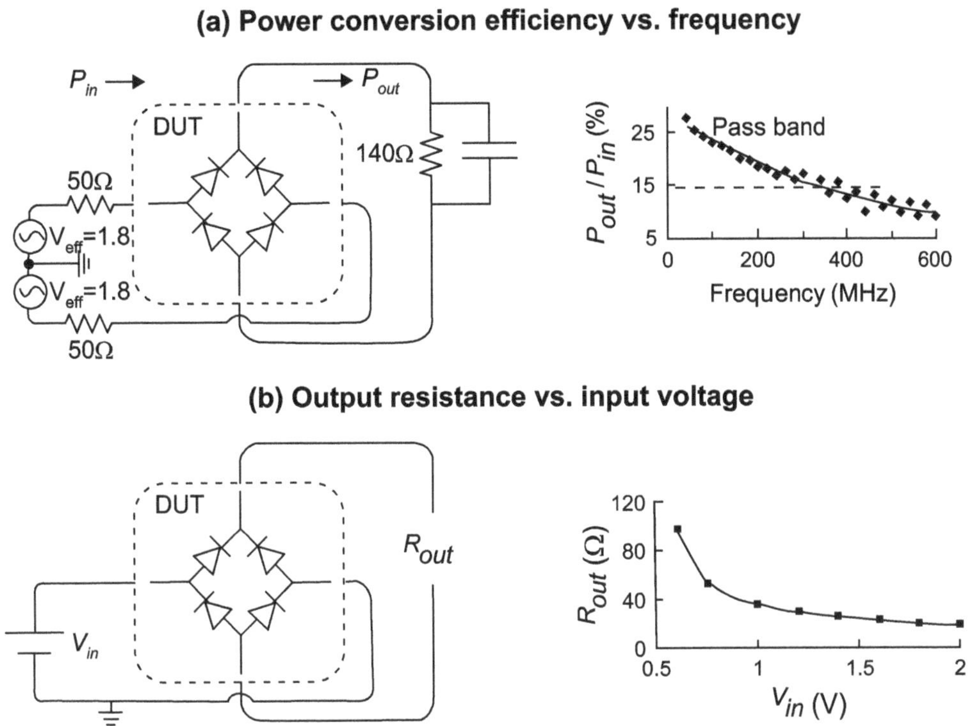

Figure 2.49 Efficiency and output resistance of enhanced NCG rectifier design. © 2007 IEEE. Reprinted, with permission, from [34]

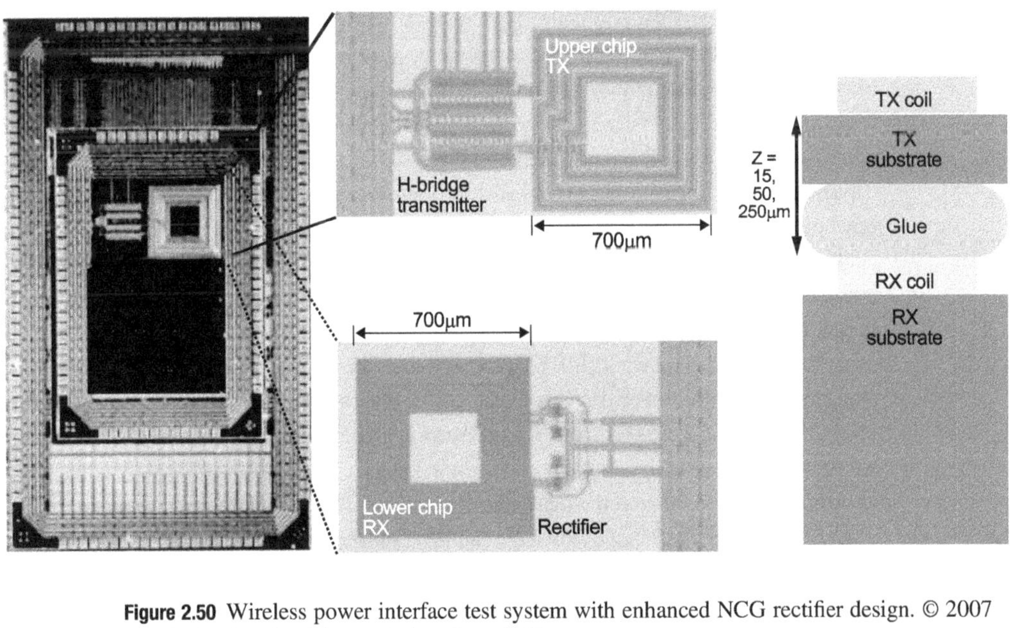

Figure 2.50 Wireless power interface test system with enhanced NCG rectifier design. © 2007 IEEE. Reprinted, with permission, from [34]

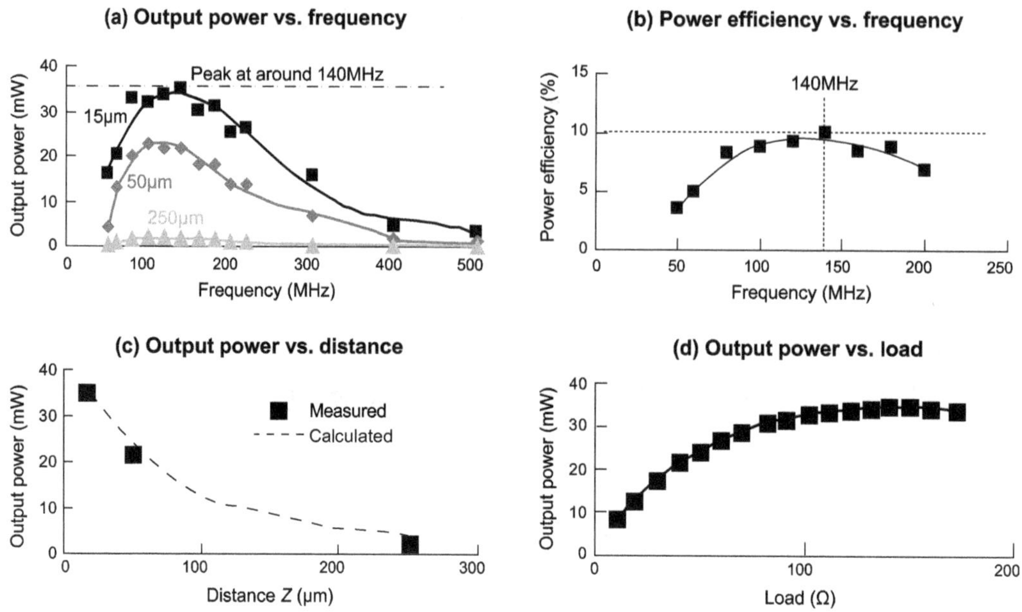

Figure 2.51 Output power and power efficiency of test system. © 2007 IEEE. Reprinted, with permission, from [34]

the rectifier is connected to a smoothing capacitor of $C_F = 100$ pF and a rheostat to vary the load.

Figure 2.51 plots the measurement results of the test system. As discussed in Section 2.1.3.2, the inductively coupled coils create a band pass filter with a peak frequency determined by the inductance and capacitance of the coils. Below the peak frequency, impedance of the inductance dominates over that of the capacitance. As a result, power transmission efficiency varies proportionally with the ratio of ωL to the coil resistance, which increases with frequency. Beyond the peak frequency, however, impedance of the capacitance dominates, and power leaks through the capacitance, resulting in decreasing power transmission efficiency with frequency. The coupled coils in the test system have a peak frequency around 140 MHz. Consequently, both the output power and power efficiency of the interface peak around 140 MHz. Furthermore, since inductance and capacitance of the coils are independent of the communication distance Z to first order, the peak frequency stays around 140 MHz as Z is varied from 15 to 250 μm. Nevertheless, as expected, the peak power delivered drops as Z increases, dropping more than 1/3 from 36 to 21 mW as Z increases from 15 to 50 μm. It is therefore important to operate around the peak frequency of the coupled coils, while keeping the communication distance short. Figure 2.51b shows that at the peak frequency of 140 MHz and $Z = 15$ μm, a maximum power transmission efficiency of 10% can be achieved.

Figure 2.51d shows that the output power increases with load and reaches its peak with a load of 140 Ω. This corresponds to the effective source resistance of the wireless power interface.

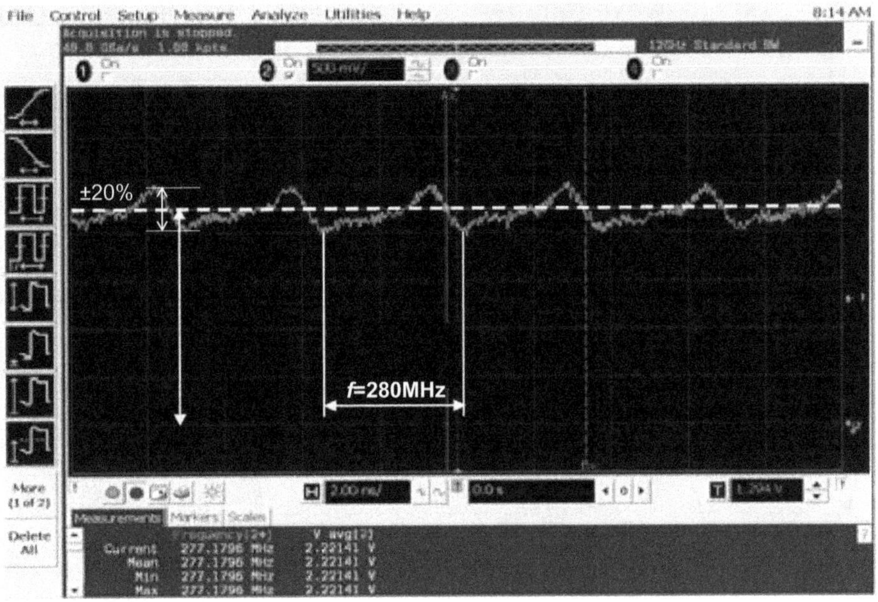

Figure 2.52 Output voltage waveform at peak frequency. © 2007 IEEE. Reprinted, with permission, from [34]

Figure 2.52 depicts the measured output voltage waveform at the peak operating frequency of 140 MHz, $Z = 15$ μm, and with a load of 140 Ω. It reveals a ripple at twice the frequency of 280 MHz and with a magnitude of ±20%, which is larger than what is generally acceptable. A solution to suppress the ripple is proposed in Section 2.4.2.3.

2.4.2.2 Half-Wave Rectifier for Efficiency

The rectifier designs discussed so far are all full-wave rectifiers that consist of two pairs of diodes conducting current in opposite directions. As a result, they utilize both phases of the AC input voltage V_{in}, as illustrated in Figure 2.53a. Since each current path consists of two diodes, V_{in} must exceed $2{*}V_{on}$ to deliver V_{out}, where V_{on} is the voltage drop across a forward-biased diode. The full-wave rectifier is therefore more suited for when V_{in} is high and V_{on} is low.

Figure 2.53b depicts a half-wave rectifier used in [38]. It utilizes only the positive phase of V_{in}, and experiences voltage drop across only one diode. As a result, the half-wave rectifier may be more efficient when V_{in} is low and V_{on} is high.

Each diode in the half-wave rectifier can be implemented by tying the gate of an NMOS to its drain as shown in Figure 2.54a. As previously noted, V_{on} of a PMOS can be reduced by an amount equal to its threshold voltage if its drain is biased against its gate by V_{th}. V_{on} of an NMOS can be similarly reduced by applying a bias voltage of V_{th} between its gate and drain, resulting in a half-wave rectifier with enhanced efficiency (Figure 2.54b).

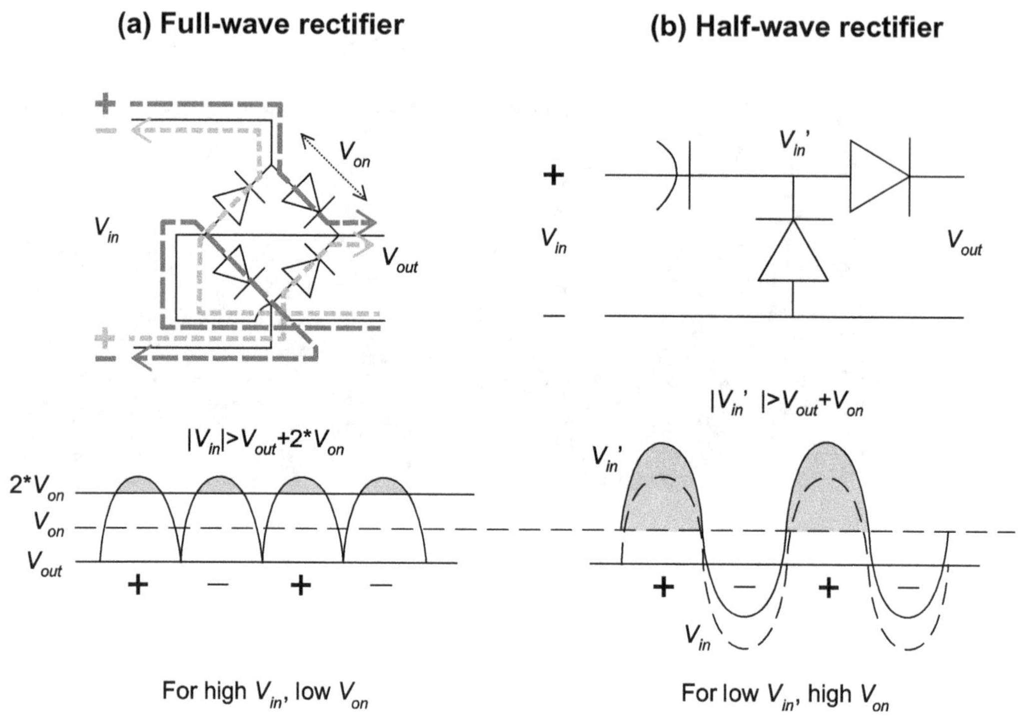

Figure 2.53 Half-wave vs. full-wave rectifier. Copyright 2008 The Japan Society of Applied Physics [38]

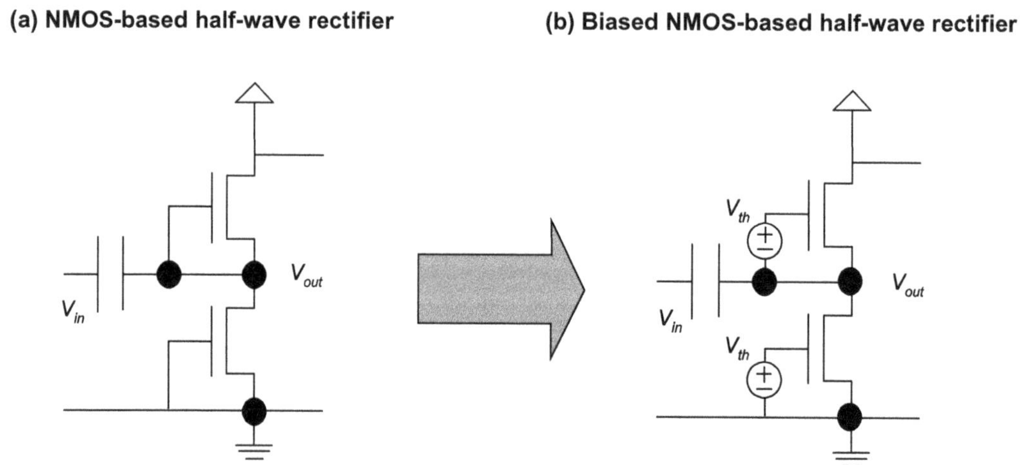

Figure 2.54 NMOS-based half-wave rectifier. Copyright 2008 The Japan Society of Applied Physics [38]

2.4.2.3 Multichannel, Multiphase for Increased Power and Reduced Ripple

The ±20% ripple measured in Figure 2.52 is more than generally acceptable supply tolerances. A larger smoothing capacitor can help to suppress it, but it is costly to build such a large capacitor on-chip. Instead, [38] proposes using the multichannel, multiphase

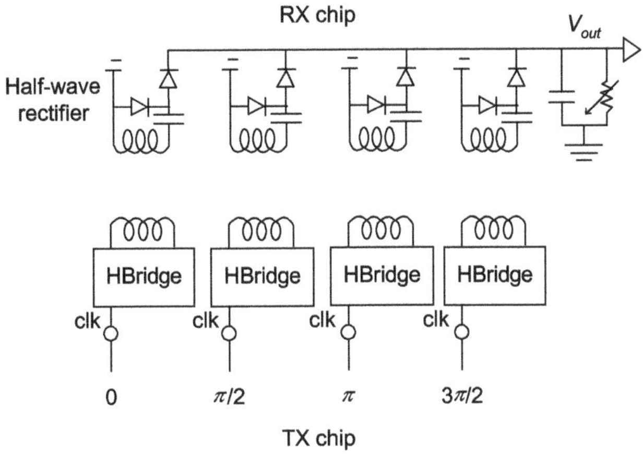

Figure 2.55 Multichannel, multiphase wireless power interface. Copyright 2008 The Japan Society of Applied Physics [38]

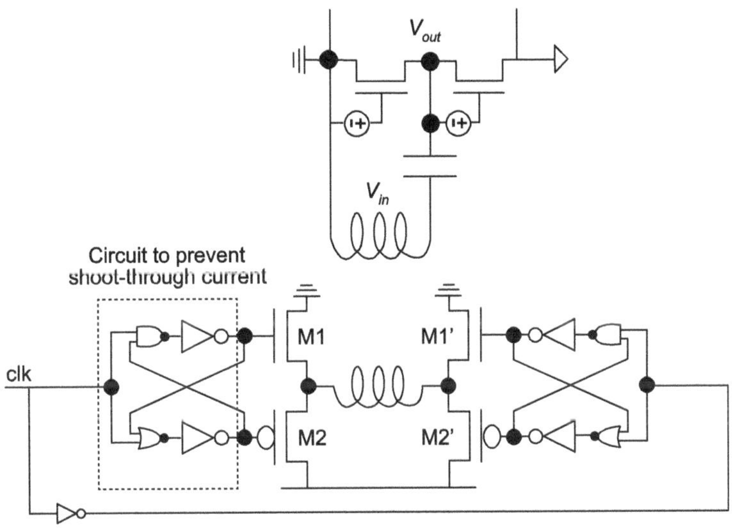

Figure 2.56 Circuit schematics per power channel. Copyright 2008 The Japan Society of Applied Physics [38]

solution in Figure 2.55, which reduces ripple and boosts the transmitted power at the same time. The interface consists of four parallel and identical power channels, each with an H-bridge transmitter, inductively coupled coils, and a biased NMOS-based half-wave rectifier. The four channels are driven by the same clock signal with a successive phase offset of 90 degrees. Currents from the four channels are fed into the same smoothing capacitor and load. As a result, they add to boost the transmitted power. At the same time, the phase offsets result in averaging of and hence reduction of the individual ripples.

Figure 2.56 depicts the circuit schematics for each channel. Even though the same H-bridge transmitter is used as in TCI, the transistors M1, M2, M1', and M2' are sized

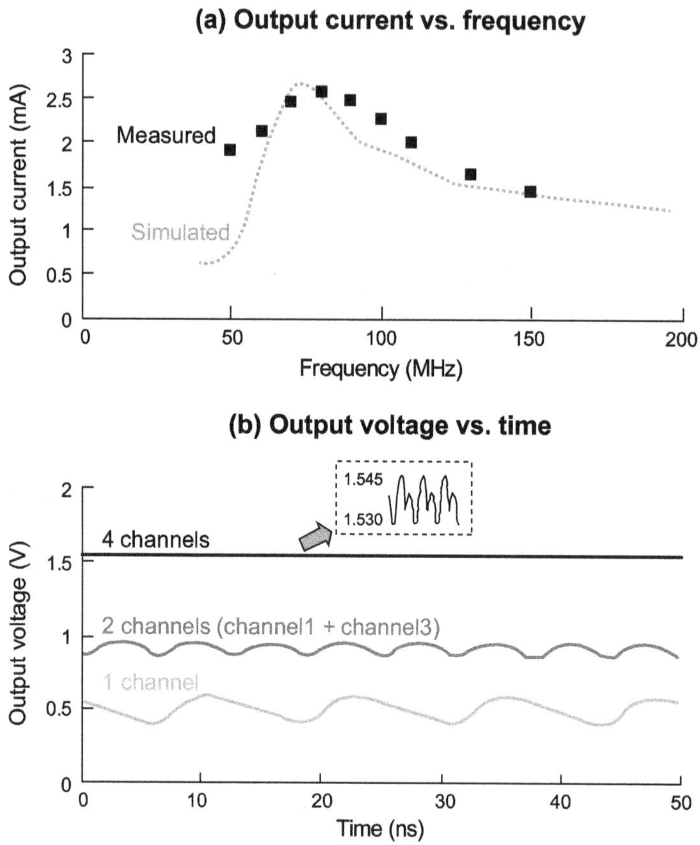

Figure 2.57 Output of four-channel, four-phase wireless power interface. Copyright 2008 The Japan Society of Applied Physics [38]

much larger than in TCI in order to deliver sufficient power. As a result, they have slower switching speed. Consequently, there is a higher risk that both the PMOS and NMOS of the same leg are momentarily on at the same time during a transition, causing large shoot-through current. The additional circuit within the dashed line uses the gate voltage of one transistor to control the switching of the other to ensure that they are not turned on simultaneously to eliminate the shoot-through current.

The proposed solution is tested in [38] and the results are summarized in Figure 2.57. Figure 2.57a depicts the average output current vs. frequency for one channel with a load of 200 Ω. The output current reaches a peak of 2.5 mA at 80 MHz. (The peak frequency is lower than before due to larger coil inductance in this test system.) Figure 2.57b illustrates how the output voltage and ripple change as more and more channels are switched on. Because the channels are connected in parallel to the load, individual currents add to each other. Consequently, the output voltage increases as more channels are turned on, although it increases less than linearly due to the phase offsets between channels. Furthermore, phase offsets help to smooth out the ripple. As a result, while the average output voltage due to one channel is 0.5 V with a 40% ripple of 0.2 V, it increases to 1.5 V with a much smaller 1% ripple of 15 mV when all four channels are switched on.

2.4.2.4 Nested Clover Coils for Reduced Interference of Data Channel

Since inductively coupled wireless power transmission uses larger coils than data transmission, it is conceivable that data channel coils can fit within power coils to reduce area penalty. However, the larger transmitter and coupling coefficient used for power transmission can result in excessive interference with data transmission. Time-interleaving power and data transmission can reduce interference, but at the expense of a lower effective data rate. Figure 2.58a depicts the nested clover coils solution proposed in [39] to minimize interference of data channel from not only power but also clock in source-synchronous data transmission. The power channel uses a single coil at both ends of the channel. By contrast, each of data and clock channels uses a differential coil that consists of two coils connected in such a way that their currents flow in opposite directions. For instance, if current flows clockwise in the upper-left data coil, it flows counterclockwise in the lower-right coil. As a result, the output voltage of the differential coil is the differential of the voltages induced in the two coils, hence the name. Data and clock differential coils are arranged as mirror images of each other. Furthermore, transmitter coils have matching shape as receiver coils.

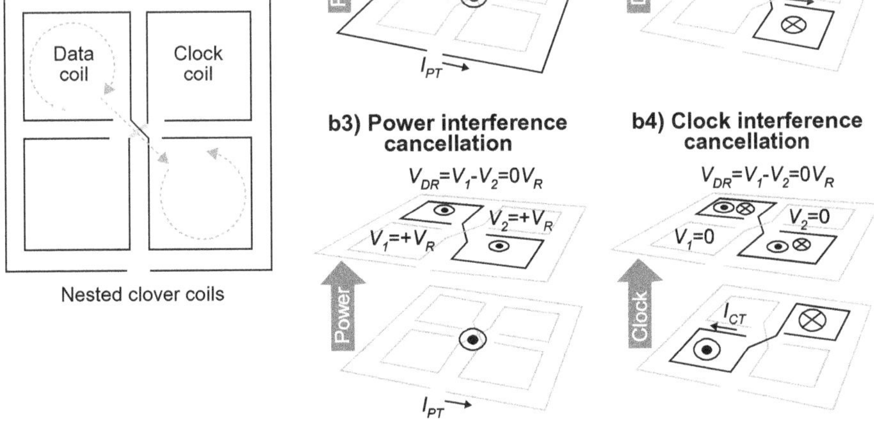

Figure 2.58 Nested clover coils for reduced interference. © 2010 IEEE. Reprinted, with permission, from [39]

The mechanism for transmission and interference cancellation is explained in Figure 2.58b. Data transmission is depicted in Figure 2.58b2. Because the data transmitter coil is a differential coil, the two halves induce voltages of opposite polarities. In other words, in the receiver differential coil, $V_1 = +V_R$ and $V_2 = -V_R$, resulting in an output voltage of $V_1 - V_2 = 2V_R$. However, because of the symmetry of the nested clover coils layout, interference from either the power or clock coils induces voltages of the same polarity in the two halves of the data receiver differential coil, resulting in cancellation at the output.

The nested clover coils design relies on the differential nature of the interference from both power and clock coils for complete cancellation. Any common mode interference, for example, as a result of misalignment between the transmitter and receiver coils, will still degrade the received data signal. Figure 2.59 plots the simulated data signal amplitude V_{SD} vs interference noise from power and clock, V_{IP} and V_{IC}, as a function of lateral misalignment Δx, for an example design where the power coil diameter is $D_P = 7$ mm and the communication distance is $Z = 0.25$ mm. The signal-to-noise ratio, SNR $= V_{SD}/(V_{IP} + V_{IC})$, stays above 20 dB for up to $\Delta x = \pm 0.2$ mm, which is about 3% of D_P and 80% of Z. The acceptable misalignment will shrink with Z, so similar analysis must be conducted to ensure that it is acceptable for your particular design.

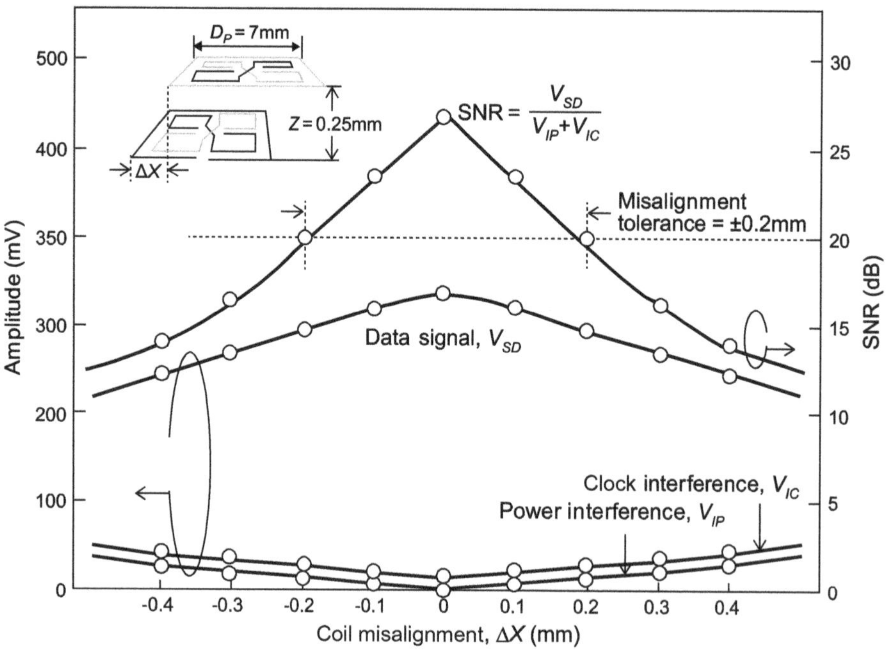

Figure 2.59 Signal-to-interference ratio (SIR) of nested clover coils vs. misalignment. © 2010 IEEE. Reprinted, with permission, from [39]

2.4.2.5 Load Tracking for Dynamic Transmit Power Control

Using simultaneous power and data transmission, it becomes possible to dynamically control power transmission in response to load changes. A load tracking circuit implemented on the receiver side of the wireless power interface can use the data channel to drive a power control circuit on the transmitter side in real time. Figure 2.60 depicts an implementation of such load tracking and power control circuits. The first stage of the load tracking circuit consists of a linear shunt-type filter to suppress ripples in the supply voltage. The output is used to generate a shunt current $I_{SHUNT,}$ which is compared against a reference current I_{REF} to determine if transmit power needs to be adjusted up or down. The outcome drives a *Pup/down* signal, which is communicated to the power control circuit using the data channel. The startup circuit serves to keep the shunt transistor initially off to avoid device lockup and slowing of the startup process.

Power control can be achieved by modulating the transmitter current. This requires adjusting either the on-resistance of the driving transistors or the supply voltage. In the former case, an increased R_{LOAD} will result in an increased on-resistance and hence reduced power transmission efficiency. In the latter case, a fast and large-capacity power converter is required that consumes large area. Because of these drawbacks, a pulse density modulation control mechanism is proposed in [39]. Under this scheme, the power control circuit uses the *Pup/down* signal to adjust the duty cycle of a mask signal *CM*. *CM* in turn controls how many cycles of input signal *Pclk* are passed through to the power transmitter. In the scenario shown in Figure 2.60c, an increase in

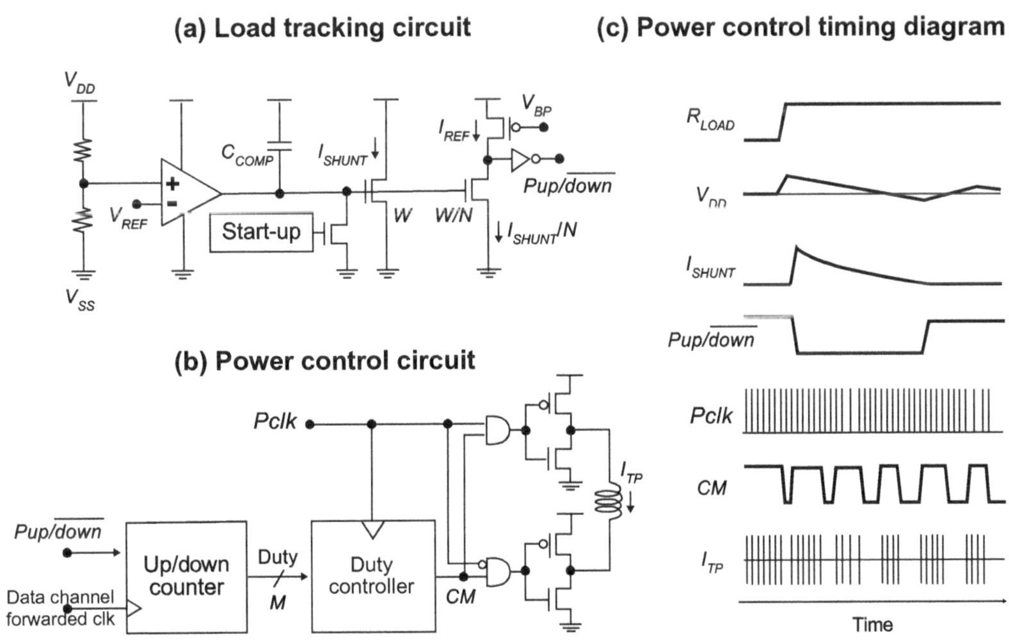

Figure 2.60 Load tracking and control circuits for dynamic transmit power adjustment. © 2010 IEEE. Reprinted, with permission, from [39]

R_{LOAD} causes V_{DD} to surge. This in turn raises I_{SHUNT}, which generates a *Pup/down* signal that directs the power control circuit to reduce power transmission. As a result, the duty cycle of *CM* is reduced, masking out a larger number of *Pclk* cycles to lower the transmitted power, until V_{DD} returns to its nominal value.

2.4.3 Example Application: A Digital Rosetta Stone

Using a wireless power interface in conjunction with TCI, it becomes possible to build a storage system that is sealed to protect it from moisture and other forms of erosion and contamination, thus enabling it to last forever – hundreds of years if not thousands. Figure 2.61 from [40] illustrates the concept of such a Digital Rosetta Stone (DRS). Mask ROM (MROM) is used as the storage medium in which data are written using an electron beam. Inductors are implemented using on-chip metal layers to create both a wireless power interface and a TCI data channel. This fully contactless interface allows the storage medium to be completely sealed in SiO_2 and SiN films to protect it from erosion. Relative humidity inside the films can be reduced to 2%, enabling a lifespan of more than 1,000 years (Figure 2.62 and [41]). The use of magnetic coupling in the contactless interface allows multiple wafers of such MROM to be stacked to create large storage capacity. For instance, stacking four 15 inch wafers fabricated in 45 nm CMOS process creates 2.5 Tb in capacity. Low-density parity check can be implemented to compensate for manufacturing defects.

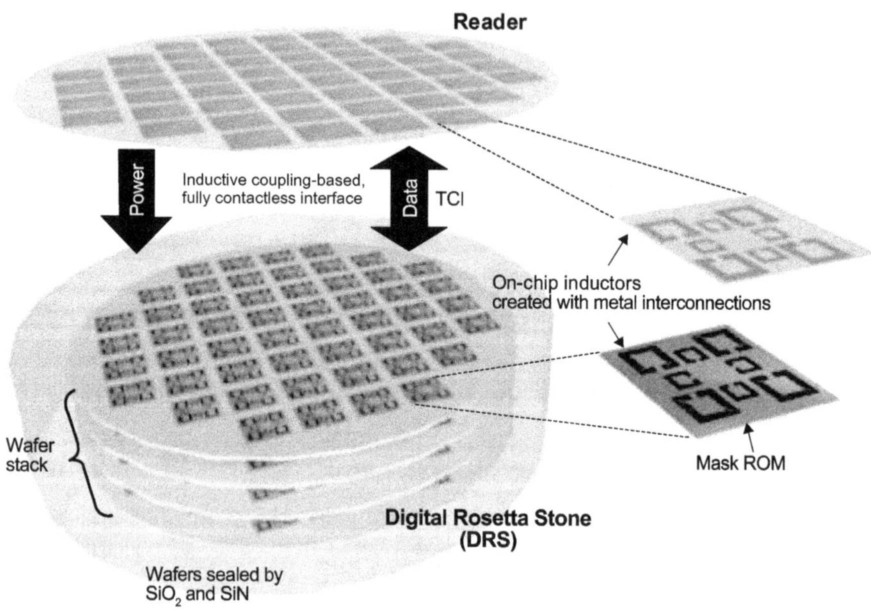

Figure 2.61 Concept of a Digital Rosetta Stone. Copyright 2009 The Japan Society of Applied Physics [40]

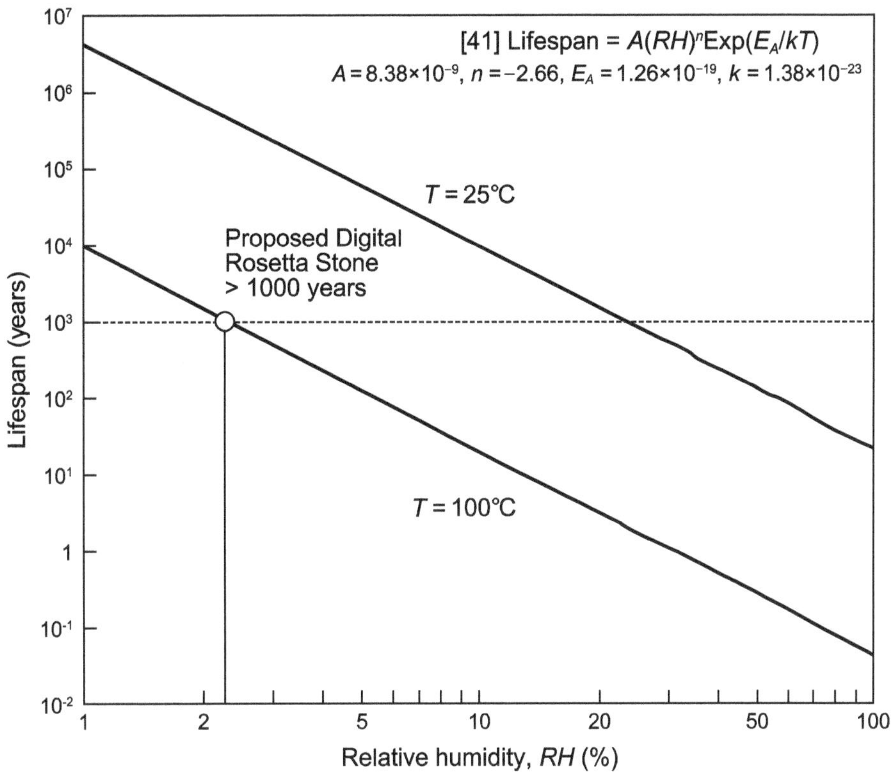

Figure 2.62 Lifespan of a silicon chip vs. relative humidity. Copyright 2009 The Japan Society of Applied Physics [40]

Since the power transmitter current is 40 dB larger than the data transmitter current, and power coils are placed in the vicinity of data coils, time-interleaved transmission is utilized to eliminate power to data interference in place of adopting nested clover coils. The reduced data rate is still acceptable for this application. The interleave scheme, consisting of four states, is controlled by a V_{SEN} signal generated by a sensor block in the power interface, as illustrated in Figure 2.63. In State I, power is transmitted from the reader to DRS which is then reset. In State II, upon suspension of power transmission, V_{SEN} activates the data interface to allow a read command to be sent from the reader to DRS. In State III, power transmission is resumed, and DRS retrieves the requested data and stores it in the register as *Mdata*. In State IV, power transmission is again suspended, which allows the data channel to be used to deliver the requested data to the reader. States III and IV are repeated until all requested data are delivered. Operation of DRS is sustained in States II and IV by energy stored in an on-chip capacitor, causing V_{DD} to drop gradually. Therefore, the data interface must be able to tolerate such V_{DD} drop. Nevertheless, effect of clock jitter due to V_{DD} drop can be minimized by using source synchronous data transmission. If nested clover coils from Section 2.4.2.4 are used to suppress power to data interference, power transmission can continue even during data transmission.

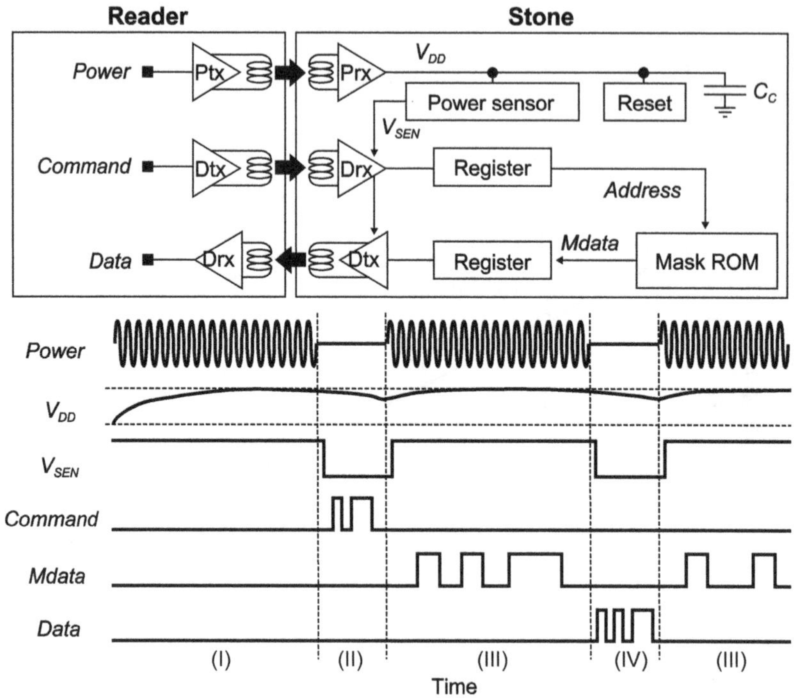

Figure 2.63 Time-interleaved transmission of power and data. Copyright 2009 The Japan Society of Applied Physics [40]

Figure 2.64 depicts the full block diagram of the DRS system. The power interface consists of four parallel channels to generate sufficient power for data retrieval. Operating frequency of 250 MHz is chosen based on the coil design to achieve high power transmission efficiency. The reset block initializes the system at startup; the clamp circuit protects the chip from voltage spikes; and the 20 nF on-chip capacitor supplies power to DRS to sustain its operation and keeps V_{DD} drop below 10% in States II and IV. The power sensor block connected to one of the power receivers is used to generate V_{SEN} to control the state transitions. The 100 MHz *Mclk* is used to retrieve data from MROM and write it to the register, while the 1 GHz *Dclk* is used to send data from the register to the reader. Initial clock cycles may not be stable, so they are masked out with Mask1 and Mask2.

A prototype system is built in a 180 nm CMOS process, with 2 mm power coils and 0.4 mm data coils, and 1 Mb of MROM storage space in a 5×5 mm^2 chip [40]. Some of the measured waveforms are depicted in Figure 2.65. At a communication distance of 0.2 mm between the reader chip and DRS chip, four parallel power channels provide a total of 56 mW of power, sufficient for the operation of the 1 Mb MROM at 100 MHz. The data channel operates at 1 Gb/s at BER $<10^{-12}$ with a V_{DD} drop $<10\%$ of 1.8 V and ripple of 33.6 mV. Since data transmission occurs only during State IV, which is 1/6 of the total read cycle of States III and IV, the effective data rate is 167 Mb/s.

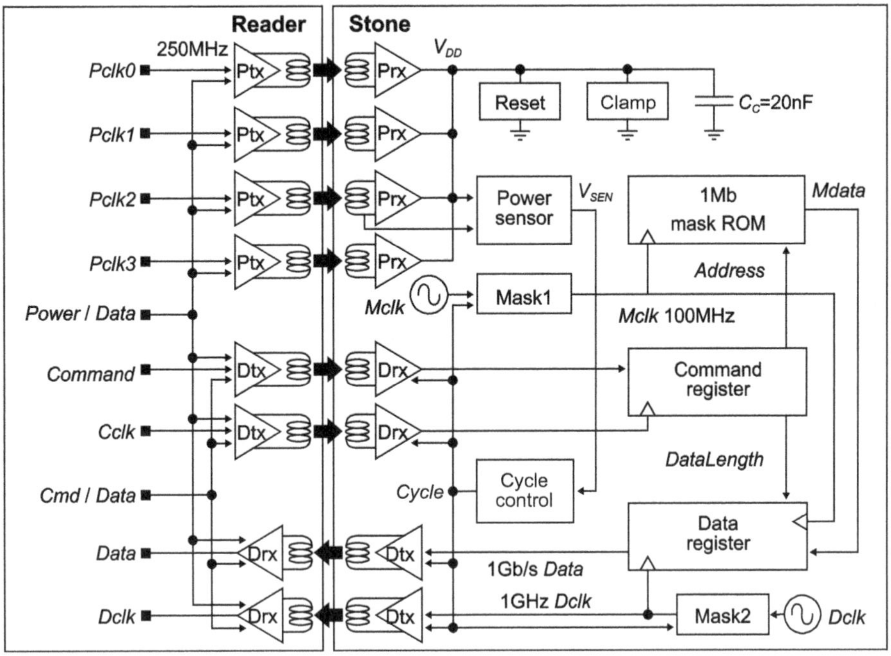

Figure 2.64 Block diagram of DRS system. Copyright 2009 The Japan Society of Applied Physics [40]

2.4.4 Closing Thoughts

The inductive coupling principle used in TCI can also be utilized to create a wireless power transmission interface. However, the coil must be designed differently in order to deliver sufficient power at sufficiently high efficiency. In addition, power transmission can be improved by optimizing the design of the rectifier used to convert the received AC voltage into DC. NMOS transistors with gates cross-connected, bias voltage to lower voltage drop across the diode, and half-wave rectifier for low input voltage and high diode voltage drop are a few such improvements. On the other hand, parallel channels with phase offsets can be used to not only boost transmission power but also smooth out ripples. When power and data channels are placed in close proximity to reduce area overhead, nested clover coils can be used to suppress interference from both power and clock to data. Finally, load tracking on the receiver side and power control on the transmitter side can be employed during simultaneous power and data transmission to dynamically adjust power transmission in response to load changes.

A completely contactless chip interface can be implemented by using inductively coupled wireless power transmission in conjunction with TCI. Such an interface improves reliability by avoiding stress in physical connections and by eliminating

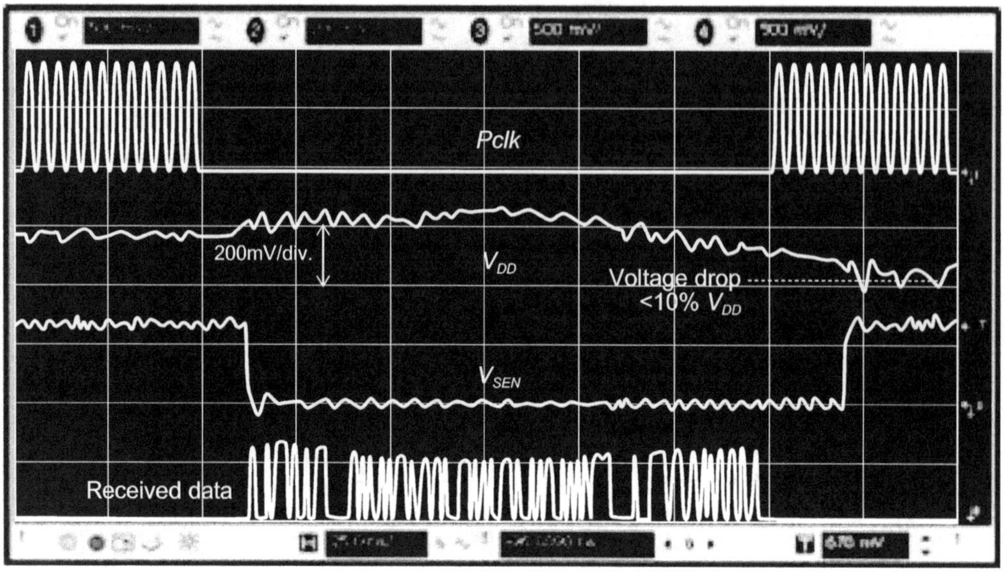

Figure 2.65 Measured prototype DRS system waveforms. Copyright 2009 The Japan Society of Applied Physics [40]

tear and wear from physical contact. It will improve the longevity of, for example, a memory card that must operate under various environmental conditions and undergo many cycles of connection and disconnection. Furthermore, elimination of physical contact enables sealing of the chip to protect it from the environment. Sealing of memory chips in SiO2 and SiN to protect them from humidity can extend their lifespan to more than 1,000 years, creating a digital equivalence of a Rosetta Stone.

Our research has not reached the stage where wireless power delivery is a viable solution ready for general commercial adoption – both the achievable power and power efficiency are not yet sufficient. Meanwhile, there are concerted efforts in the industry to develop a commercially attractive solution as exemplified by the Wireless Power Consortium (Qi) for multiple end markets and the Power Matters Alliance (PMA) for the mobile device market, both using inductive coupling-based technology. In the meantime, our recommended solution for a high-performance chip interface is to couple wireless TCI data interface with a wired power delivery technology. Lower-cost systems can use the time-tested wirebonding technology, while higher-cost systems can benefit from the better performance of TSV. By eliminating wired connections from the data interface, some of the major problems suffered by a conventional wired interface can be alleviated. For instance, a 3D flash memory chip stack with TCI for data and wirebonding for power has an order-of-magnitude fewer wires than the conventional wired solution, making it possible to wirebond a stack with an order-of-magnitude more chips to dramatically increase capacity. On the other hand, a 3D DRAM chip stack with TCI for data and TSV for power is smaller in chip area and lower cost than a conventional 2.5D solution due to smaller interface area and elimination of the interposer. The TSV yield problem can be circumvented by adding redundant power and ground TSVs.

Nevertheless, there is ongoing research on a lower-cost alternative to TSV as a wafer-level wired power delivery solution, which is the subject of the next section.

2.5 Highly Doped Silicon Via (HDSV)

For high-performance 2.5/3D IC integration, the current solution of choice for power delivery through the silicon substrate is through-silicon-via (TSV). The use of plated copper in TSV achieves a very low-resistance path for efficient power delivery. Unfortunately, due to additional processing steps that are outside of the conventional CMOS fabrication process, TSV remains a costly solution. On the other hand, although silicon is not a good electrical conductor, its conductivity can be improved by orders of magnitude through doping. Therefore, using deep and heavy doping (10–100 times the normal dose) of focused regions, parallel conductive paths can be created to enable power delivery across the chip without etching and plating via holes [42–43]. The technology is known as highly doped silicon via (HDSV) since the doped conduits function like plated vias, even though they are not actual physical vias. Because the required doping process, including annealing, can be integrated into the conventional IC manufacturing process without a large cost adder, HDSV potentially offers a low-cost alternative to TSV power delivery.

2.5.1 HDSV Fabrication

HDSVs are fabricated using a "via-first" process where highly doped regions to be used as power vias are first created using ion implantation. This is then followed by the normal front end of line (FEOL) process to fabricate individual devices and back end of line (BEOL) process to fabricate metal wirings. After contact pads are added to the HDSVs, the wafer is thinned and stacked.

Two types of impurities are possible for creating the highly doped regions, resulting in either a P-HDSV or an N-HDSV. For a p-type substrate, a P-HDSV consists of a highly doped p-well (p++well) inside a normal n-well which provides isolation from the underlying p-substrate, while an N-HDSV consists of a highly doped n-well (n++well) directly buried in the p-substrate.

A simplified fabrication process flow is depicted in Figure 2.66. Because P-HDSV requires forming an n-well first before the p++well, it naturally requires a few more steps and a larger area than N-HDSV. After implantation and before annealing, a high-density plasma (HDP) oxide layer is formed over the substrate surface to prevent out-diffusion during annealing. Based on technology computer-aided design (TCAD) simulation, the key process parameters to achieve a vertical resistivity of $<0.0002\ \Omega\text{m}$ are summarized in Table 2.19. With this resistivity, the vertical resistance of a $20 \times 20\ \mu\text{m}^2$ HDSV through a $2\ \mu\text{m}$ thick substrate is less than $1\ \Omega$. The

Table 2.19. HDSV fabrication process parameters.

		P-HDSV		N-HDSV
Process	Parameter	N-well	P++well	N++well
Implant	Energy	50 keV	50 keV	50 keV
	Dosage	$1e^{15}/cm^2$	$1e^{16}/cm^2$	$1e^{15}/cm^2$
	Dopant	Phosphorous	Boron	Phosphorous
Anneal	Temperature	1,200 °C	1,200 °C	1,200 °C
	Duration	4 hours	2 hours	2 hours

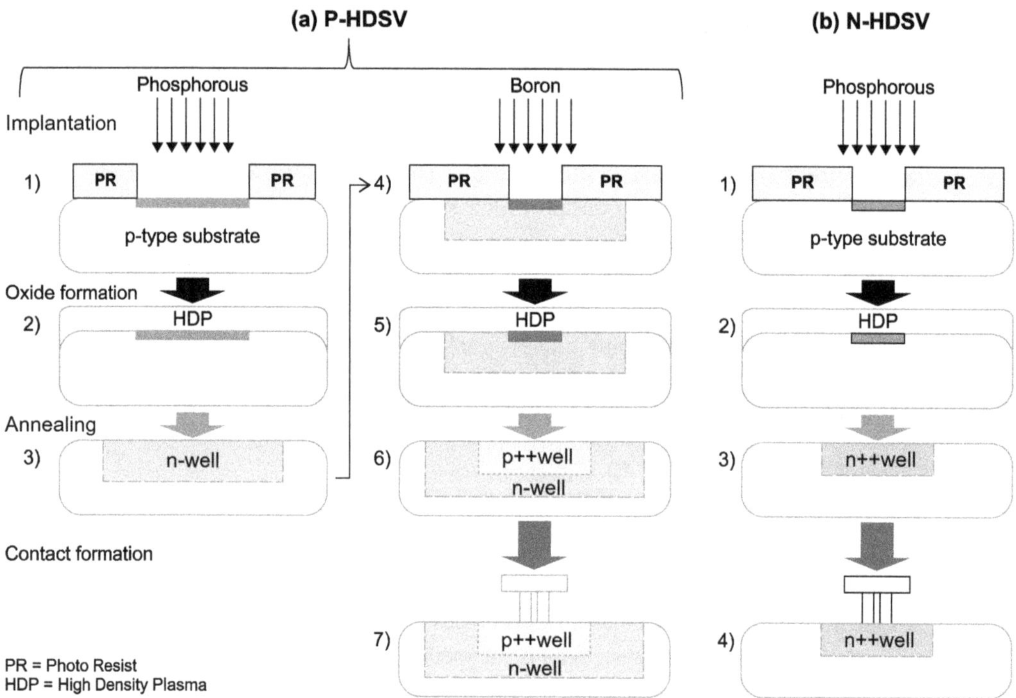

Figure 2.66 HDSV formation.

implantation dosage is up to an order of magnitude higher than in a normal process, but it is still achievable using conventional implant equipment.

Figure 2.67 illustrates the process for creating wafer-on-wafer stacking using HDSV. Chemical mechanical polishing (CMP) of the backside silicon substrate of the upper wafer exposes the highly doped p++ or n++ well. Meanwhile, CMP of the front side of the lower wafer exposes the copper contact pads and prepares their surface for bonding. The wafers are then stacked, and a room-temperature wafer bonding machine is used to apply pressure to create solid intermetallic bonding by diffusion to establish an electrical connection between the matching wells. The process can be repeated to stack additional wafers underneath.

2.5.2 HDSV Characterization

A prototype system has been built to characterize both the P-HDSV and N-HDSV fabricated using the process described in Section 2.5.1, and the results are summarized in [43]. Figure 2.68 depicts the simulated doping concentration as a function of depth. A highly doped concentration of $>10^{18}$ cm^{-3} is achieved for a depth of >3 μm for both HDSV types, which is sufficient for the characterization of a 2 μm thick substrate.

Proper wafer-on-wafer bonding after the CMP process in Figure 2.67 requires the exposed surface to be smooth. However, surface roughness characterization shows

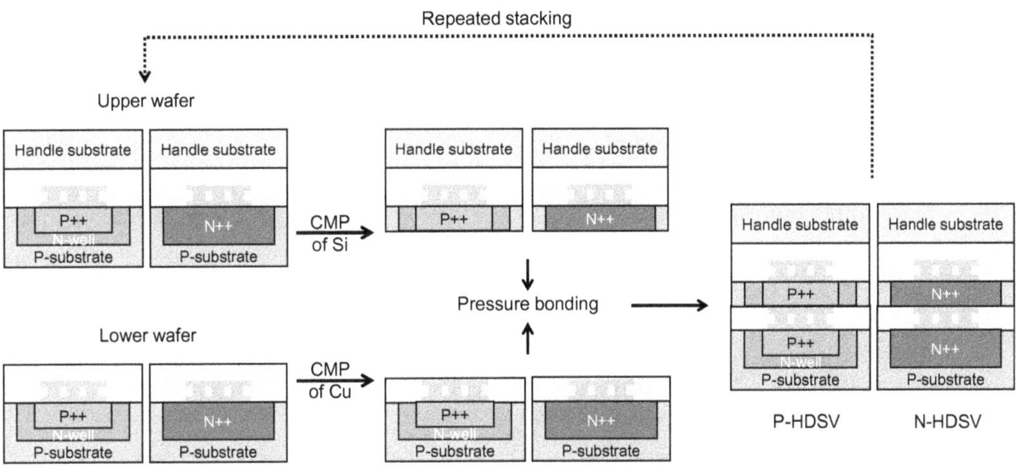

Figure 2.67 Wafer-on-wafer stacking to establish HDSV connection. Copyright 2019 The Japan Society of Applied Physics [43]

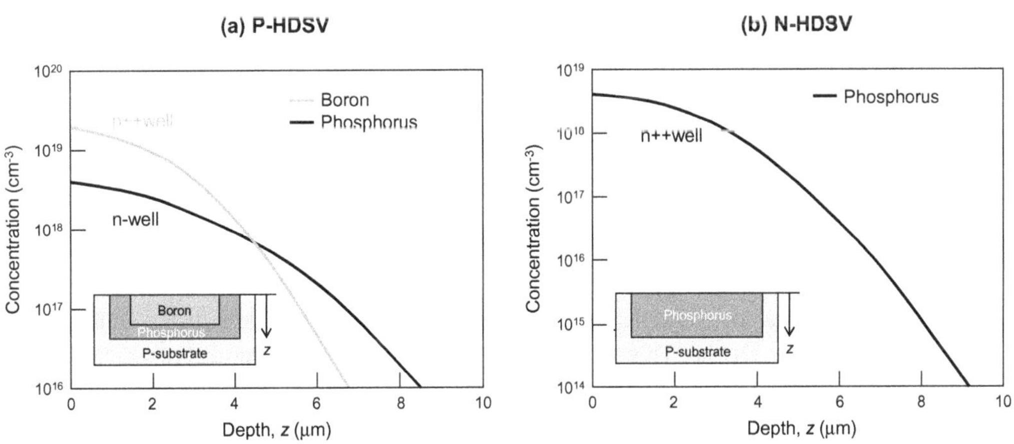

Figure 2.68 Doping concentration vs. depth. Copyright 2019 The Japan Society of Applied Physics [43]

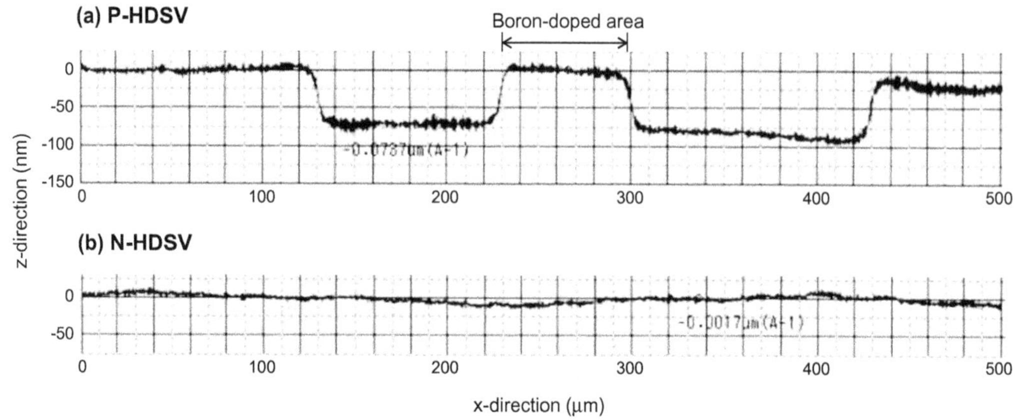

Figure 2.69 Surface roughness comparison after the CMP process. Copyright 2019 The Japan Society of Applied Physics [43]

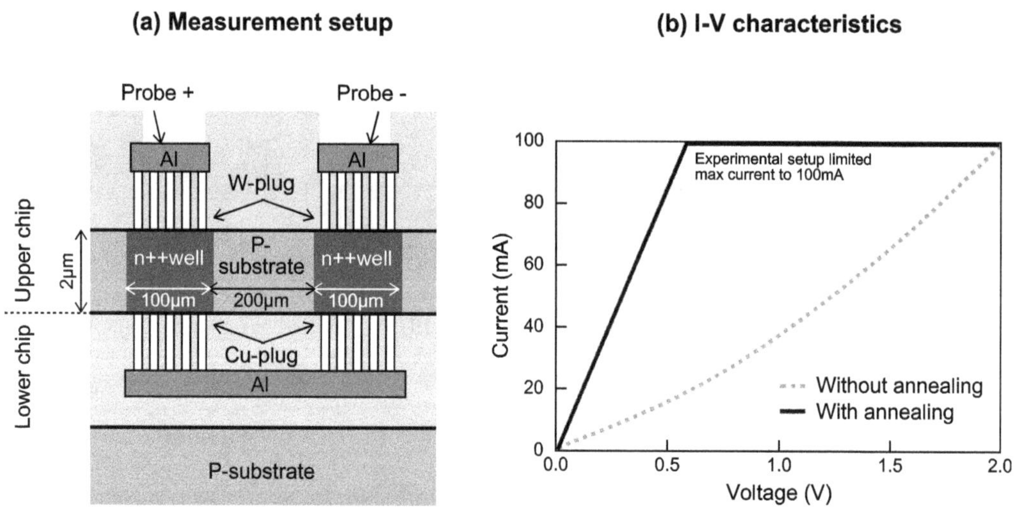

Figure 2.70 Measured I-V characteristics of N-HDSV. (b) Copyright 2019 The Japan Society of Applied Physics [43]

significant difference between the two HDSV types. The result of a comparative study using the conventional CMP process is shown in Figure 2.69. The measured height variations across the polished surface are more than 90 nm for P-HDSV and less than 2 nm for N-HDSV. This can be attributed to slowed CMP removal rate due to boron dopant as described in [44]. Given these data, while the N-HDSV surface is sufficiently smooth for proper bonding, the P-HDSV surface is not. Consequently, subsequent characterization is focused on N-HDSV.

Current and voltage (I-V) characteristics of the N-HDSV are measured and plotted in Figure 2.70. Using the setup shown in Figure 2.70a, the measured resistance is that of the entire path of probe contact, HDSV, wiring through the lower chip, HDSV, and probe contact, which is a little more than the resistance of two HDSVs in series. Each probe contact includes an aluminum pad and multiple tungsten (W) plugs, while wiring through the lower chip includes copper (Cu) plugs connecting the two ends of the wiring to the chip interface. When measured right after wafer bonding, the current path exhibits nonohmic/nonlinear I-V characteristics as shown by the dotted line in Figure 2.70b. At 2 V, the resistance is 20 Ω. But after additional annealing, the characteristics become ohmic with a resistance of 5.4 Ω as shown by the solid line (before current reaching the maximum value supported by the experimental setup). It is postulated that the change is due to the formation of silicide at the copper–silicon interface through annealing. Nevertheless, since the estimated resistance of one N-HDSV is 22 m Ω, there is significant room (two orders of magnitude) for further contact resistance reduction. In fact, it is observed that the postbonding annealing process created holes in the plugs that are void of copper. Therefore, to realize the full potential of HDSV, further optimization of the fabrication and bonding process is necessary. In addition to optimization of the postbonding annealing process, optimization of the plug material such as replacing copper with cobalt or nickel may also help to reduce contact resistance.

The N-HDSV prototype is used for power delivery in a 3D SoC test system in [43]. The system consists of eight SRAM dice stacked on top of a processor die. Each SRAM die consists of 24 vaults, and each provides storage to one core on the processor die, enabling parallel processing for DNN computation. Communication across the stack is achieved through TCI. The same test system was previously constructed using TSV for power delivery as described in [45]. In this implementation, HDSV is used instead. More detail of the processing system is provided in Section 4.1.2.

Since the current consumption of each vault is estimated at 50 mA, to meet the power supply ($V_{DD} + V_{SS}$) IR drop specification of 100 mV, the maximum individual resistance of V_{DD} and V_{SS} across eight SRAM dice needs to be within $100/50/2 = 1$ Ω. As a result, effective resistance of the parallel HDSVs across one die should be $<1/8 = 125$ m Ω. Since the measured resistance of one N-HDSV is approximately $5.4/2 = 2.7$ Ω, the power delivery network of each die can be constructed with 22 N-HDSVs each for V_{DD} and V_{SS}. Since each HDSV is 100×100 μm^2, and the pitch is 108 μm, the total HDSV area overhead amounts to 11%. If further process optimization is able to reduce contact resistance such that the total HDSV resistance approaches that from simulation, it is possible to reduce this area overhead down to below 0.1%, making HDSV a viable power delivery solution.

2.5.3 Closing Thoughts

Table 2.20 provides a comparison of key attributes between TSV and HDSV as a means of power delivery in 3D IC integration, using simulated resistivity for HDSV. While TSV provides orders-of-magnitude lower resistivity, HDSV provides significant cost advantage and logistical simplicity. In one study [46], HDSV was found to

Table 2.20. HDSV vs. TSV in power delivery. Copyright 2019 The Japan Society of Applied Physics [43].

Power delivery technology	TSV	HDSV
Added wafer processing	Via etch and fill	Implant and anneal
Manufacturing logistics	New model	Conventional model
Cost adder	A few 10s of %	A few %
Order of resistivity	10 nΩm	10–100 μΩm

incur a 7% cost adder, as opposed to the 23% incurred by TSV. Nevertheless, the technology is still a work in progress. In particular, significant process optimization is necessary to lower contact resistance so as to achieve resistivity that approaches simulated results. Furthermore, wafer thinning technology needs to further advance in order to deliver thickness of a few microns in mass production to make HDSV a practical solution.

The achievable thickness in wafer thinning is limited due to the loss of Gettering effect when wafers are thinned with backgrinding. Gettering effect occurs when Gettering sites, which are defects and distortions formed in the crystal structure of bulk silicon and its backside surface, capture and remove impurities that can otherwise cause metal contamination. As a result, loss of Gettering effect results in increased metal contamination and hence loss in yield and performance especially due to increased leakage. Fortunately, significant strides have been made in recent years to address the problem, resulting in rapid advances in wafer thinning. The Gettering Dry Polish wheel achieves thickness of a few microns by forming Gettering sites during grinding while maintaining physical strength [47]. Meanwhile, [48] demonstrates thinning of 300 mm wafer to a 4 μm thickness, and reports no degradation in 40 nm, 2 Gb DRAM retention characteristics and distribution. Consequently, there does not seem to be major technical hurdles to achieving the thickness required for mass production of HDSV.

2.6 Wideband DRAM Interface

With DRAM memory access continuing to be a key performance bottleneck of state-of-the-art computing systems, the DRAM industry has turned to 2.5/3D integration to narrow the performance gap, in the form of Wide I/O, Hybrid Memory Cube (HMC), and high bandwidth memory (HBM). In particular, HBM has become the solution of choice in high-end systems, including the latest GPU computing platforms, achieving total memory interface bandwidth of 341 GB/s in 2018 [49]. HBM employs TSVs to connect a 3D stack of DRAM core dice to a base controller chip, which is in turn connected to the processor chip through a silicon interposer in a 2.5D configuration to keep interconnect length short and hence interface bandwidth high and power consumption low (Figure 2.71). The massive number of parallel TSVs used in the DRAM stack allows the per-pin data rate to remain relatively low, making it a practical

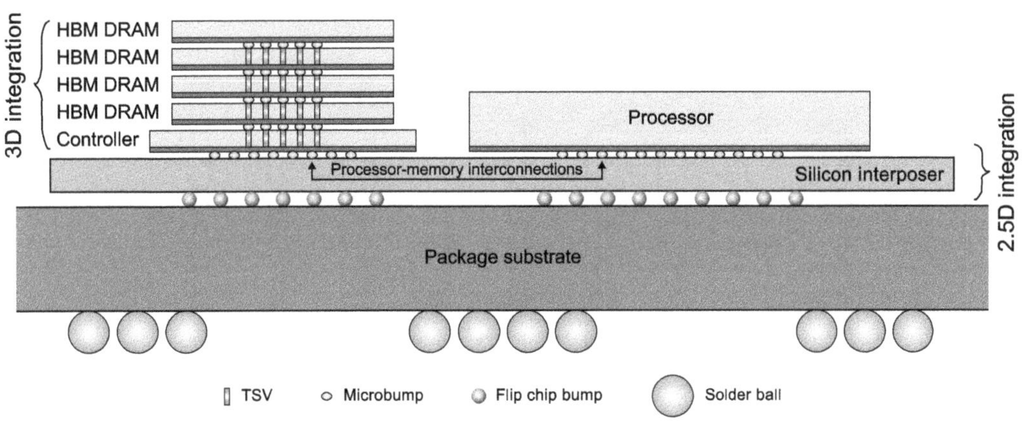

Figure 2.71 Construction of an HBM DRAM memory system.

solution for the slower DRAM process while keeping power consumption manageable. Nevertheless, power consumption and cost challenges remain. Given the better power efficiency and lower cost of TCI, replacing TSV with TCI can overcome some of these challenges to enable further performance leaps using the same wideband DRAM interface architecture.

As noted in Chapter 1, the size of a TSV remains large compared to device feature size. Furthermore, since it creates a hole through the silicon substrate, no device can be placed within the keep out region, making TSV area overhead large. By contrast, devices including DRAM cells and sense amps can be placed underneath TCI coils. This not only minimizes area overhead of the interface but also enables coils to be placed in more optimal positions to reduce signal line length and propagation delay. Furthermore, by using constant current transceivers such as the standard transceiver, TCI induces much lower simultaneous switching noise (SSN) than TSV. Smaller area overhead and lower SSN means that more parallel TCI channels can be implemented, resulting in higher interface bandwidth. Meanwhile, TCI enables the elimination of ESD, reducing capacitive loading on the channel. Besides, the weak magnetic coupling used by TCI means that as more DRAM dice are stacked, the effective loading on the channel increases much slower than the linear increase experienced by TSV. Consequently, TCI can achieve a higher per-pin data rate and power efficiency than TSV especially in a deep die stack. Finally, since TCI is fabricated within the conventional manufacturing process, it has better reliability than TSV, which still suffers from yield loss due to the added mechanical processing for its fabrication.

2.6.1 Prototype TCI Wideband DRAM System

2.6.1.1 Architecture

Figure 2.72a depicts the block diagram of a prototype TCI wideband DRAM system completed in 2018. It consists of a 3D stack of four identical DRAM dice in a 25 nm

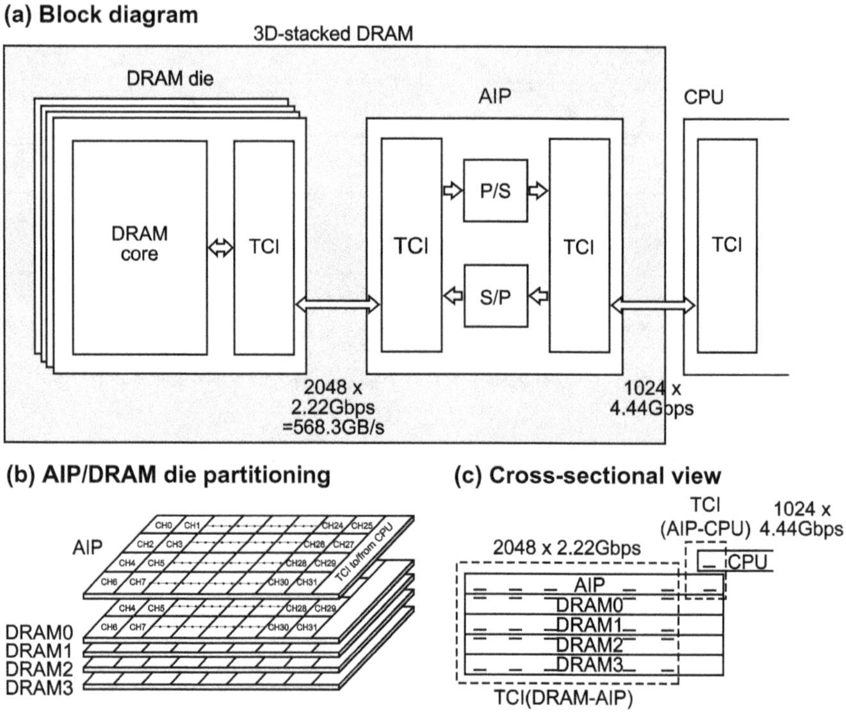

Figure 2.72 Structure of a prototype TCI wideband DRAM system.

DRAM process and an active interposer (AIP) in a 40 nm logic process. Each DRAM die consists of a 9 Gb memory core and a TCI interface, where 1 Gb is used for ECC and the remaining 8 Gb for storage, resulting in a total system storage capacity of 4 GB. The AIP includes both a TCI interface to the DRAM stack and another to the CPU, serial-to-parallel and parallel-to-serial convertors bridging the two TCI interfaces running at different data rates, as well as FIFO buffers, and control logic. The AIP–DRAM interface consists of 2,048 parallel TCI DQ links each running at 2.22 Gb/s for an interface bandwidth of 568.3 Gb/s, divided into 32 parallel channels of 64 links each (Figures 2.72b and 2.73a, b). Since the AIP is fabricated in a logic process, it supports a higher data rate of 4.44 Gb/s for the AIP–CPU interface, reducing the number of DQ links on the CPU side to 1,024. The AIP is placed on top of the DRAM stack, and the entire stack underneath the flip-chip bonded CPU in a 2.9D configuration, resulting in a small footprint and low profile (Figure 2.72c). Elimination of the silicon interposer results in additional cost saving compared to HBM.

While TSV is used to provide power delivery through the 3D stack, a bumpless TSV technology is used to minimize chip-to-chip gap to achieve a small 12 μm communication distance across each chip, resulting in a maximum communication distance of 48 μm (Figure 2.73d). Removing the air gap between adjacent chips also improves thermal conductivity to facilitate heat removal from the chip stack.

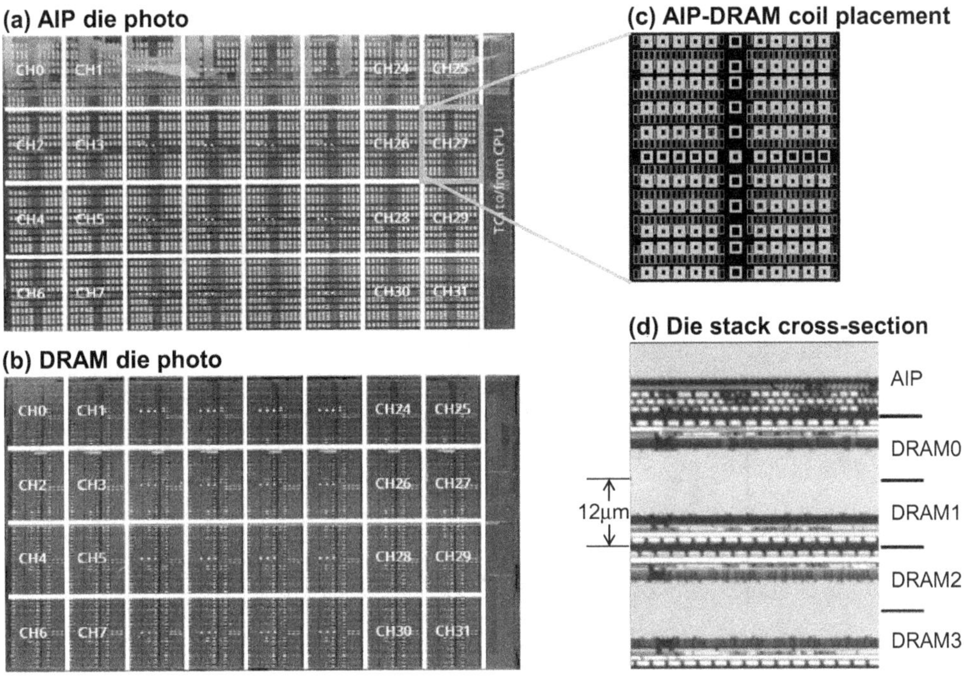

Figure 2.73 Die photos and 3D stack cross section.

Placement of TCI coils in the AIP–DRAM interface is depicted in Figure 2.73c. Coils are laid out using only one metal layer and placed over memory cells to minimize area overhead.

2.6.1.2 Transceiver Design

Figure 2.74 illustrates the transceiver design for the TCI interfaces. The standard H-bridge transmitter is used for clock while a pulse transmitter is used for DQ. The same standard hysteresis comparator receiver is used with both transmitter types. The pulse transmitter drives current only during a transition and is thus lower power. However, it requires a larger coil due to weaker induced voltage and induces more switching noise. Since clock switches continuously, the power saving from using a pulse transmitter is not as significant as it is for DQ. Therefore, the standard H-bridge transmitter is chosen for the clock for its benefits over a pulse transmitter.

Since inductive coupling strength varies proportionally with communication distance, to achieve comparable voltage margin, a larger transmitter current is needed for communication between AIP and a DRAM farther away in the stack. As a result, both transmitter types are equipped with binary-weighted current adjustment capability controlled by a 4-bit enable signal. After 3D integration, the current strength of individual transmitters is adjusted based on the location and hence communication distance of the destination chip, and the settings are saved in registers.

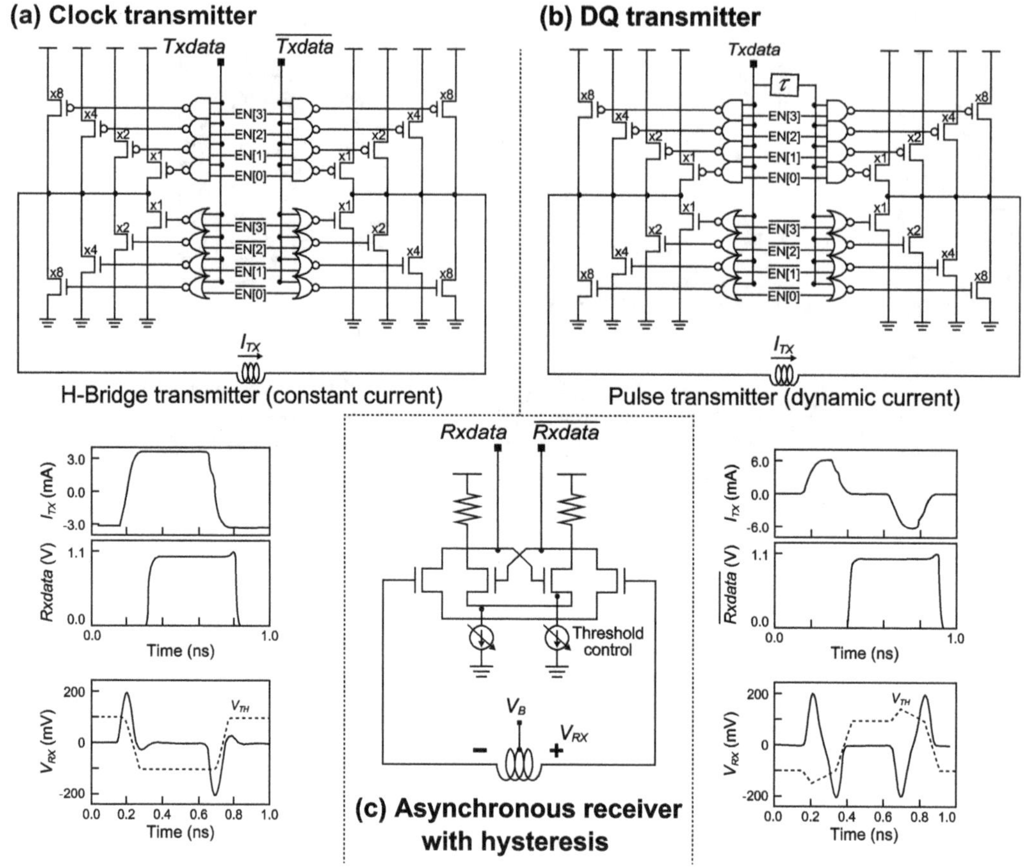

Figure 2.74 Transceiver design for a wideband DRAM interface.

The prototype system operates successfully running the DRAM–AIP and AIP–CPU interfaces at 2.22 Gb/s and 4.44 Gb/s respectively. Since this was an initial prototype, the interface was designed to run at a conservative data rate. As noted in [11], we expect TCI using DRAM transistors to be able to reach a data rate of 8 Gb/s, so there is a lot of performance headroom. Furthermore, successful operation of the prototype demonstrates that the magnetic field from TCI does not appreciably interfere with DRAM operation even when its coils are placed above memory cells and circuits.

2.6.2 Case Study: Area Saving vs. TSV-HBM

The HBM stacked DRAM from [49] is used as a baseline for a case study in [42] to quantify the area saving achievable by converting the data interface within the stack from being TSV-based to TCI. The area and layout of the baseline DRAM and a comparable hypothetical TCI-stacked DRAM are shown in Figures 2.75a and 2.75b respectively. The hypothetical TCI-stacked DRAM is created by replacing only the

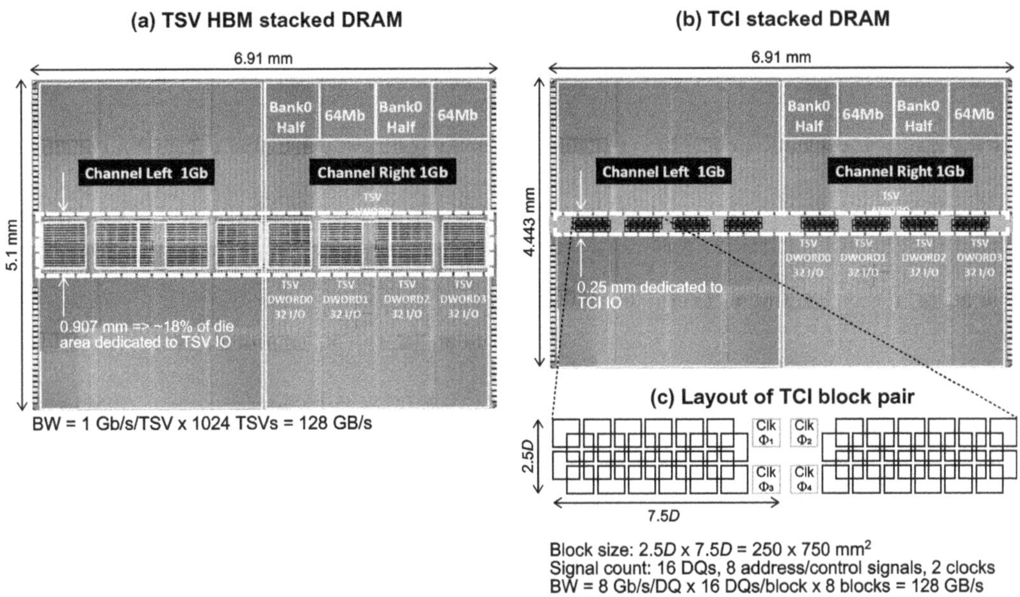

Figure 2.75 Die size comparison between TSV- HBM and TCI-stacked DRAM.

dedicated TSV area with a dedicated TCI area that produces the same interface bandwidth of 128 GB/s. The dedicated TCI area consists of four identical block pairs, and the layout of each is depicted in Figure 2.75c. It adopts the 4PDM overlapping coil placement scheme prototyped in [11]. Assuming that each DRAM die is thinned to 10 μm, stacking four DRAMs on top of the AIP to provide the same storage capacity as [49] would result in a maximum communication distance of 40 μm. Consequently, a coil diameter of 100 μm ($D = 2.5Z$) is chosen, resulting in a TCI block size of 250 × 750 μm. With a per coil data rate of 8 Gb/s, eight TCI blocks each with 16 DQs achieve the same total bandwidth of 128 GB/s as the baseline DRAM. Since the area of the DRAM is reduced from $6.91 \times 5.1\ \text{mm}^2$ to $6.91 \times 4.443\ \text{mm}^2$, an area saving of 13% is achieved. As previously noted, there is additional cost saving due to improved yield from elimination of TSV yield loss, plus elimination of the 2.5D interposer.

2.6.3 Closing Thoughts

Table 2.21 summarizes the comparison between the prototype TCI wideband DRAM from Section 2.6.1 and the HBM2 results from [49] published in 2018. The DRAMs used are comparable parts, but TCI enables twice as many I/Os as HBM2, resulting in 66% higher total interface bandwidth using a slower DRAM process. Furthermore, as noted in Section 2.6.1, TCI should have plenty of headroom to scale the data rate and hence interface bandwidth.

Table 2.21. TCI wideband DRAM vs. HBM.

		HBM2 [49]	TCI
Year achieved		2018	2018
DRAM	Pin data rate	2.667 Gb/s	2.22 Gb/s
	Process	20 nm	25 nm
	Nominal VDDC/VDDQ/VPPE	1.2/1.2/2.5 V	1.2/1.2/2.5 V
DRAM stack bus width		1,024	2,048
Interface bandwidth		341 GB/s	568 GB/s

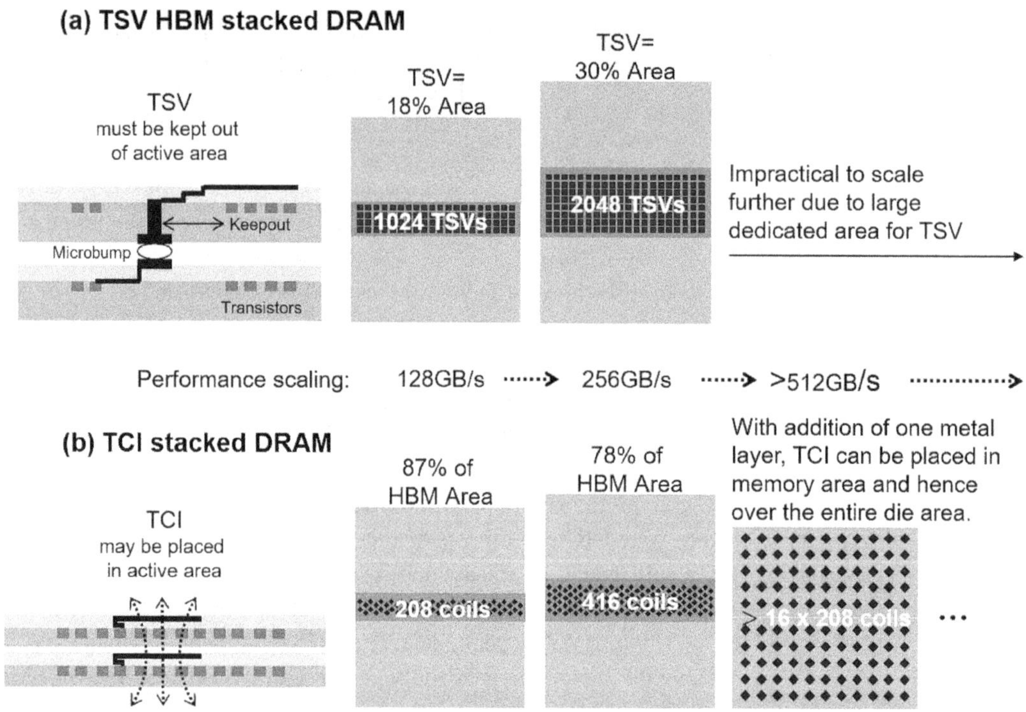

Figure 2.76 Performance scaling comparison between TSV HBM and TCI DRAM.

Meanwhile, Figure 2.76 compares TCI wideband DRAM with HBM in bandwidth performance scaling at a constant per pin data rate. Extending the comparison in Section 2.6.2, doubling the interface bandwidth to 256 GB/s without changing data rates will result in a TCI area saving of 22% over HBM where DQ alone will require a dedicated TSV area of 30% of the die without further optimization. Moreover, it will be challenging to further double the HBM data rate to 512 GB/s because the area overhead will become impractically large without drastic improvement in TSV area efficiency. Meanwhile, while more exhaustive testing is required to qualify TCI coils placed over DRAM cells for mass production, such placement freedom can potentially

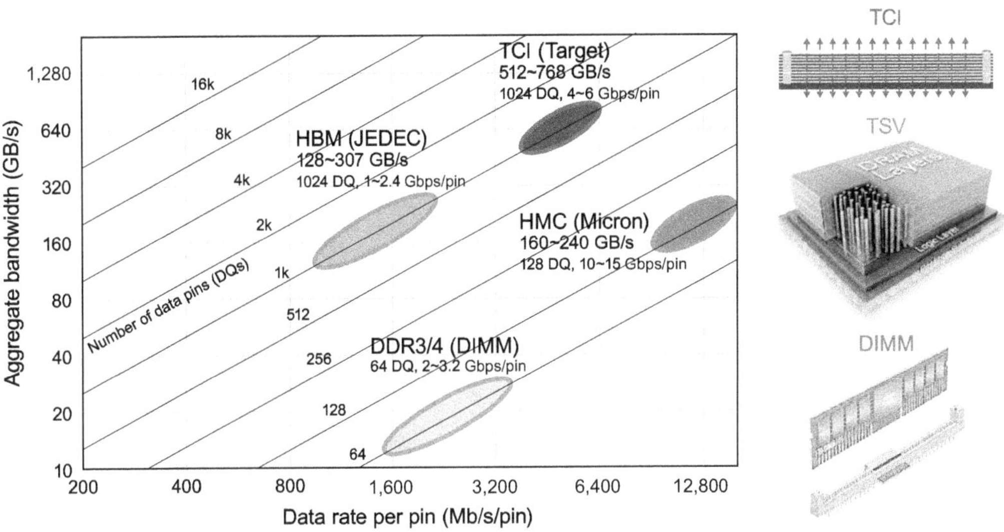

Figure 2.77 Memory performance roadmap comparison.

enable TCI coils to be placed over the entire chip area, resulting in bandwidth scaling well beyond 512 GB/s.

With such performance headroom, we expect TCI wideband DRAM to deliver a strong performance roadmap against competing DRAM technologies. Figure 2.77 compares the performance roadmaps among TCI, HBM, HMC, and DDR. With a high-end target of 768 GB/s, we expect TCI to deliver more than twice the bandwidth of HBM, three times the bandwidth of HMC, and 30 times the bandwidth of DDR4. Furthermore, it is expected to deliver significant cost saving through better yield and the elimination of the 2.5D silicon interposer used by HBM. While the current proposal is for power to be delivered by TSV, when HDSV becomes ready for mass production, it can be used in place of TSV to achieve another level of cost reduction.

2.7 Stacked Flash Memory

With the falling cost and rising capacity of NAND flash memory, solid state drive (SSD) is fast becoming a higher-performance replacement for hard disk drive (HDD) as a storage solution. In addition to the higher speed, the elimination of mechanical parts for data access also makes SSD lower power and more reliable than HDD. However, SSD is still lacking in capacity compared to HDD. As a result, there is continuous technology development to enable more and more NAND chips to be stacked. However, it is extremely difficult to stack a large number of NAND chips in one package using conventional wirebonding technology for interconnection due to the prohibitively large number of bond wires required. Even just an eight-chip stack would require about 200 bond wires. By applying TCI to the NAND chip stack, the

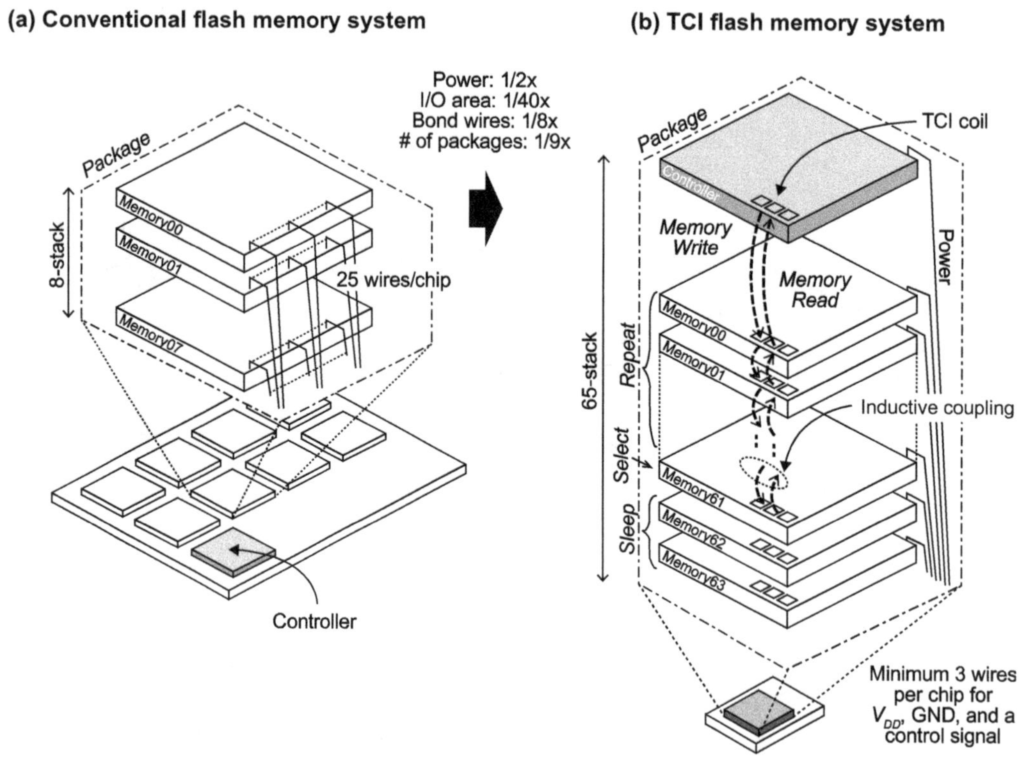

Figure 2.78 TCI vs. wirebond as interconnect for a flash memory system. © 2009 IEEE. Reprinted, with permission, from [50]

number of bond wires can be significantly reduced to enable 128 or more chips to be stacked. Figure 2.78 illustrates the improvement to a flash memory system by converting the signal interface from wirebond to TCI and when a controller chip is stacked on top of 64 NAND chips in one package. At minimum, only three bond wires per chip are required to deliver V_{DD}, ground and a control signal. Using conventional stacking technology, the same capacity would require eight packages of eight-stack NAND chips plus a separate controller package, where each NAND chip requires 25 bond wires for power delivery and signals. Consequently, TCI enables almost an order-of-magnitude reduction in number of bond wires and form factor. Furthermore, the elimination of ESD protection results in a 1/40× reduction in I/O area and 50% reduction in power.

2.7.1 Stacking and Communication Schemes

While high-speed signal connections are provided by TCI, to keep cost down, wirebonding is still used for power delivery and low-speed control signals such as reset and chip ID. As a result, when chips are stacked, the upper chip must be offset horizontally relative to the lower chip to expose the bond pads of the lower chip.

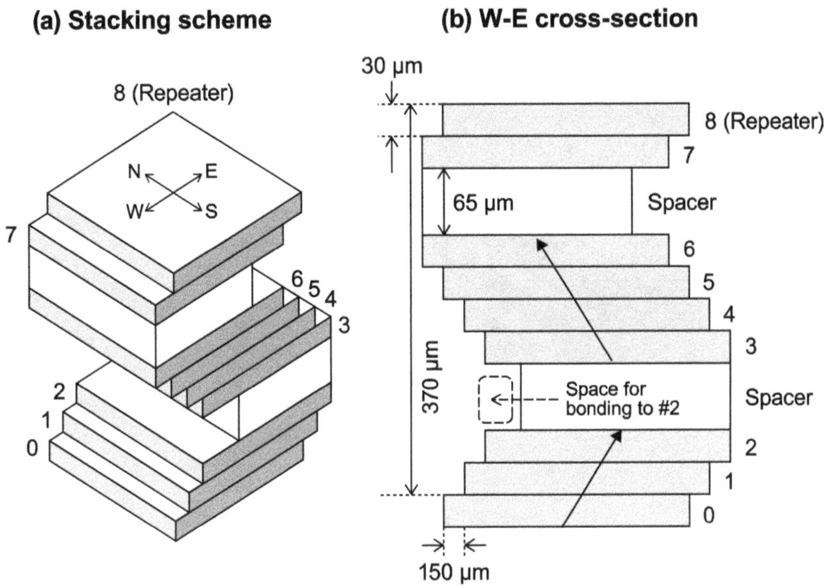

Figure 2.79 Conventional terraced stacking scheme for a TCI flash memory system. © 2010 IEEE. Reprinted, with permission, from [51]

However, to maintain mechanical stability, the sliding direction must be reversed periodically. Figure 2.79 shows the conventional terraced stacking scheme described in [51], where the sliding is in the east direction for memory chips #0–2, but the west direction for chips #3–6. Additionally, between chips #2 and 3, where sliding direction is reversed, a spacer is inserted to create space for bonding to chip #2. Consequently, the overall profile of the stack as well as maximum TCI communication distance is increased due to the spacer.

To eliminate the spacer, a spiral terraced stacking scheme shown in Figure 2.80 is proposed in [51]. In the conventional terraced scheme, the sliding direction is changed by 180 degrees at a time. By contrast, in the spiral terraced scheme, the sliding direction is changed by 90 degrees at a time, resulting in a spiral staircase. In the implementation in Figure 2.80, the spiral is counterclockwise. First, chip #1 is slid in the south direction relative to chip #0. Then the sliding direction changes 90 degrees counterclockwise and chips #2 and 3 are slid in the east direction relative to chip #1. Next, chips #4 and 5 are slid in the north direction relative to chip #3, followed by chips #6 and 7 sliding in the west direction relative to chip #5. Finally, the sliding direction changes another 90 degrees counterclockwise and chip #8 is slid in the south direction relative to chip #7, putting it in the same horizontal location as chip #0. In this manner, the spiral pattern is repeated at every eighth chip. Other variations of spiral terraced stacking are possible, such as changing sliding direction clockwise or sliding in a diagonal direction (northwest, etc.) and so forth. More examples can be found in [52].

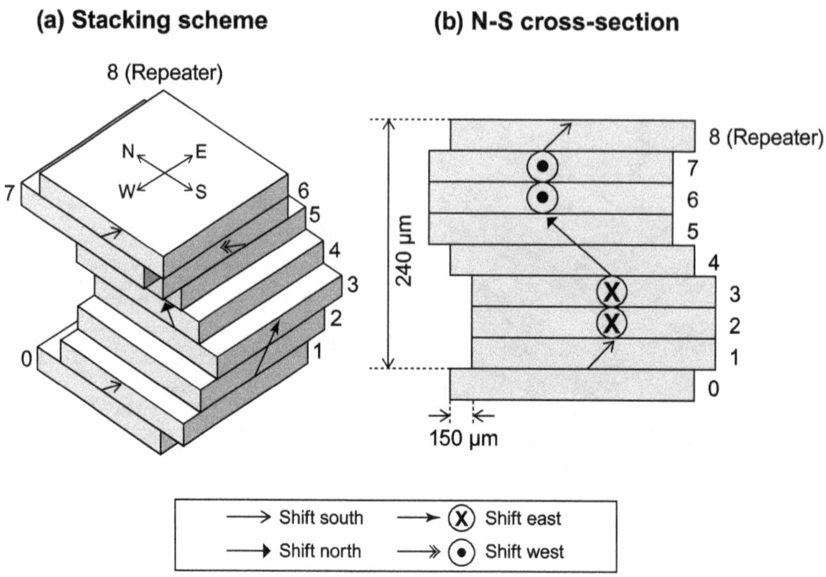

Figure 2.80 Spiral terraced stacking scheme for a TCI flash memory system. © 2010 IEEE. Reprinted, with permission, from [51]

By eliminating the spacer, spiral terraced stacking results in a much lower profile and shorter TCI communication distance. For instance, if the thickness of a chip and a spacer including bonding adhesive are 30 and 65 μm respectively, the total height of a 129 die stack (128 flash memory chips + controller) using the schemes in Figures 2.79 and 2.80 are 5.95 and 3.87 mm respectively, corresponding to a difference of 2.08 mm and a 35% reduction using spiral terraced stacking. Furthermore, maximum TCI communication distance across a repeating pattern of eight chips decreases from 370 to 240 μm.

To minimize coupling between TCI coils and bit and word lines of the memory array, [51] places the coils diagonally with edges at 45 degrees relative to chip edges. The resulting coil alignment across the stack of eight chips in Figures 2.79 and 2.80 are shown in Figure 2.81a and b respectively. Since the stacking pattern repeats at every eighth chip for both schemes, the coil of repeater chip #8 is directly above and aligned with the coil on chip #0 in both cases.

Coupling coefficient k of the TCI coil pair is simulated between the repeater chip #8 and each of the eight chips underneath in Figures 2.79 and 2.80 and plotted in Figure 2.81a and b respectively. Due to coil alignment, the farthest chip #0 does not have the worst k. Because of its shorter communication distance, the spiral terraced stacking scheme has generally stronger coupling coefficients. In particular, the worst-case coupling coefficient is improved significantly from 0.07 to 0.104 (+48.6%), while the coupling coefficient to the farthest chip #0 is increased from 0.083 to 0.132 (+59%).

With TCI, direct communication between the controller and a large number of stacked memory chips is not viable because the rapid decay of transmitted energy over

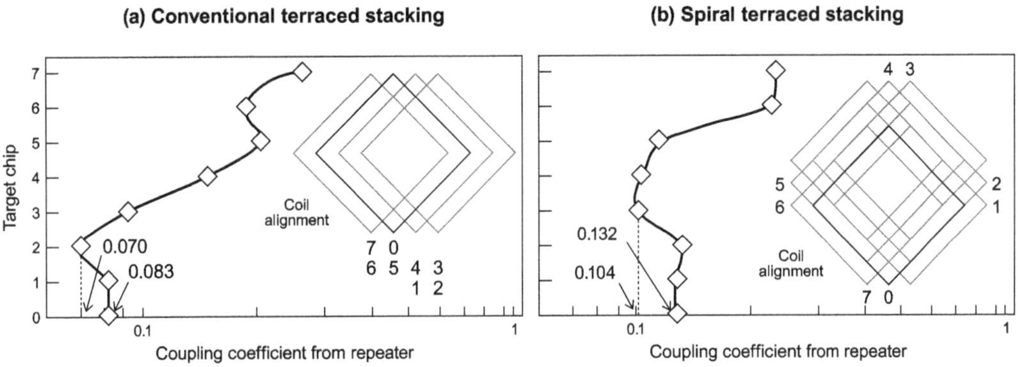

Figure 2.81 Coil alignment and coupling coefficient from TCI coil on repeater chip #8. © 2010 IEEE. Reprinted, with permission, from [51]

distance would necessitate an impracticably large coil diameter. Instead, a relayed communication approach is adopted, where intervening chips act as repeaters to propagate the signal to the target receiver far from the transmitter in the stack. The more often transmission is repeated, the shorter the communication distance between repeaters and the fewer chips the magnetic field has to penetrate, leading to a smaller coil diameter and lower transmission power per transmitter. The trade-off is a larger number of repeaters and hence active transceivers that are required to complete a transmission. Transmission power per transmitter and total transmission power as a function of how often transmission is repeated in a (64 + 1)-chip stack are measured in [51]. The results, plotted in Figure 2.82, indicate that repeating transmission at every eighth chip ($N_P = 8$) yields the minimum total power.

2.7.2 Area Efficiency Optimization

Figure 2.83 illustrates the coil alignment and sequence of events to relay transmission through three adjacent repeater chips n, $n + 1$ and $n + 2$, for read (left-hand side) and write (right-hand side) respectively. On read, signal received by RX_n of chip n is sent to TX_n to relay to chip $n + 1$, which in turn relays it to chip $n + 2$, and the process repeats until the signal reaches the controller chip at the top. The receiver coil of chip $n + 1$ is shown to be offset from the transmitter coil of chip $n + 2$. This is because when TX_{n+2} is transmitting, even though the intended target is RX_{n+3}, its magnetic field also extends downward and links to RX_{n+1}, RX_n, and so forth. Placing RX_{n+1} with a horizontal offset from TX_{n+2} is an effective countermeasure against this unintended coupling or crosstalk. This offsetting scheme as demonstrated in [1] results in up to 47% reduction in transmission power for the same BER compared to the solution in [50] using shields, enabling 1.6 Gb/s data transmission across an eight-chip stack at BER < 10^{-12} in a 180 nm process. Nevertheless, the offsetting scheme results in an effective area overhead of more than three times the coil size per channel and hence a significant reduction in TCI I/O area efficiency.

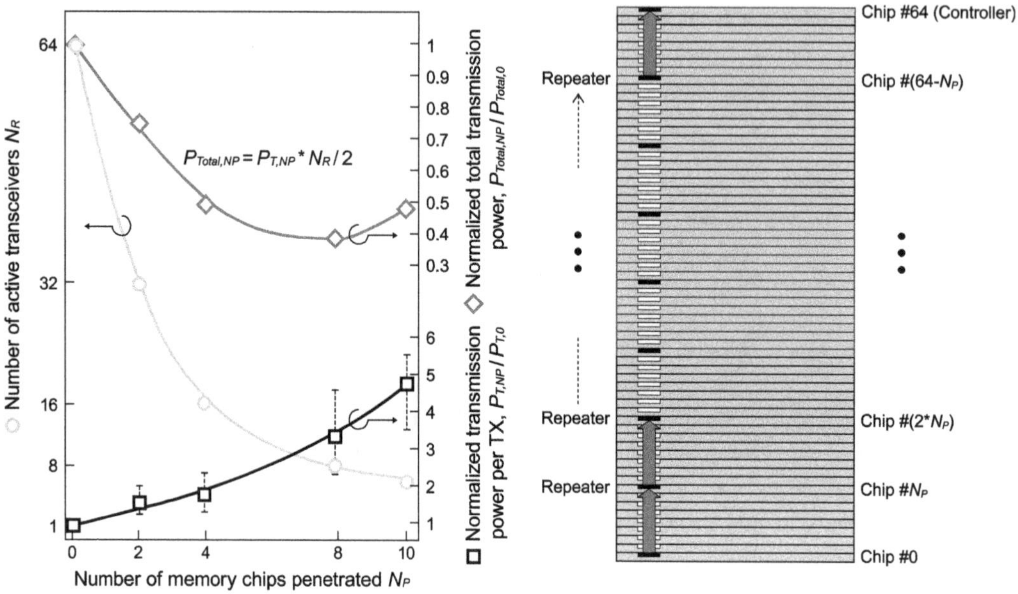

Figure 2.82 Measured transmission power vs. number of repeaters. © 2010 IEEE. Reprinted, with permission, from [51]

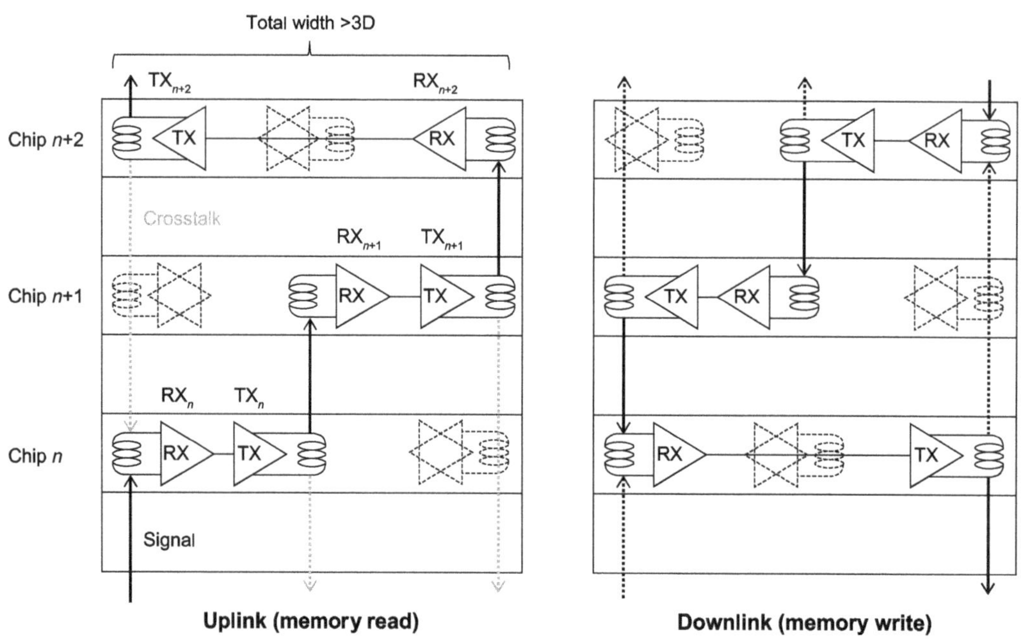

Figure 2.83 Coil alignment in relayed transmission. © 2009 IEEE. Reprinted, with permission, from [1]

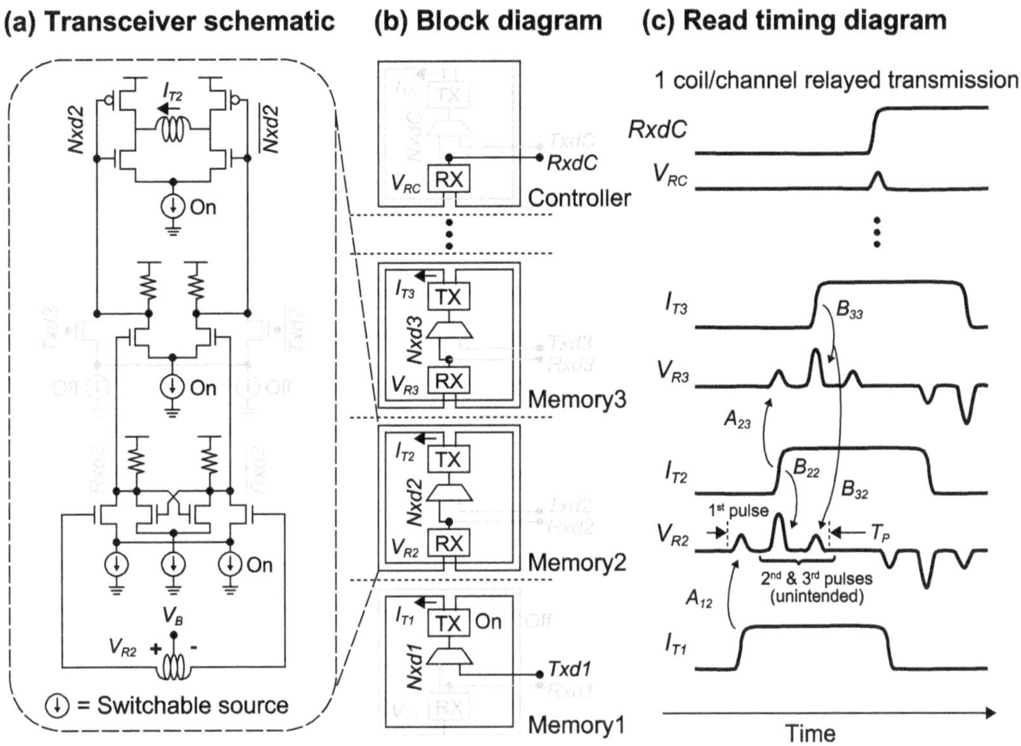

Figure 2.84 1× coil area per channel design for NAND chip stacking. © 2011 IEEE. Reprinted, with permission, from [53]

Figure 2.84a and b depicts a 1× coil area per TCI channel design proposed in [53] to minimize TCI area overhead. The transceiver is an asynchronous design with a common mode logic (CML) buffer. The current sources are switchable to allow the transceiver to be programmed as a transmitter, receiver, or repeater. The transmitter coil is placed concentrically inside the receiver coil of the same transceiver. Furthermore, coils from different chips in the same channel are all concentrically aligned. Hence the area overhead of a channel is reduced from 3 to 1× the coil area. The scenario depicted in Figure 2.84 is a data access (read) from Memory1, with intervening chips programmed to be repeaters to relay the transmission. Gray shading indicates that a signal or circuit element is disabled and inactive. Referring to the timing diagram in Figure 2.84c, signal A_{12} is transmitted from Memory1 to Memory2 by using a step in I_{T1} to induce a pulse in V_{R2}. The received signal is then amplified by the CML buffer and relayed to Memory3 as signal A_{23} through a step in I_{T2} that induces a pulse in V_{R3}. Meanwhile, the same I_{T2} step also induces a pulse in V_{R2} in its own receiver through self-crosstalk B_{22}. In fact, due to the close proximity of the transmitter and receiver coils of the same transceiver, this second, unintended pulse in V_{R2} is larger than the first, intended pulse. Nevertheless, since the receiver includes a hysteresis comparator whose input threshold has been changed upon the arrival of the first pulse to only detect a subsequent pulse of opposite polarity (Figure 2.2b), the second, unintended pulse that is of the same polarity as the initial pulse is safely ignored. The large second pulse does shift the operating voltages in the comparator and degrade its sensitivity. By designing the time constant of the internal circuits to be small, the comparator can quickly recover from this noise disturbance.

The pulse in V_{R3} caused by A_{23} leads in turn to a step in I_{T3} to relay the signal to Memory4. This step also induces a large pulse inV_{R3} through self-crosstalk B_{33}. At the same time, it induces a small, third pulse in V_{R2} through crosstalk B_{32}. This pulse is again ignored by the hysteresis comparator as long as the next data bit does not arrive until after the third pulse has faded. Therefore, if the bit time is set to be larger than T_P in Figure 2.84c, crosstalk due to placing all the coils concentrically can be ignored. The circuits in [53] are designed to operate with low-swing signaling and achieve $T_P = 0.4$ ns, resulting in a maximum data rate of 2.5 Gb/s.

In a conventional source synchronous SerDes architecture, each data link is accompanied by a parallel clock link to provide the clock for retiming in the receiver. To further reduce the TCI area overhead, [53] employs a clock data recovery (CDR) scheme to eliminate the individual clock links. However, this necessitates distribution of a global reference clock to all the chips in the stack. Furthermore, since this reference clock is always on to support random memory access, its power consumption must be minimized. To this end, [53] adopts coupled resonation to create the reference clock link using the circuit shown in Figure 2.85a. The coupled-resonation-based clock distribution network consists of an LC oscillator to generate a local clock in each chip, and it uses inductive coupling between coils to synchronize the individual clocks in frequency and phase. The coupled coils are designed with a high Q-factor for efficient coupling and low power consumption. The output clock of each coupled resonator is used to provide transmission timing (Txc) as well as the reference frequency to the local, replica PLL used by the receiver. Figure 2.85b depicts the simulated impedance Z_{CR} of the coupled resonators in a four-chip stack. The coupled resonators can oscillate at four different frequencies but are in phase only at the first peak. Since the necessary condition for oscillation is

$$Z_{CR} > \frac{1}{g_m}, \tag{2.29}$$

where g_m is the transconductance of the driving circuit, the resonators are designed such that (2.29) is satisfied across process variations only at the first peak to ensure that all resonators oscillate only at this peak but not the others. As a result, they are guaranteed to oscillate in phase by design. Since the peaks have decreasing impedance as frequency increases, this can be achieved by designing $1/g_m$ to lie between the first and second peak. The large difference in impedance between the first two peaks ensures a robustness design with tolerance of more than $\pm 50\%$ in g_m.

Figure 2.85c depicts the simulated lock range of the resonators, demonstrating a large range in the tail current I_{TAIL} of the driving circuit as long as the coupling coefficient k is sufficiently large. Additionally, coupling between the resonators reduces phase noise by a factor of $(1 + k)^{0.5}$ compared to a standalone LC oscillator.

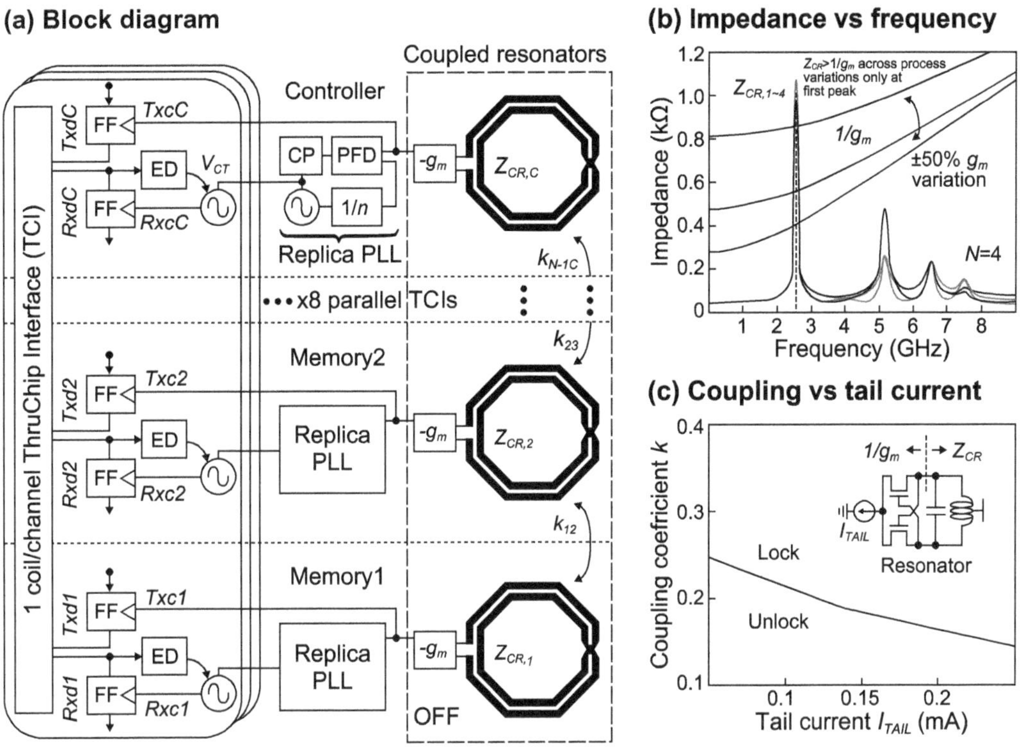

Figure 2.85 Coupled-resonation-based reference clock distribution. © 2011 IEEE. Reprinted, with permission, from [53]

By sharing the coupled resonator and replica PLL on each chip among parallel data links, power consumption and area overhead required for clocking are significantly reduced compared to the conventional source synchronous scheme on a per-data-link basis.

A test chip demonstrating the 1× coil area per channel and coupled-resonation-based clock distribution was fabricated in a 180 nm CMOS process. Sixteen chips are stacked and communicate through eight parallel TCI data links. Figure 2.86 shows the evaluation results of four of the local clocks in the coupled-resonation-based clock distribution network. Figure 2.86a demonstrates that the four clocks successfully achieve coupled resonance for I_{TAIL} = 0.21–0.39 mA, even though variation in resonance frequency between the clocks is >10% for smaller I_{TAIL}. Furthermore, phase discrepancy is ≤0.01% and root mean square (RMS) timing jitter is <2.5 ps, which is <0.6% of the clock period at 2.4 GHz, resulting in BER < 10^{-12}.

Figure 2.86b shows the evaluation results of lock range as a function of resonator capacitance variation. Locking is achieved for up to a variation of ±17.5%, which is more than sufficient to achieve robustness against the typical ±10% variation of metal–insulator–metal (MIM) capacitor in mass production. Within ±10% capacitance variation, frequency shift is <0.03%.

Table 2.22. Improvement from 1× coil per channel + coupled resonation clock. © 2011 IEEE. Reprinted, with permission, from [53].

		Previous [51]	Latest [53]	Improvement
Test chip process		250 nm CMOS embedded flash	180 nm CMOS embedded flash	–
Clocking		Source synchronous	Coupled resonator + CDR	–
Data rate (Gb/s)		2.0	2.4	1.2×
# of channels		1	8	–
Aggregate BW (Gb/s/channel)		2.0	19.2	9.6×
# of coils		6/channel	1/channel + clock	–
Coil diameter (mm)		1.1	Data: 0.9, clock: 0.6	–
Total layout area (mm^2)		7.3	7.0	0.96×
Area efficiency ($Gb/s/mm^2$)		0.27	2.7	10×
Energy consumption / chip (pJ/b)	Data	0.9	0.8	0.89×
	Clock	0.9	0.1	0.11×
	Total	1.8	0.9	0.5×

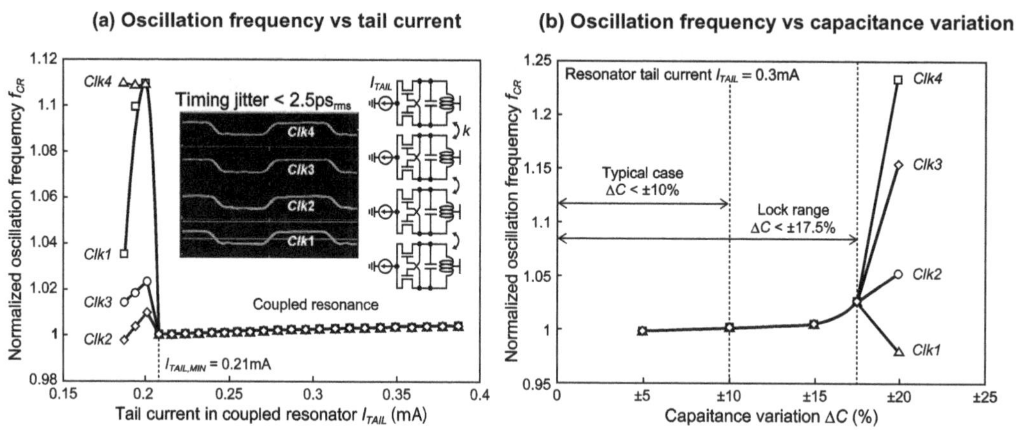

Figure 2.86 Experimental evaluation of coupled-resonation-based clock distribution. © 2011 IEEE. Reprinted, with permission, from [53]

Table 2.22 summarizes the performance of the test system in comparison with that from previous work [51] that utilizes a source-synchronous clock. In the latter case, each data link is accompanied by a clock link, resulting in a layout area of 3 × 2 = 6 × coil area per channel. The new design implements eight channels in less area than the old design implementing only one channel. Together with an increase of per-channel data rate from 2 to 2.4 Gb/s/channel, the aggregate bandwidth is increased by 9.6×, and area efficiency 10×. Furthermore, clock energy consumption is reduced by 9×, resulting in a 50% reduction in total energy consumption per bit per chip.

2.7.3 Closing Thoughts

By improving area and energy efficiencies while drastically reducing the number of bond wires, the adoption of TCI to replace wirebonding as the data interface enables a large number of flash memory chips to be stacked in one package. With each chip being thinned to 30 µm, 128 memory chips and one controller chip add up to only 3.9 mm in height. This results in drastic reduction in form factor of a flash memory stack and hence enables much larger capacity SSD to make it a viable high-performance replacement of HDD for a wider range of applications.

Nevertheless, since power delivery and low-speed control signals still require wirebonding, the memory chips must be stacked in a way to expose the required bonding pads. By using a spiral terraced stacking scheme, spacers required by conventional terraced stacking can be eliminated, resulting in lower stack height and shorter TCI communication distance. Meanwhile, area and energy efficiencies must be optimized to limit coil diameter and power required for transmission across a large number of chips with long communication distance. By using a relayed transmission scheme, communication distance and hence coil size can be limited to achieve high area and energy efficiencies. Further improvement in area efficiency is achieved by placing all transmitter and receiver coils of the same data channel concentrically, both within the same chip and across chips, resulting in an effective area overhead of only $1\times$ coil area per channel. Careful timing design is necessary though to ensure that the hysteresis receiver is immune to the increased crosstalk. Additional improvement in area and energy efficiencies is also achieved by replacing source synchronous clocking with CDR together with a coupled resonator to deliver the required global reference clock.

2.8 3D Network-on-Chip (NoC)

Network-on-chip (NoC), the application of router-based computer networking technology to on-chip communication, has the benefit of reducing on-chip routing complexity and area, as well as interconnect power. The benefit is gaining significance as more and more modules have to be interconnected on-chip in order to create an SoC with rich functionality. More recently, the concept has been extended in the third dimension in the form of 3D NoC [54]. Compared to monolithic NoC, 3D NoC has the added benefit of interconnecting chips (also referred to as "planes") of different functionalities that are made in their respective, optimal processes, such as processor, memory, accelerator, and sensor chips. Furthermore, by mixing and matching components from a set of building block chips, 3D SoC with different specifications can be constructed to serve a wide range of applications as well as to meet new system requirements that are not anticipated at the time of development of the building blocks. For instance, 3D NoC can be used to build chip multiprocessors (CMPs) consisting of different numbers and types of processor core and memory chips. Figure 2.87 illustrates the basic structure of a 3D NoC. In this particular example,

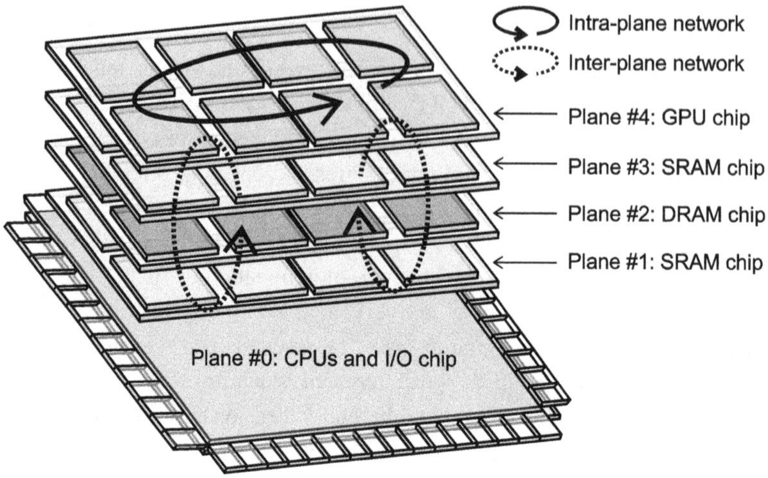

Figure 2.87 Basic structure of a 3D NoC. © 2014 IEEE. Reprinted, with permission, from [55]

one CPUs-and-I/O chip, two SRAM chips, one DRAM chip, and one GPU chip are interconnected through an interplane network to form a system-in-package (SiP). Meanwhile, communication within each chip is achieved through an intraplane network. The interplane network must be compatible with the intraplane networks implemented in individual chips for proper system operation.

Similar to other 3D ICs, 3D NoC can be implemented using either wired connection technology such as microbump and TSV, or wireless connection technology such as TCI. TCI offers a lower cost and more reliable solution than TSV. Furthermore, when wireless power delivery becomes available for general adoption such that interplane connections can be completely contactless, TCI will enable changing component chips of the 3D NoC even postassembly out in the field. With the example 3D NoC in Figure 2.87, TCI enables easy variation of the numbers and types of processing and memory chips to hit different system performance points without changing the interplane network construction. Such flexibility is of increasing value given the exponentially rising NRE costs of new chip development. In this section, we will discuss the application of TCI to 3D NoC, including different network design options, their respective shortcomings, and solutions.

2.8.1 Interplane Network Design for TCI 3D NoC

The main requirements for the TCI interplane network are as follows:

- Independence from intraplane networks on individual chips. For maximum postfabrication mix-and-match flexibility, the interplane network should function correctly regardless of the particular intraplane network implementations in the individual chips. For instance, one chip may use a 2D mesh or star topology, while another may not have an on-chip network at all. The only requirement is that the intraplane networks are deadlock-free.

- Ease of addition, removal, and interchange of chips.
- Deadlock-free.

If the larger size of a TCI coil compared to a microbump or TSV connection is a concern, the interplane network design may limit the use of TCI to high-speed data communication and not for slow control signals such as handshake signal for flow control to minimize the number of coils required.

2.8.2 Unidirectional Ring Network with Bubble Flow Control

Figure 2.88 depicts a simple unidirectional ring network design that is independent of intraplane network implementations and enables easy modification to the chip stack. Incoming packets to a plane are passed to either router on the next plane in the forward direction, or the core if the current plane is the destination. As a result, packet routing does not require consultation of routing tables. Consequently, changes to the chip stack can be made without modification of the routing tables in individual chips. Individual chips are only required to comply with the following rules:

- Each chip must support two inductively coupled links, one upward (uplink), the other downward (downlink). Each link must include a data line that transmits in the forward direction, and it may include a credit line that transmits in the opposite

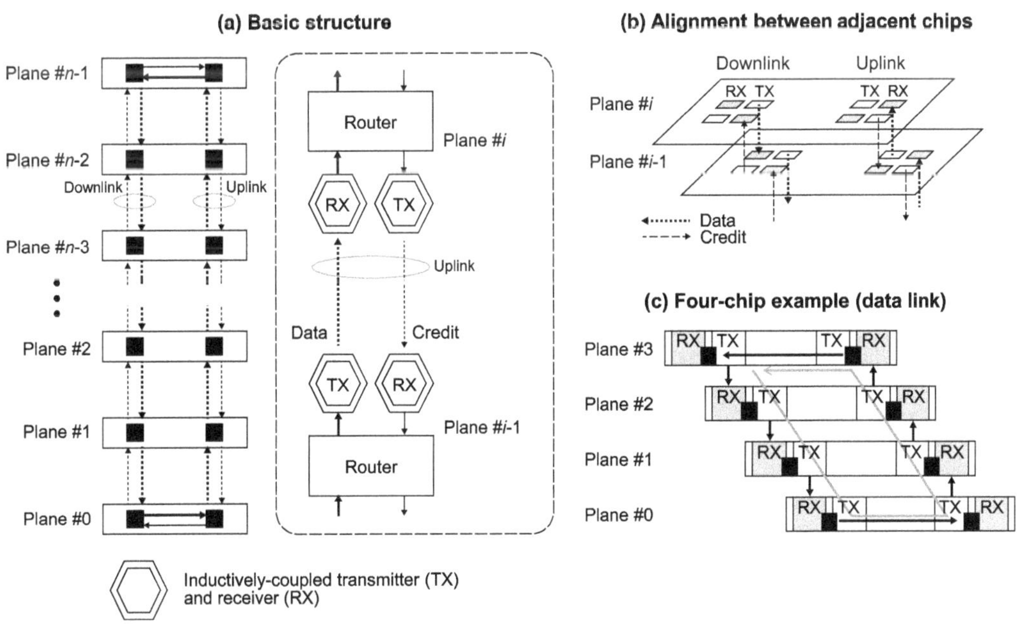

Figure 2.88 A unidirectional ring network. © 2014 IEEE. Reprinted, with permission, from [55]

direction. Individual transmitter and receiver coils are placed in prespecified locations such that adjacent chips can be mounted with matching transmitter and receiver coil pairs aligned simultaneously for both the uplink and downlink. An example implementation is shown in Figure 2.88b, where alignment is made by a lateral offset between the adjacent chips. This implementation has the added advantage of providing space for wirebonding if needed for power, clock, and control and configuration signals.
- All cores on each chip must be connected to either the uplink or downlink through its intraplane network.
- The uplink and downlink of both the top and bottom chips must be connected together through their respective intraplane networks to complete the ring, as shown in Figure 2.88c.

This unidirectional ring structure has inherent cyclic dependency and is therefore prone to structural deadlocks. To avoid deadlocks, a bubble flow control approach is adopted where all routers must obey the following flow control rules [56–57]:

- A packet can be forwarded to the next plane only if there is empty buffer space for at least one packet in the receiving router.
- A packet can be injected from an intraplane network into the ring only if there is empty space for at least two packets in the vertical input buffer of the ingress router.
- A packet can be moved out of the ring into an intraplane network only if there is empty space for at least one packet in the horizontal output buffer of the egress router. Otherwise, the packet must continue to flow through the ring, which is known as miss-routing.

The flow control mechanism is illustrated in Figure 2.89. Following the prescribed rules, planes #4 and #5 can both forward their respective packets to the next plane up. Furthermore, both planes #4 and #5 may inject packets into the ring, but plane #6 may not. Meanwhile, packets destined for planes #4 and 5 can be moved out of the ring, while those destined for plane #6 cannot and must loop through the ring once before the next attempt to exit.

These rules ensure that packet movement in and out of the ring only proceeds when space to accept it is guaranteed. As a result, a packet can continue to move along the ring unimpeded and no deadlocks will ever occur. Since this flow control does not require any virtual channel (VC) buffers, it works regardless of how many VC buffers are implemented in the individual chips. Therefore, unlike a conventional VC-based deadlock avoidance approach, it places no restriction on what chips can be included in the stack since chips with any number of VCs can be accommodated. However, when miss-routing occurs, the packet has to go through the entire ring at least once before reaching its destination, resulting in significant degradation of network performance. To alleviate such problem, a deep horizontal output buffer should be implemented.

This unidirectional ring network design with bubble flow control is used to implement a scalable 3D heterogeneous multicore processor prototype system known as Cube-1 in [58]. Figure 2.90 shows a partial block diagram of the prototype system.

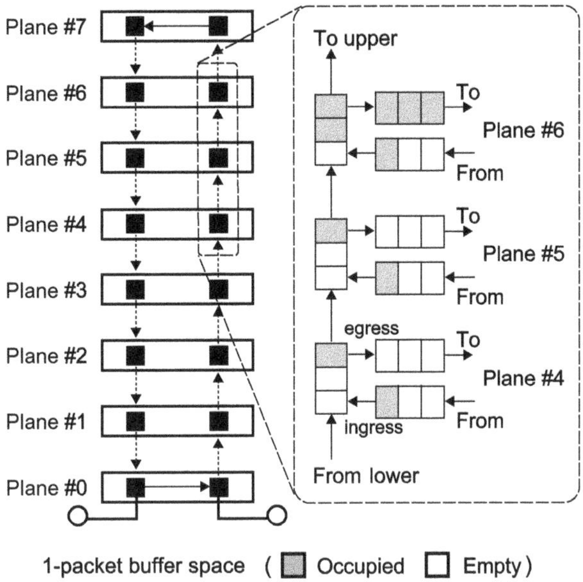

Figure 2.89 A vertical bubble flow network. © 2014 IEEE. Reprinted, with permission, from [55]

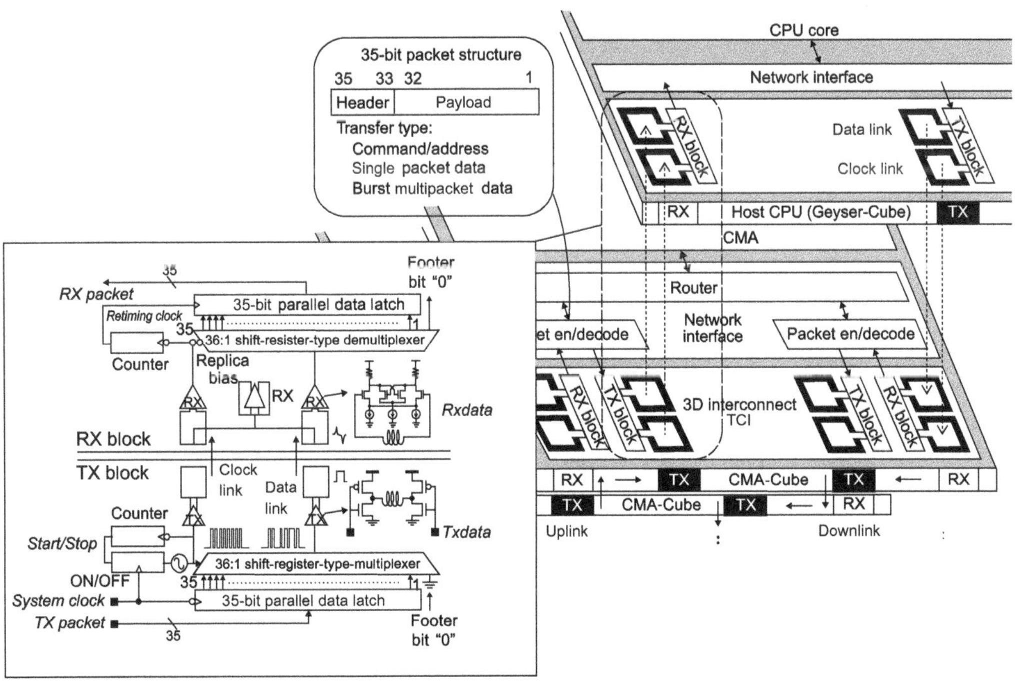

Figure 2.90 Block diagram of a scalable 3D heterogeneous multicore processor. © 2013 IEEE. Reprinted, with permission, from [58]

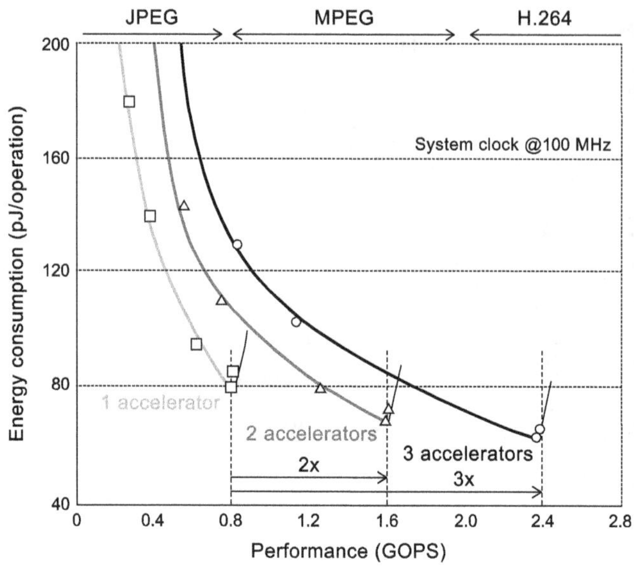

Figure 2.91 Simulated streaming processing energy efficiency vs. number of accelerators. © 2013 IEEE. Reprinted, with permission, from [58]

It consists of a general-purpose host CPU known as Geyser-Cube stacked on top of one or more multicore accelerator chips known as CMA-Cubes. It is targeted for multimedia streaming processing applications that generally have enough parallelism to benefit from simultaneous multicore processing. As a result, performance can be scaled up by increasing the number of accelerators without degrading energy efficiency. In fact, Figure 2.91 shows decreasing energy consumption per operation as performance scales up by adding accelerators.

The TCI interface on Geyser-Cube consists of one data-clock channel pair for uplink and another for downlink, while that on CMA-Cube consists of two data-clock channel pairs for each link, so that CMA-Cube can communicate with both the chip above and the chip below, if any. The TCI transceiver consists of the standard H-bridge driver and hysteresis comparator receiver as well as a receiver replica bias generation circuit. The transceiver is also connected to a packet en/decoder to pack/unpack and serialize/deserialize command, address, and parallel data bits into a single data stream for burst transmission. The parallel clock is generated by a local ring oscillator as described in Section 2.2.4.2. The TX coils of a data-clock pair on one plane can be aligned with the RX coils on an adjacent plane by a lateral offset between the two planes to establish a link. A coil diameter of 240 μm is used in the prototype to cover a communication distance range of 50–90 μm, since each chip is thinned down to 40–80 μm and the bonding epoxy glue is 10 μm thick. By thinning the chip further to 20 μm, which was already possible in 2013, the coil diameter can be reduced to $\leq$90 μm, and 16 chips can be stacked in a standard JEDEC 1.4 mm package.

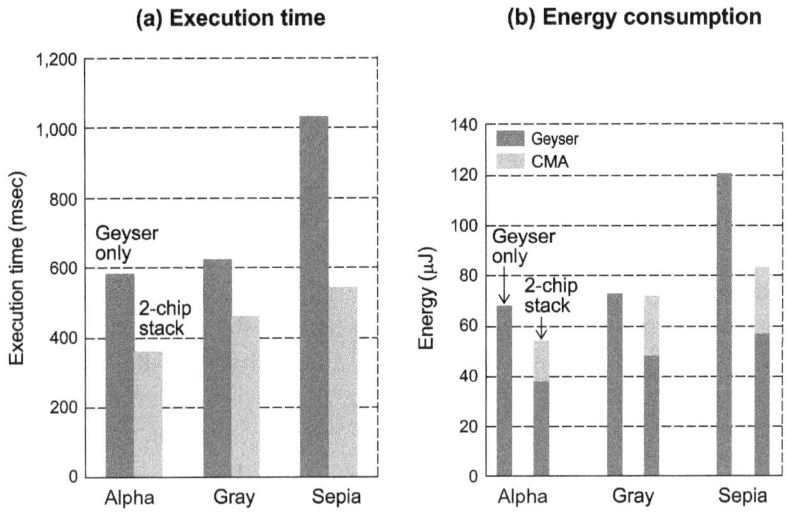

Figure 2.92 Measured application performance of a two-stack Cube-1. © 2013 IEEE. Reprinted, with permission, from [58]

A two-stack (Geyser-Cube + 1 CMA-Cube) prototype system is used to evaluate the performance improvement of the accelerator. Figure 2.92 depicts the results in terms of execution time and energy consumption in processing an image with 100,240 pixels using three simple filter applications – alpha blender (Alpha), gray filter (Gray), and sepia filter (Sepia). Results from using a standalone Geyser-Cube CPU for processing are also shown as the baseline for comparison. When the accelerator is used for processing, the Geyser-Cube CPU is used only for managing I/O and data movement. The accelerator is able to reduce execution time by about 30–50%. Improvement in energy consumption, however, is not as significant. This is because the Geyser-Cube spends a lot of time being idle. More aggressive power management to reduce idle power can be expected to further improve energy efficiency.

2.8.3 Dynamically Reconfigurable Bidirectional Ring Network with Bubble Flow Control

A unidirectional link generally suffers from low scalability in throughput and communication latency. The adoption of TCI enables high throughput since it supports data rates as high as the transistor speed allows. For instance, a TCI channel fabricated in a 65 nm process is simulated to support a data rate as high as 30 Gb/s per channel [55].

To improve communication latency, the TCI interplane network can be modified to dynamically support a bidirectional link. This can be achieved without increasing the number of coils by replacing both the TCI transmitters and receivers in the unidirectional ring network with dynamically reconfigurable transceivers. As explained in Section 2.1.3, there are design requirement differences between a transmitter and a receiver coil, and so generally they are optimized differently. But in this implementation, each coil must be designed to be both an effective transmitter and receiver. Finally, the routers on individual chips must be equipped with both an input and output port for each connection to both the intraplane and interplane networks. This is generally not a restrictive requirement since most on-chip routers have both input and output in every connection. For each connection to a TCI link, both the input and output ports of the router are connected to the same transceiver. Only one of the two ports is active at any given time based on whether the transceiver is configured as a transmitter or receiver.

With these modifications to the unidirectional ring network, communication direction can be dynamically reversed by reconfiguring a TX–RX pair as RX–TX pair, effectively creating a bidirectional ring (Figure 2.93). To ensure that the pair of communicating transceivers are reconfigured in synchronization, the credit line is used by the current RX to send a reconfiguration request to the current TX, and the data line is then used by the current TX to respond. Reconfiguration occurs only if the request is acknowledged by the current TX. The overhead required for reconfiguration will result in some reduction in network throughput, but the support for bidirectional communication will significantly shorten latency.

Performance of both the TCI unidirectional and bidirectional ring network designs in conjunction with the vertical bubble flow deadlock avoidance approach

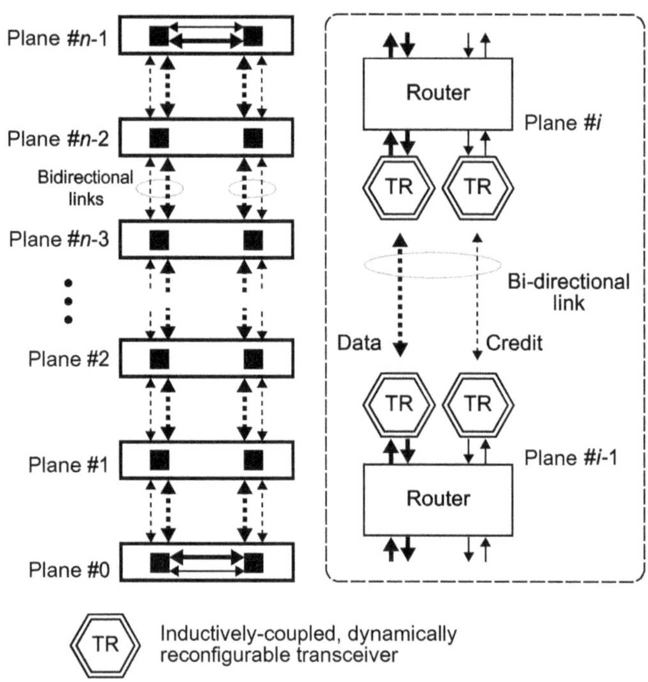

Figure 2.93 A bidirectional ring network using dynamically reconfigurable transceivers. © 2014 IEEE. Reprinted, with permission, from [55]

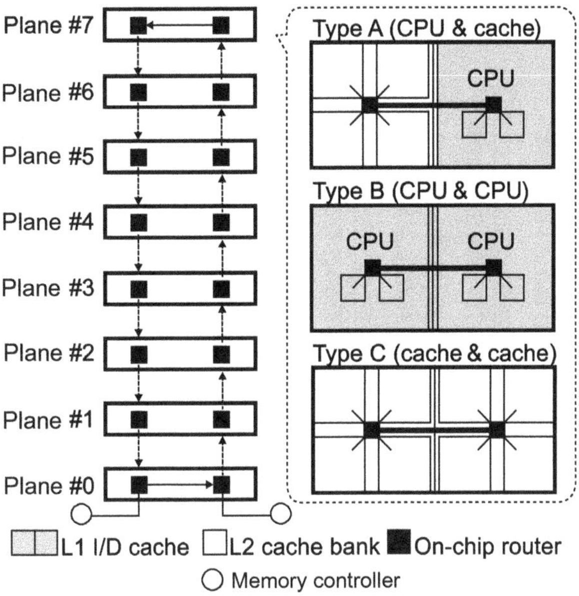

Figure 2.94 CMP for performance evaluation of TCI ring network interplane designs. © 2014 IEEE. Reprinted, with permission, from [55]

implemented in the CMP in Figure 2.94 is evaluated in [55]. The CMP consists of multiple 3D-integrated chips, each of which can have a different combination of numbers of CPU cores and shared L2 cache banks. Ten programs from the OpenMP implementation of the NAS Parallel Benchmarks [59] are used to quantify application performance. A vertical broadcast bus interplane network is used as a baseline for evaluating the network designs. The broadcast bus enables any plane to communicate directly with any other plane. However, each plane must wait for its turn to transmit when all other planes listen. For simplicity, an 8-cycle transmission window is assigned to each plane periodically, whether it has data to transmit or not.

The vertical bubble flow control implementation, Bubble (15-flit), has a 15-flit FIFO buffer for each input port. It is compared against two VC-based deadlock avoidance approaches, 6-VC (18-flit) and 6-VC (30-flit). Both implementations have six virtual channels for each input port, with each virtual channel having a 3-flit and 5-flit FIFO buffer in 6-VC (18-flit) and 6-VC (30-flit) respectively.

The application benchmark results are depicted in Figure 2.95 for an eight-plane CMP. Execution times are normalized by that of the vertical broadcast bus. The results show that both the unidirectional and bidirectional ring designs outperform the vertical bus. In conjunction with Bubble (15-flit), on average unidirectional ring is 15% while bidirectional ring 32.5% faster than the vertical bus. Furthermore, bidirectional ring is faster than unidirectional ring across the board, and 20.7% faster on average using Bubble (15-flit). Meanwhile, Bubble (15-flit) is faster than both VC-based implementations almost across the board and for both ring designs, due to

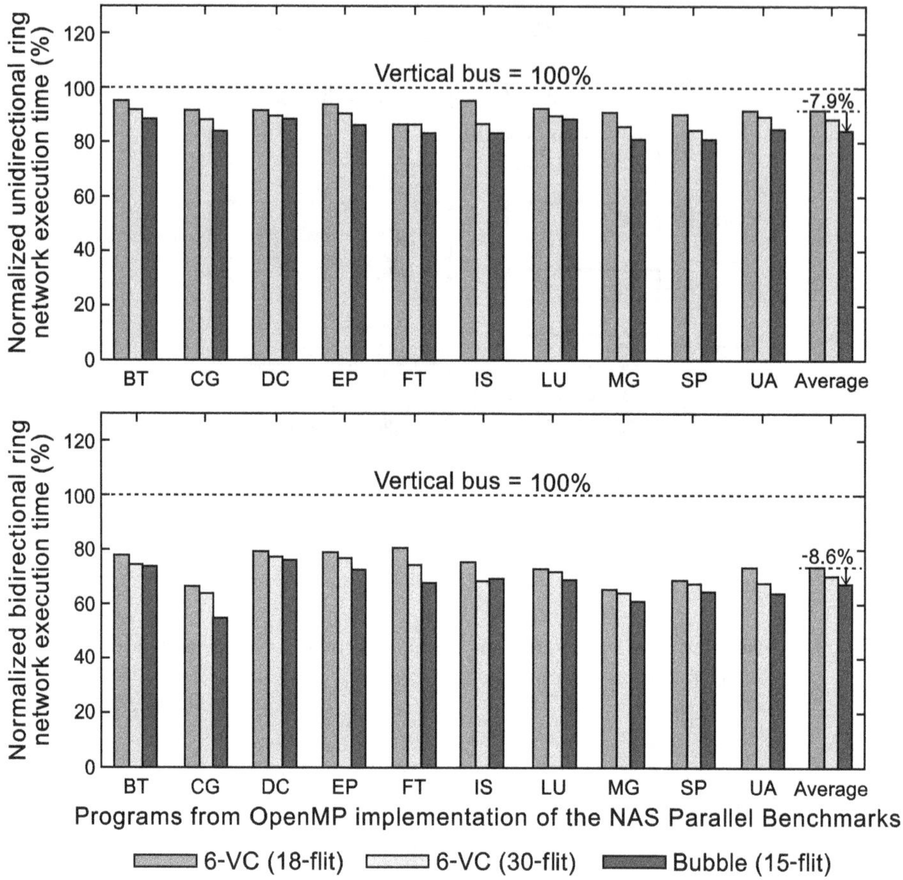

Figure 2.95 Benchmark results of TCI uni- and bidirectional ring networks. © 2014 IEEE. Reprinted, with permission, from [55]

a deeper 15-flit buffer compared to a 3- and 5-flit buffer respectively in the VC implementations.

2.8.4 Escalator Network with Piggyback Flow Control

The prototype 3D heterogeneous multicore processor Cube-1 built with a unidirectional ring network and bubble flow control described in Section 2.8.2 is found to suffer from several shortcomings. First, because of the lateral offset required in stacking the chips, uplink TCI coils on one plane can come close to the horizontal location of and interfere with downlink coils on another plane, depending on the amount of offset and the separation between uplink and downlink coils within the same plane. As the number of chips increases, it becomes difficult to avoid this

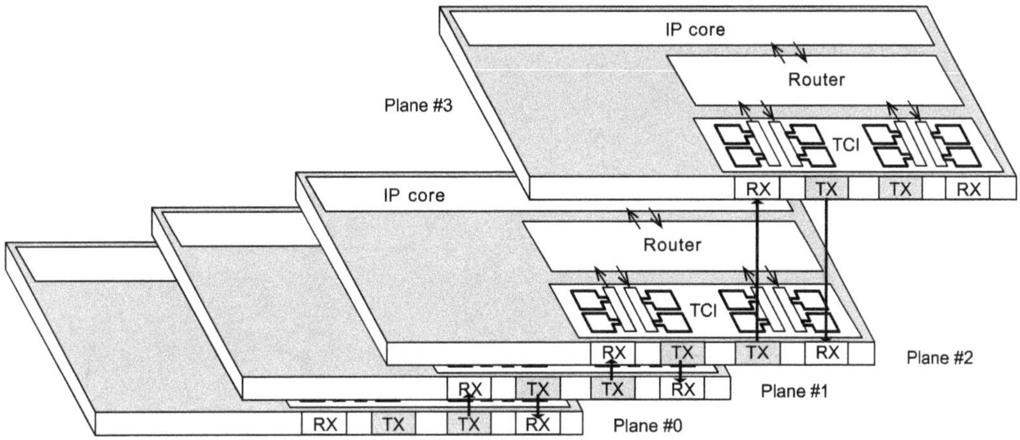

Figure 2.96 Construction of an escalator network [60].

interference. As a result, it is found to be difficult to stack more than three chips in Cube-1.

Second, the bubble flow control approach requires enough bubbles to be in the network for smooth data flow. As a result, a router can insert packets into the network only within certain time window, which limits injection rate.

Third, as previously noted, a unidirectional ring network can suffer from large latency due to a large number of hops to reach a destination plane, especially when miss-routing occurs in a large chip stack.

Latency can be improved by using the dynamically reconfigurable bidirectional ring described in Section 2.8.3, but it requires a credit line to synchronize reconfiguration, which increases the number of coils.

To circumvent these shortcomings, an escalator network together with piggyback flow control is proposed in [60] and used to build the second prototype, known as CUBE-SOTB, where SOTB refers to the fabrication process known as silicon on thin box [61]. Construction of the escalator network is illustrated in Figure 2.96. Unlike both the unidirectional and dynamically reconfigurable bidirectional ring networks, the TCI block in the escalator network uses full-duplex, bidirectional channels to support both uplink and downlink data transmission. Specifically, each block consists of two such channels to communicate with both the plane above and the one below. As a result, the router connected to each TCI block is required to have three pairs of input/output ports to connect to the two channels and the core respectively. Similar to Cube-1, coupled TX–RX pairs are aligned by a lateral offset before a chip is stacked. But unlike Cube-1, since there is only one TCI block on each plane, there is no concern of a TCI block on one plane being shifted to the horizontal proximity of another block on another plane, thus avoiding interplane coil interference. Packets are routed by comparing the destination chip ID with current chip ID. If they are the same, packets are routed to the core. If the destination chip ID is smaller, packets are routed

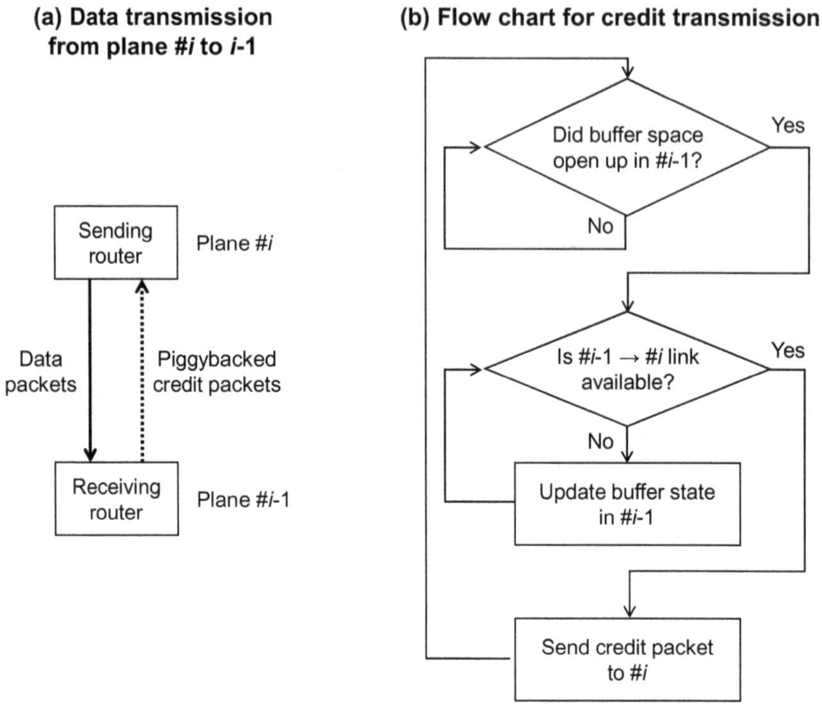

Figure 2.97 Piggyback flow control mechanism [60].

to the downlink; otherwise, they are routed to the uplink. Consequently, packets are moved either up or down the stairs, hence the name "escalator network."

In place of a bubble flow control approach, a credit-based flow control is implemented by piggybacking flow control information on data transmitted in the opposite direction, taking advantage of the bidirectionality of the TCI channels. Figure 2.97 illustrates the flow control mechanism for data packet transmission from plane #*i* to *i*–1. When a buffer space in the router on plane #*i*–1 opens up since the last time a credit is sent to plane #*i*, the router checks to see if the data link to plane #*i* is available. If yes, it uses the link to send a credit packet to plane #*i*. Otherwise, it waits until the link becomes available while continuously updating its buffer state. Consequently, there are two main shortcomings in this control mechanism. First, the sharing of the same link by data and credit packets degrades network throughput. Second, data packet transmission has to wait both for buffer space on the receiving plane to open up and for the link in the opposite direction to become available for piggybacking. This wait can be long when a large data packet is being transmitted in the opposite direction.

Nevertheless, evaluation results from [60] show that degradations from both shortcomings are small. The degradation in throughput is on the order of 3–4%. Meanwhile, increase in latency is negligible for common applications.

Performance of a four-chip implementation of Cube-1 and CUBE-SOTB are simulated and compared in [60]. While the escalator network with piggyback flow control does not require VCs, the input port of each buffer in CUBE-SOTB includes

Table 2.23. Transmission performance of CUBE-SOTB vs. CUBE-1.

	Normalized throughput		Normalized zero-load latency	
Traffic	Uniform	Bi-reverse	Uniform	Bi-reverse
CUBE-1: unidirectional ring + bubble	100%	100%	100%	100%
CUBE-SOTB: escalator + piggyback without VCs	126%	93%	75%	82%
CUBE-SOTB: escalator + piggyback with VCs	159%	128%	–	–

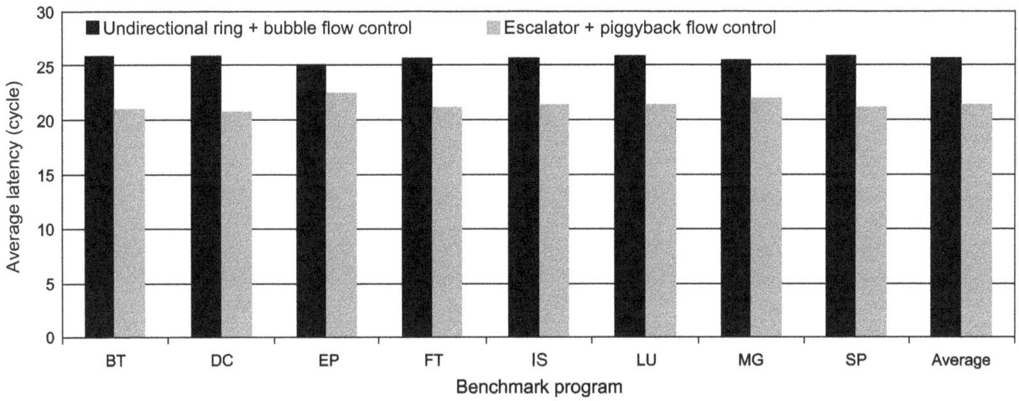

Figure 2.98 Application performance comparison between Cube-1 and CUBE-SOTB [60].

eight VCs as required by the operating system. As a result, application performances are compared between three networks, namely the unidirectional ring network with bubble flow control in Cube-1, and the escalator network with piggyback flow control in CUBE-SOTB with and without VCs. The packet transmission performance comparison in terms of throughput and zero-load latency using uniform and bit-reverse traffic is summarized in Table 2.23. The escalator network with piggyback flow control and no VCs, while delivering 7% less throughput under bit-reverse traffic, provides 26% more throughput under uniform traffic, with 18–25% lower zero-load latency than the unidirectional ring network with bubble flow control. Furthermore, addition of VCs results in 28–59% improvement in throughput.

Application performance is also simulated and compared using eight programs from the OpenMP implementation of the NAS Parallel Benchmarks as shown in Figure 2.98. The results confirm latency improvement of the escalator network with piggyback flow control at the application level of 10–20%. Consequently, in addition to avoiding interplane coil interference, this improved network alleviates the latency shortcomings of the unidirectional ring network with bubble control as well.

2.8.5 Closing Thoughts

Compared to wired 3D integration technology such as microbump and TSV, TCI offers added flexibility in mixing and matching chips to create a 3D NoC. Without the

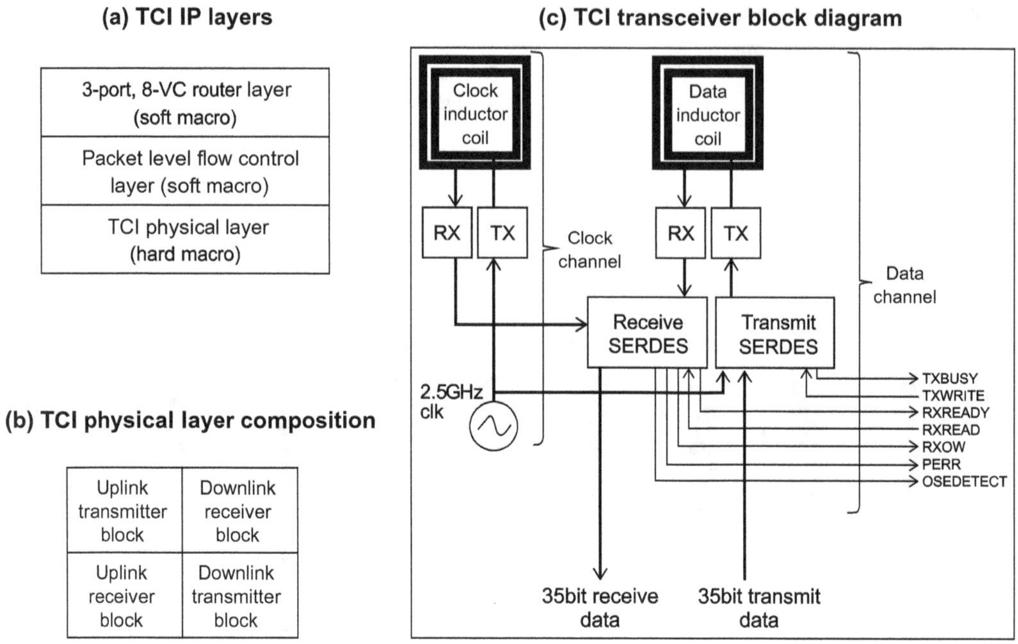

Figure 2.99 TCI IP for 3D NoC [60].

need for a physical connection, TCI can be used in conjunction with an interplane network design that is agnostic regarding the intranetwork topology of individual chips to build a flexible and scalable 3D NoC. The three interplane network designs discussed in this section – unidirectional and dynamically reconfigurable bidirectional ring network with bubble flow control, and escalator network with piggyback flow control – do not require any advance knowledge of the intra-plane network design of the individual chips and do not rely on virtual channels in their routers. Any two chips with matching TCI blocks can be stacked together to establish the interplane network. This can in theory be done not only postfabrication of individual chips but post–system assembly in the field by the user, when wireless power delivery becomes feasible and the interplane connection becomes completely contactless.

To facilitate adoption of TCI in 3D NoC, a TCI IP has been developed as described in [60]. Figure 2.99 depicts the composition of the TCI IP. It consists of three layers – a router layer and a link control (packet level flow control) layer each implemented in a soft macro, and a physical layer provided as a hard macro (Figure 2.99a). The physical layer is composed of four transceiver blocks for downlink transmit and receive and uplink transmit and receive respectively (Figure 2.99b). The four blocks are created by configuring the same transceiver block physical design shown in Figure 2.99c for either transmit or receive. While the transceiver block is dynamically reconfigurable between transmit and receive, their configurations are fixed in the physical layer of the IP. An escalator network step is formed by laterally shifting and stacking one chip over another to align their respective TCI IPs. In this manner,

3D NoCs consisting of different building block chips can be integrated into a heterogeneous system connected through an escalator interplane network to address diverse computational needs.

2.9 Postscript: Parallel TCI Research

The potential of TCI as a better alternative in system cost and performance to TSV for 3D IC integration has motivated parallel research efforts at other institutions. While it is not intended to be an exhaustive and comprehensive discussion, this section introduces a sample of such parallel efforts.

2.9.1 Coil Design Methodology

The closed-form equations in Section 2.2.1, in particular those in Table 2.5, relate TCI electrical performance parameters to its physical design parameters in an intuitive manner. While they are approximate, they provide insight into how to optimize the interface design, especially the coil design, for power and area efficiency. However, the use of such equations may not represent the best methodology in terms of speed and accuracy in identifying the optimal coil design. Meanwhile, the motivation of [62] and [63] is to generate an optimal coil design methodology that is automated and fast while achieving high accuracy. It achieves its goals by accomplishing the following:

- Creating semi-empirical equations to compute coil electrical parameters from its physical design parameters
  - o Semi-empirical equations are closed-form equations modified with empirically determined correction factors. Being closed-form equations, they enable fast extraction of electrical parameters. At the same time, the correction factors result in improved accuracy.
- Reducing the size of the search space by observing that the fill factor of the optimal solution can be predetermined, and that the RX coil can be optimized first instead of simultaneously with the TX coil, without losing accuracy in the optimal design
  - o Fill factor is a measure of how much of the area within the coil diameter is used for coil wiring. By predetermining the fill factor of the optimal design, the search space can be reduced to that around this fill factor value only.
- Using dynamic programming to reuse computation results from previous search iterations

Implicit in the definition of coil physical design parameters in Figure 2.7 is that the wiring width and spacing are uniform across the entire coil. Given the reduction in computation time achieved by [62] and [63], the research expands the solution space to include nonuniform coil designs where wiring width and spacing are varied incrementally at each turn. By adopting a scalable inductor model based on [64], nonuniform coil designs can be quickly analyzed so as to minimize the added computation time due to the expanded solution space.

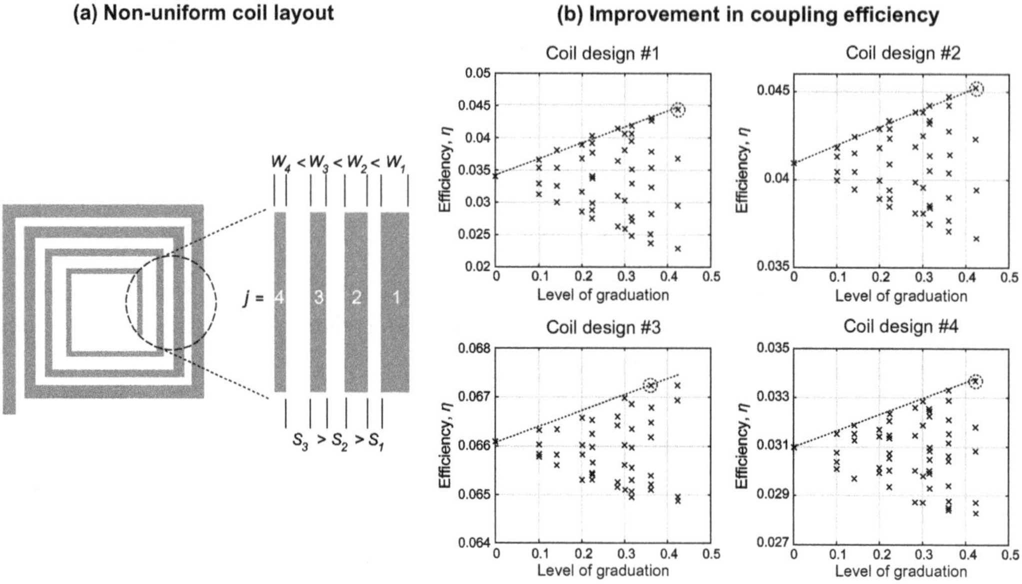

Figure 2.100 Nonuniform coil design and resultant performance improvement. © 2019 IEEE. Reprinted, with permission, from [63]

Figure 2.100a depicts the concept of a nonuniform coil design where the wiring width decreases and spacing increases at each turn as the coil spirals inward. Figure 2.100b quantifies the improvement in coupling efficiency, defined as V_R/V_T, as the level of graduation increases for four different base example layouts. Level of graduation is a measure of how much the wiring width and spacing change incrementally at each turn and is computed by

$$\text{Level of graduation} = \sqrt{\left(\frac{w_n - w_1}{n}\right)^2 + \left(\frac{s_{n-1} - s_1}{n}\right)^2}, \tag{2.30}$$

where w_1 and w_n are the width of the first and last turn, s_1 is the spacing between the first and second turn, and $s_{n\text{-}1}$ the spacing between the second-to-last and last turn respectively. For each level of graduation, different permutations of w_1, w_n, s_1, $s_{n\text{-}1}$ are evaluated, resulting in multiple data points in the graphs. A level of graduation = 0 corresponds to a uniform coil design. For this set of evaluation, nonuniform coil design achieves a maximum improvement of 30% and average improvement of 14% in coupling efficiency compared to a uniform design. Such improvement enables the coil area to be reduced for a given coupling efficiency.

2.9.2 Application of TCI to Heterogeneous 3D IC Integration

Heterogeneous 3D IC integration is the stacking of disparate chips, where disparate may refer to the nature of the chips as being digital or analog, and/or the process technology with which the chips are fabricated. It can be employed to create a small form-factor, low-cost, low-power intelligent device such as an IoT device, a lab-on-a-chip, or a neuromodulator where digital processing is combined with sensing and actuation functionalities. While a digital processing chip may be fabricated in an advanced process for performance, a sensor chip that does not require speed is more likely manufactured in a mature process to keep the cost down. Heterogeneous 3D integration of such a combination of disparate chips may therefore require the chip interface to support communication across a wide gap in process technology. In such applications, TCI offers the advantages of front-to-back integration compared to capacitive and microbump integration to expose the sensor chip to the environment, as well as the advantages of low cost, high reliability, and the elimination of level shifters compared to TSV.

The application of TCI is investigated in [65] to develop a closed-loop neuromodulator, which is an implantable device that analyzes neural signals in real time and then responds with customized neural stimulation [66]. Given its function, a closed-loop neuromodulator requires the integration of two disparate chips – an analog chip for amplification, filtering, and analog-to-digital conversion of the neural signals, and a digital processing chip to detect biomarkers from the signals, such as those foretelling an oncoming seizure. Furthermore, given its implantable nature, it must be low-power, small form factor, and highly reliable, which is why TCI represents an attractive 3D integration solution.

Figure 2.101 depicts the construction of the closed-loop neuromodulator investigated in [65]. It consists of a digital processing silicon chip made in a 65 nm LP CMOS process stacked front-to-back on top of an analog chip fabricated in a 180 nm SiGe BiCMOS process, which in turn is stacked on top of a substrate with an electrode array for sensing neural signals. While the bottom chip is bonded to the substrate using

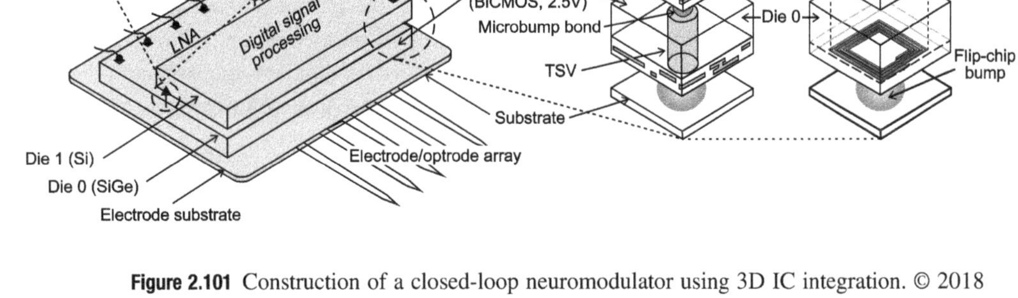

Figure 2.101 Construction of a closed-loop neuromodulator using 3D IC integration. © 2018 IEEE. Reprinted, with permission, from [65]

flip chip bumps, the front-to-back integration of the two chips is implemented using either TSV or TCI. The investigation compares the two implementations in terms of power consumption. For fair comparison, both implementations are realized using the same total area and operated to deliver the same total bandwidth. Since a single TSV is smaller than a single TCI coil, multiple TSVs are implemented within the same area as a TCI coil, thus allowing the TSV implementation to run at a lower data rate for lower power. However, TSV requires a level shifter to convert between the 2.5 V domain of the BiCMOS chip and the 1.2 V domain of the LP CMOS chip, resulting in added power. The investigation evaluates nine different variants of the TSV implementation using different combinations of TSV diameter and pitch and signal-to-ground ratio. By eliminating the level shifters and the associated power consumption, the TCI implementation achieves the lowest power, which is 37% less power on average than the TSV implementations.

The low power advantage of TCI for 3D integration of disparate chips through the elimination of level shifters is therefore validated in [65]. Additional power reduction is possible by optimizing the transceiver design to take advantage of the process asymmetry as proposed in [67], which investigates the integration of a 65 nm processing chip with a 350 nm sensor chip. Specifically, because the supply voltage of the 350 nm process is large, the transistors in the sensor chip transceiver are sized for minimum current to achieve low power. Meanwhile, transistors in the 65 nm processing chip transceiver are sized up to compensate for the weaker sensor chip transceiver. As a result, the processing chip transmitter drives a large current to enable communication with a sensor chip receiver that is sized for minimum sensitivity. On the other hand, the processing chip receiver is sized for maximum sensitivity so that the sensor chip transmitter can be implemented with minimum drive current. In other words, the optimization makes the better performing chip work harder to enable large reduction in power consumption of the slower chip to achieve overall interface power reduction.

2.10 Postscript: The Phenomenon of Coupled Resonation

The phenomenon of coupled resonation, which is made use of in the distribution of a global reference clock described in Section 2.7.2, is a widely observed natural phenomenon [68]. An excellent visual example can be found in the collective synchronization of the ticking of a set of mechanically coupled metronomes. Independently ticking metronomes, when placed on a common platform that allows mechanical vibration from one metronome to be coupled to another, eventually transition into ticking in synchronization, regardless of their initial phase relationship (Figure 2.102 and [69]).

The same phenomenon manifests itself in different forms in different domains. In mammals, thousands of cells inside the suprachiasmatic nucleus located in the hypothalamus of the brain are synchronized to create the biological clock that regulates the rhythm of the body such as the sleep cycle. Meanwhile, fireflies have been observed to gather in mangrove forests in East Asia and flash light in unison. In the domain of

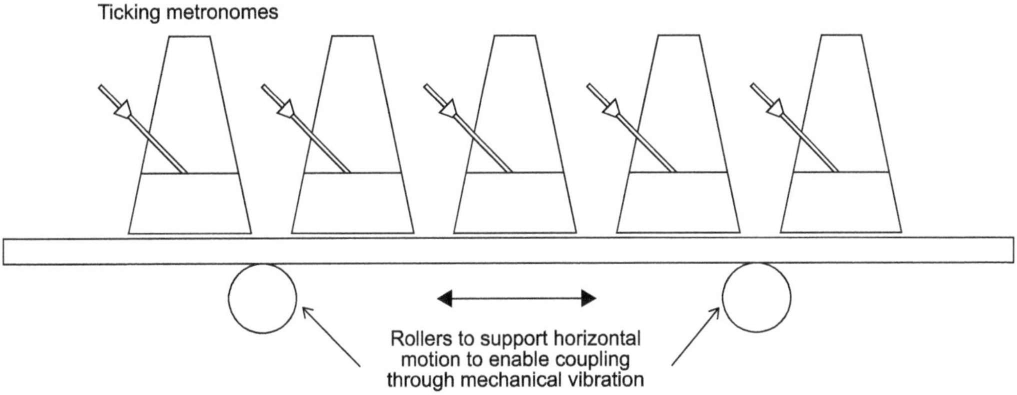

Figure 2.102 Collective synchronization of ticking metronomes [68].

inanimate objects, superconductors achieve superconductivity when a certain condition makes individual electrons move in lockstep, bringing electrical resistance to zero. On the other hand, a laser is generated by causing individual atoms to emit photons of the same frequency and phase in unison. Even manmade objects exhibit collective synchronization, such as generators connected to the same power grid. Their rotational speeds are synchronized through energy flowing from high-speed to low-speed generators. All these manifestations of the same fundamental phenomenon suggest that nature has the tendency to move toward a rhythmic harmony. On the other hand, the second law of thermodynamics states that the total entropy of an isolated system can never decrease over time. Since entropy is a measure of disorder, this law suggests that nature has the tendency to move toward chaos. Thus, the second law of thermodynamics seems to be at odds with the natural tendency of collective synchronization. So which is nature's true intention?

The first law of thermodynamics states that energy must conserve. In other words, while energy can convert from one form to another, for example from optical to electrical in a solar panel, the total sum of energy must remain constant. But the second law stipulates that as energy conversion occurs, the amount of free (usable) energy can only decrease. Put differently, the overall quality of energy judged by whether it can be used to perform work and be useful to human being can only degrade. Therefore, the universe is moving from a world of motion to stillness, and from life to death. This is the idea of heat death first proposed by Scots-Irish mathematical physicist William Thomson (also known as Lord Kelvin) in 1851, which suggests that motion will cease and the universe die when all energy has been converted to heat.

But in reality, structure can form even while free energy decreases. The two may not be as inconsistent as it appears, since structures can result in energy efficiency, limiting the decrease in free energy. For example, when a candle burns, it generates heat, which in turn creates a flame. Therefore, as entropy increases (heat is generated), a structure (flame) is formed. The flame vanishes when heat generation stops and thermal equilibrium is reached. The energy that exists in the system that includes the candle and its environment is shared and spreads out to stop further heat transfer. The system then reaches a metastable state where the atoms and molecules within maintain their state. If the candle is lit again, it resumes the generation of entropy as well as the flame. The two are generated concurrently. In other words, chaos and structure are not mutually exclusive.

The scientific study of collective synchronization is challenging since it is a nonlinear behavior. In a linear system, the effect of the sum is the sum of the individual effects. Mathematically, a linear system can be characterized by the following equation:

$$f(\mathrm{a}x_1 + \mathrm{b}x_2) = \mathrm{a}f(x_1) + \mathrm{b}f(x_2), \tag{2.31}$$

where f(x) denotes the output of the system when the input is x, and a, b are constants.

In other words, interaction between individual elements can be ignored (Figure 2.103a). Therefore, a complicated, large-scale system can easily be analyzed by breaking it down into small pieces and analyzing them individually. The behavior of the whole system can then be predicted by adding the behaviors of the individual pieces. For example, molecules in gas state move independently and can be analyzed using linear statistical mechanics.

By contrast, molecules in solid and liquid states interact strongly with each other, and hence linear analysis is not applicable. In other words, when material undergoes a phase change from gas to liquid or solid, it exhibits fundamentally new behavior due to interaction between its constituents that cannot be predicted through linear extrapolation. Likewise, in radio engineering, when a sinusoidal wave is input into a

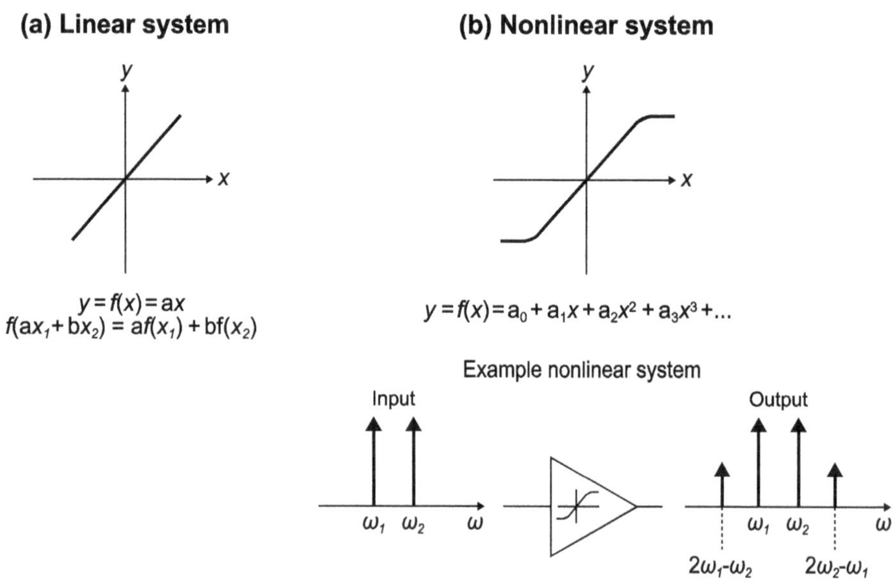

Figure 2.103 Linear vs. nonlinear system [68].

Figure 2.104 Distinguished researchers of collective synchronization. Norbert Wiener: by Konrad Jacobs – https://opc.mfo.de/detail?photo_id=4520, CC BY-SA 2.0 de, https://commons.wikimedia.org/w/index.php?curid=6090164; Arthur Winfree: by Gene Selkov – Cropped from Image on Flickr, CC BY 2.0, https://commons.wikimedia.org/w/index.php?curid=59260097; Charles Peskin: By L. B. Peskin. Robert Adler: Fair use, https://en.wikipedia.org/w/index.php?curid=9551099, photo trimmed

nonlinear system, the output contains components at new frequencies that did not exist in the input (Figure 2.103b). Analyzing such complicated nonlinear behavior requires the use of computer simulation based on mathematical models of the system.

Collective synchronization, where the individual actions of a group of elements reach synchronization collectively through mutual interaction, is such a nonlinear behavior. Its analysis has been undertaken by distinguished researchers from various disciplines (Figure 2.104). Analysis was first undertaken by Massachusetts Institute of Technology mathematician Norbert Wiener, who was also a pioneer of cybernetics, which is the study of the control and communication mechanism common to living organisms and machines [70]. Wiener postulated that alpha wave is the master clock of the brain that brings individual neurons with uncorrelated rhythms into unison, resulting in frequency entrainment (synchronization to a single rhythm). Unfortunately, he passed away in 1964 before he could prove his hypothesis. A year later, Cornell University student Arthur Winfree, who later became a theoretical biologist at the University of Arizona, established the mathematical technique for tackling the problem. Then in 1975, New York University applied mathematician

Charles Peskin proposed a model for the natural pacemaker cells of the heart [71]. In Peskin's model, the voltage oscillation of the cell membrane is modeled with an oscillator circuit, where the membrane is modeled with a capacitor, and its leakage a resistor. It is a nonlinear model where the cycle starts with the capacitor being gradually charged up, leading to a rising voltage, which in turn decreases the charging rate. And when its voltage reaches a certain threshold, the capacitor discharges rapidly and its voltage drops to zero, starting the next cycle. The sudden discharge corresponds to the firing of the pacemaker cell. The individual oscillators are of equal strength and coupled. But they interact with each other only during firing. In other words, when one oscillator fires and creates an impulse, coupling causes the voltages of other oscillators to jump up above the threshold so they fire too, in unison. Unfortunately, the mathematical techniques available at the time were not adequate for handling such an impulse response of a large number of oscillators. Therefore, Peskin simplified the problem by limiting his study to only two identical oscillators with weak coupling, and succeeded in proving that the two oscillators would always reach synchronization.

Later, Winfree took Peskin's model and further abstracted it to create a mathematical model for coupled oscillators. In Winfree's model, two oscillators rotate at the same speed in the same direction. There is an interacting force in play between the oscillators, which can be a pulling or pushing force. It is a nonlinear force determined by the relative location or phase of the two oscillators. It alters the speeds of the oscillators and synchronizes them either in phase or out of phase. Specifically, when the force is a pulling force, the two oscillators will synchronize in phase; otherwise, they will synchronize out of phase. If the initial speeds of the oscillators are different, they will stabilize at a phase difference that is a small deviation from either 0 or 180 degrees. The magnitude of the deviation is a function of the magnitude of the interacting force. A strong force leads to a small deviation. But if the force is weaker than a certain threshold, synchronization will not occur.

Next, Winfree developed the equations that govern the relationship between the speeds of individual oscillators when the number of oscillators increases beyond two. In his model, the instantaneous speed of an oscillator is determined by three factors. The first factor is the characteristic frequency of the oscillator. The second is the overall sensitivity of the oscillator to the influence of the group, which is a function of its location within the group. The third is overall influence exerted by the group. Since the governing equations are simultaneous nonlinear equations, they must be solved by simulation. Given the initial locations of the oscillators, the equations can be used to determine their individual instantaneous speeds, which in turn allow computation of their locations at the next instant. By repeating the calculation, the behavior of the oscillators can be predicted.

After repeating his simulation while varying the values of the sensitivity and coupling coefficients of the system, Winfree made a few observations. First, there are times when collective synchronization occurs, and times when it is broken. Second, collective synchronization does not require any core oscillators. But the most important observation of them all is that, as the oscillators become more and more homogeneous and similar, there is a critical point beyond which they exhibit collective synchronization. This is similar to when water is cooled to a certain temperature it changes phase into ice. Phase transition occurs when there is a sudden reversal in the balance between two forces, one creating structure, the other destroying it. If the spread in the frequency distribution of the oscillators can be thought of as temperature, collective synchronization induced by mutual coupling is like the freezing of water molecules when a certain temperature is reached, resulting in the creation of structure at the macroscopic level.

Winfree's work established a bridge between nonlinear and statistical mechanics. He published the book *The Geometry of Biological Time* in 1980 and paved the way for subsequent research by numerous distinguished scientists [72].

Kyoto University physics professor Yoshiki Kuramoto improved on Winfree's model and proposed the Kuramoto model, which serves to provide exact analytical solutions without simulation [73].

On the other hand, Cornell University applied mathematics professor Steven Strogatz, while advancing his research on the synchronization of pulse-coupled biological oscillators [74], published his "Small Worlds" theory with his student Duncan Watts [75], and extended the study of nonlinear science into the area of societal network [76].

In 1978, Zenith engineer Robert Adler published his analysis of the phenomenon of frequency entrainment of oscillator circuits, which represents the first attempt at controlling synchronization in the field of engineering [77]. More recently, University of California at Los Angeles electrical and electronic engineering professor Behzad Razavi expressed injection locking and frequency entrainment in graphs and analyzed the collective synchronization of two oscillators [70, 78].

The first successful attempt at creating collective synchronization of more than two circuit oscillators was made by our research team in 2007, when Takayuki Shibasaki integrated four oscillators onto one chip, coupled their outputs through transmission lines, and successfully caused collective synchronization [79]. He also developed the conditions for synchronized oscillation and the analytic formulae to determine the oscillation frequency and amplitude, and confirmed the operation through both simulation and experimentation using a test chip. This paved the way for our successful development of the coupled resonation-based global clock distribution network across stacked chips described in Section 2.7.2.

Figure 2.105 depicts the two modes of resonance of two coupled LC oscillators. An LC oscillator is created by connecting a coil in parallel with a capacitor (Figure 2.105b). Its standalone, uncoupled resonance frequency is given by

$$f(L) = \frac{1}{2\pi\sqrt{LC}}, \tag{2.32}$$

while its Q-factor, which defines the sharpness of the resonance, is given by

$$Q = \frac{\omega L}{R} = \frac{\sqrt{L/C}}{R}. \tag{2.33}$$

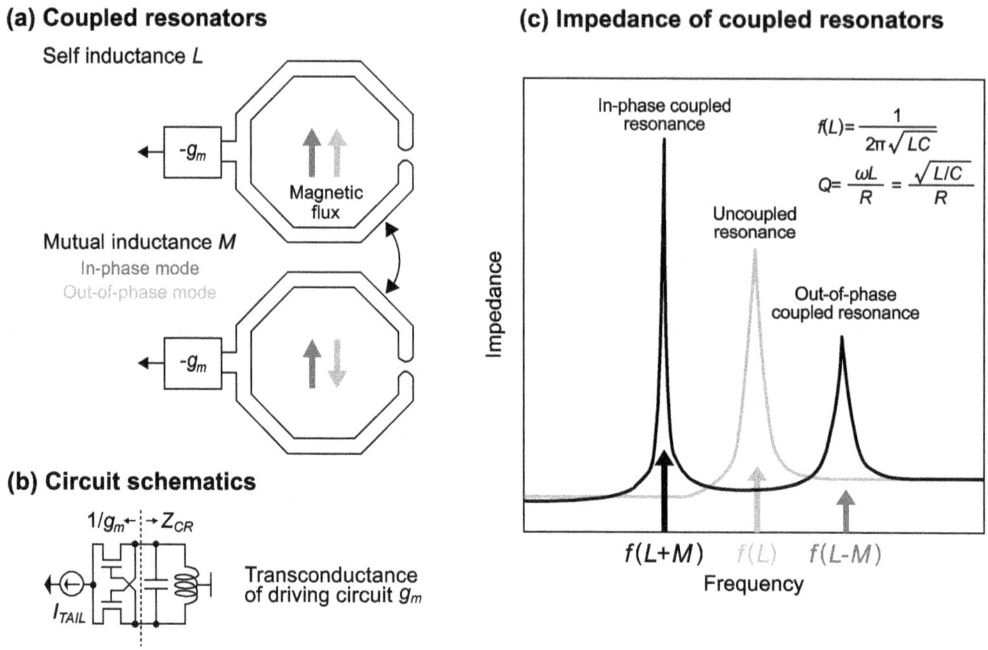

Figure 2.105 Coupled resonation of two LC oscillators [68].

At resonance, energy in the system converts back and forth between magnetic energy stored in the coil and electric energy stored in the capacitor, while a small amount of energy is lost as current flows through the parasitic resistance. By driving the oscillator with a circuit that has lower output impedance than the input impedance of the oscillator to compensate for the energy loss, resonance of the oscillator will persist (Figure 2.105b).

When two such LC oscillators are coupled magnetically through their respective coils, the mutual inductance M alters the resonance frequency. Specifically, magnetic field from one coil can add constructively or destructively with that of the other. When the magnetic fields add constructively (creating a pulling force), the effective inductance of the coil increases by M. From (2.32) and (2.33), resonance frequency decreases while Q-factor increases. This results in the two coils resonating in phase with input impedance defined by the left peak in Figure 2.105c that has a lower frequency and a sharper, higher peak than uncoupled resonance defined by the middle peak. Conversely, when the magnetic fields add destructively (creating a pushing force), the effective inductance of the coil decreases by M, resulting in the two coils resonating 180 degrees out of phase with input impedance defined by the right peak, which has a higher frequency but lower peak. By driving the coupled resonators with an output impedance that is lower than the high impedance of the in-phase mode but higher than the low impedance of the out-of-phase mode, the resonators can be pushed toward the in-phase mode to settle into resonating in phase. This is made easy by the fact that g_m naturally decreases and hence $1/g_m$ increases as frequency goes up.

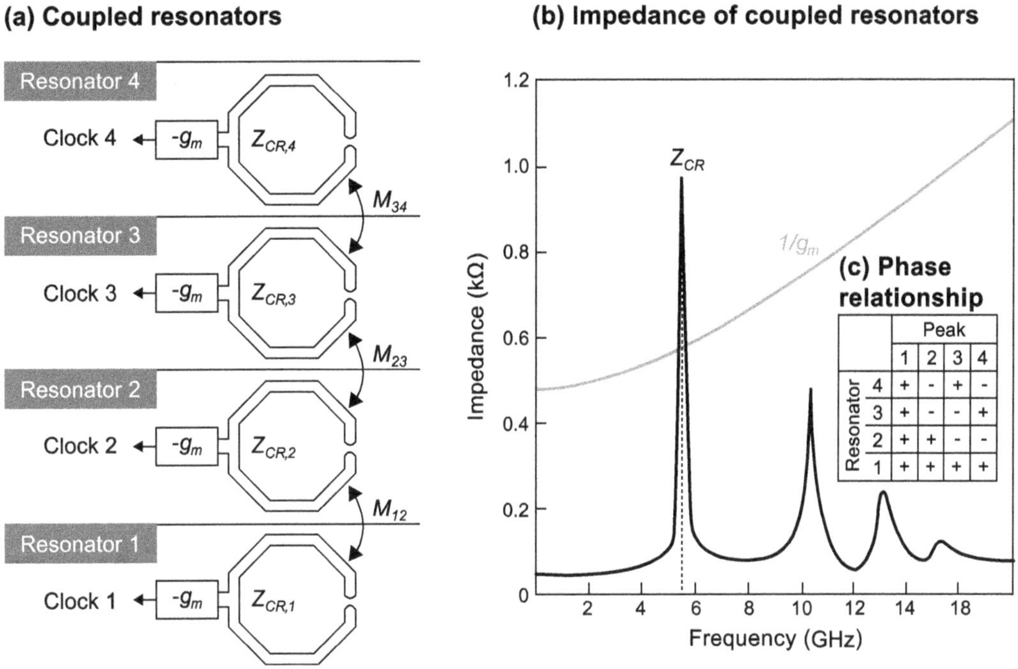

Figure 2.106 Coupled resonation of four LC oscillators [68].

When more than two oscillators are coupled, additional combinations of constructive and destructive coupling result in additional coupled resonance modes. Figure 2.106 depicts the coupled resonance of four LC resonators. The first and the sharpest peak occurs around 5.5 GHz. All four resonators resonate in phase at this frequency. The second sharpest peak occurs around 10.5 GHz, where resonator 1 resonates in phase with resonator 2, and resonator 3 with resonator 4, but the two pairs are 180 degrees out of phase relative to each other. The third sharpest peak occurs around 13.5 GHz, where resonator 2 resonates in phase with resonator 3, and resonator 1 with resonator 4, but the two pairs are 180 degrees out of phase relative to each other. The last peak occurs around 15.5 GHz, where resonator 1 resonates in phase with resonator 3, and resonator 2 with resonator 4, but the two pairs are 180 degrees out of phase relative to each other. It can thus be seen that as the number of destructively coupled adjacent resonator pairs increases, overall destructive coupling within the system increases, leading to increased resonance frequency and decreased sharpness. By designing $1/g_m$ to be only lower than the first peak, all resonators can be driven to resonate in phase, with no relative delay despite the physical distance in between. Furthermore, as explained in Section 2.7.2, the power consumption is small, primarily determined by the amount of parasitic resistance, and the tolerance of variations in capacitance across resonators is high, making it suitable for mass production.

In this manner, the natural phenomenon of collective synchronization, where similar, individual resonators are brought to synchronization through certain coupling force, has been applied to electronics. In particular, similar, individual LC oscillators coupled through mutual inductance have been used to create a system capable of collective synchronization. Furthermore, by understanding the properties of individual synchronization modes of the system, it can be nudged toward a particular mode, resulting in the successful creation of a set of in-phase, reference clocks to provide tightly controlled relative timing for digital circuits at low power and with high tolerance of manufacturing tolerances. This represents an excellent example of learning from nature and taking advantage of its tendency toward structure to develop an elegant solution to a challenging technical problem.

References

[1] M. Saito, Y. Sugimori, Y. Kohama, Y. Yoshida, N. Miura, H. Ishikuro, and T. Kuroda, "47% power reduction and 91% area reduction in inductive-coupling programmable bus for NAND flash memory stacking," *2009 IEEE Custom Integrated Circuits Conference*, pp. 449–452, Sep. 2009.

[2] N. Miura, T. Sakurai, and T. Kuroda. (2007, Feb.). Crosstalk countermeasures for high-density inductive-coupling channel array. *IEEE Journal of Solid-State Circuits. 42*(2), pp. 410–421.

[3] N. Miura, "積層チップ間誘導結合通信の高密度高速化に関する研究 [High density, high speed inductive coupling interface for 3D chip integration]," in Japanese, Ph.D. dissertation, Dept. of Electronics and Electrical Engineering, Keio Univ., Yokohama, Kanagawa Prefecture, Japan, 2007.

[4] K. Niitsu, Y. Sugimori, Y. Kohama, K. Osada, N. Irie, H. Ishikuro, and T. Kuroda, "Interference from power/signal lines and to SRAM circuits in 65nm CMOS inductive-coupling link," *2007 IEEE Asian Solid-State Circuits Conference*, pp. 131–134, Nov. 2007.

[5] Japan Ministry of Internal Affairs and Communications. (1990, Jun.). 電波防護指針 [Radio radiation protection guidelines]. In Japanese. [Online]. Available: www.tele.soumu.go.jp/resource/j/material/dwn/guide38.pdf

[6] N. Miura, D. Mizoguchi, M. Inoue, K. Niitsu, Y. Nakagawa, M. Tago, M. Fukaishi, T. Sakurai, and T. Kuroda. (2007, Jan.). A 1 Tb/s 3 W inductive-coupling transceiver for 3D-stacked inter-chip clock and data link. *IEEE Journal of Solid-State Circuits. 42*(1), pp. 111–122.

[7] T. Miyata, "誘導結合を用いた三次元積層SRAM間無線通信 [An inductive-coupling wireless interface between 3D-stacked SRAM module]," in Japanese, Master's degree dissertation, Department of Electronics and Electrical Engineering, Keio University, Yokohama, Kanagawa Prefecture, Japan, 2018.

[8] M. Saito, N. Miura, and T. Kuroda, "Asynchronous pulse transmitter for power reduction in ThruChip Interface," *2011 International Conference on Solid State Devices and Materials*, pp. 1075–1076, Sep. 2011.

[9] N. Miura, Y. Kohama, Y. Sugimori, H. Ishikuro, T. Sakurai, and T. Kuroda. (2009, Mar.). A high-speed inductive-coupling link with burst transmission. *IEEE Journal of Solid-State Circuits. 44*(3), pp. 947–955.

[10] Y. Take, N. Miura, and T. Kuroda. (2011, Nov.). A 30Gb/s/link 2.2Tb/s/mm^2 inductively-coupled injection-locking CDR. *IEEE Journal of Solid-State Circuits. 46*(11), pp. 2552–2559.

[11] A. R. Junaidi, Y. Take, and T. Kuroda, "A 352Gb/s inductive-coupling DRAM/SoC interface using overlapping coils with phase division multiplexing and ultra-thin fan-out wafer level package," *2014 Symposium on VLSI Circuits*, pp. 29–30, Jun. 2014.

[12] K. Kumagai, C. Yang, H. Izumino, N. Narita, K. Shinjo, S. Iwashita, Y. Nakaoka, T. Kawamura, H. Komabashiri, T. Minato, A. Ambo, T. Suzuki, Z. Liu, Y. Song, S. Goto, T. Ikenaga, Y. Mabuchi, and K. Yoshida, "System-in-silicon architecture and its application to H.264/AVC motion estimation for 1080HDTV," *2006 IEEE International Solid State Circuits Conference*, pp. 430–431, Feb. 2006.

[13] C. C. Lin, J. W. Chen, H. C. Chang, Y. C. Yang, Y. H. O. Yang, M. C. Tsai, J. I. Guo, and J. S. Wang. (2007, Jan.). A 160K gates/4.5 KB SRAM H.264 video decoder for HDTV applications. *IEEE Journal of Solid-State Circuits. 42*(1), pp. 170–182.

[14] N. Miura, D. Mizoguchi, M. Inoue, K. Niitsu, Y. Nakagawa, M. Tago, M. Fukaishi, T. Sakurai, and T. Kuroda, "A 1Tb/s 3W inductive-coupling transceiver for inter-chip clock and data link," *2006 IEEE International Solid State Circuits Conference*, pp. 424–425, Feb. 2006.

[15] N. Miura, H. Ishikuro, K. Niitsu, T. Sakurai, and T. Kuroda. (2008, Jan.). A 0.14pJ/b inductive-coupling transceiver with digitally-controlled precise pulse shaping. *IEEE Journal of Solid-State Circuits. 43*(1), pp. 285–291.

[16] N. Miura, H. Ishikuro, T. Sakurai, and T. Kuroda, "A 0.14pJ/b inductive-coupling inter-chip data transceiver with digitally-controlled precise pulse shaping," *2007 IEEE International Solid-State Circuits Conference*, pp. 358–359, Feb. 2007.

[17] T. Ezaki, K. Kondo, H. Ozaki, N. Sasaki, H. Yonernura, M. Kitano, S. Tanaka, and T. Hirayarna, "A 160 Gb/s interface design configuration for multichip LSI," *2004 IEEE International Solid-State Circuits Conference*, pp. 140–141, Feb. 2004.

[18] A. Fazzi, R. Canegallo, L. Ciccarelli, L. Magagni, F. Natali, E. Jung, P. L. Rolandi, and R. Guerrieri, "3D capacitive interconnections with mono- and bi-directional capabilities," *2007 IEEE International Solid-State Circuits Conference*, pp. 356–357, Feb. 2007.

[19] N. Miura, Y. Kohama, Y. Sugimori, H. Ishikuro, T. Sakurai, and T. Kuroda, "An 11Gb/s inductive-coupling link with burst transmission," *2008 IEEE International Solid-State Circuits Conference*, pp. 298–299, Feb. 2008.

[20] J. S. Kim, C. S. Oh, H. Lee, D. Lee, H. R. Hwang, S. Hwang, B. Na, J. Moon, J. G. Kim, H. Park, J. W. Ryu, K. Park, S. K. Kang, S. Y. Kim, H. Kim, J. M. Bang, H. Cho, M. Jang, C. Han, J. B. Lee, K. Kyung, J. S. Choi, and Y. H. Jun, "A 1.2V 12.8GB/s 2Gb mobile Wide-I/O DRAM with 4×128 I/Os using TSV-based stacking," *2011 IEEE International Solid-State Circuits Conference*, pp. 496–498, Feb. 2011,.

[21] K. Osada, M. Saen, Y. Okuma, K. Niitsu, Y. Shimazaki, Y. Sugimori, Y. Kohama, K. Kasuga, I. Nonomura, N. Irie, T. Hattori, A. Hasegawa, and T. Kuroda, "3D system integration of processor and multi-stacked SRAMs by using inductive-coupling links," *2009 Symposium on VLSI Circuits*, pp. 256–257, Jun. 2009.

[22] S. Kawai, H. Ishikuro, and T. Kuroda, "A 4.7Gb/s inductive coupling interposer with dual mode modem," *2009 Symposium on VLSI Circuits*, pp. 92–93, Jun. 2009.

[23] C. Cheng, K. Shiba, M. Hamada, and T. Kuroda, "2.5D integration using inductive-coupling TSV-less miniature interposer achieving 317Gb/s/mm^2, 1.2pJ/b data transfer," *2019 International Conference on Solid State Devices and Materials*, pp. 517–518, Sep. 2019.

[24] J. Poulton, W. J. Dally, X. Chen, J .G. Eyles, T. H. Greer, S. G. Tell, and C. T. Gray, "A 0.54pJ/b 20Gb/s ground-referenced single-ended short-haul serial link in 28nm CMOS for

advanced packaging applications," *2013 IEEE International Solid-State Circuits Conference*, pp. 404–405, Feb. 2013.

[25] A. Shokrollahi, D. Carnelli, J. Fox, K. Hofstra, B. Holden, A. Hormati, P. Hunt, M. Johnston, J. Keay, S. Pesenti, R. Simpson, D. Stauffer, A. Stewart, G. Surace, A. Tajalli, O. T. Amiri, A. Tschank, R. Ulrich, C. Walter, F. Licciardello, Y. Mogentale, and A. Singh, "A pin-efficient 20.83Gb/s/wire 0.94pJ/bit forwarded clock CNRZ-5-coded SerDes up to12mm for MCM packages in 28nm CMOS," *2016 IEEE International Solid-State Circuits Conference*, pp. 182–183, Feb. 2016.

[26] T. O. Dickson, Y. Liu, S. V. Rylov, B. Dang, C. K. Tsang, P. S. Andry, J. F. Bulzacchelli, H. A. Ainspan, X. Gu, L. Turlapati, M. P. Beakes, B. D. Parker, J. U. Knickerbocker, and D. J. Friedman, "An 8×10-Gb/s source-synchronous I/O system based on high-density silicon carrier interconnects," *2011 Symposium on VLSI Circuits*, pp. 80–81, Jun. 2011.

[27] M. S. Lin, C. C. Tsai, C. H. Chang, W. H. Huang, Y. Y. Hsu, S. C. Yang, C. M. Fu, M. H. Chou, T. C. Huang, C. F. Chen, T. C. Huang, S. Adham, M. J. Wang, W. W. Shen, and A. Mehta, "An extra low-power 1Tbit/s bandwidth PLL/DLL-less eDRAM PHY using 0.3V low-swing IO for 2.5D CoWoS application," *2013 Symposium on VLSI Circuits*, pp. 16–17, Jun. 2013.

[28] B. Dehlaghi, and A. C. Carusone, "A 20 Gb/s 0.3 pJ/b single-ended die-to-die transceiver in 28 nm-SOI CMOS," *2015 IEEE Custom Integrated Circuits Conference*, pp. 25.1.1–25.1.4, Sep. 2015.

[29] K. Shiba, C. Cheng, M. Hamada, and T. Kuroda. (2020, Feb.). 2.5D integration using inductive-coupling TSV-less miniature interposer achieving 317 Gb/s/mm2, 1.2 pJ/b data-transfer. *Japanese Journal of Applied Physics. 59*(SGGL06), pp. 1–6.

[30] S. Hasegawa, J. Kadomoto, A. Kosuge, and T. Kuroda, "A 1 Tb/s/mm^2 inductive-coupling side-by-side chip link," *42nd European Solid-State Circuits Conference*, pp. 469–472, Sep. 2016.

[31] M. Kumagai, N. Uchiyama, E. Ohmura, R. Sugiura, K. Atsumi, and K. Fukumitsu, "Advanced dicing technology for semiconductor wafer - Stealth Dicing," *2006 IEEE International Symposium on Semiconductor Manufacturing*, pp. 215–218, Sep. 2006.

[32] Toray. Super high accuracy flip chip bonder OF2000. [Online]. Retrieved on Feb. 3, 2020. Available: www.toray-eng.com/products/semicon/sem_005.html

[33] T. Kuroda, "Near-field coupling integration technology," *10th IEEE/ACM International Symposium on Networks-on-Chip*, Aug. 2016.

[34] Y. Yuan, Y. Yoshida, and T. Kuroda, "Non-contact 10% efficient 36mW power delivery using on-chip inductor in 0.18-μm CMOS," *2007 IEEE Asian Solid-State Circuits Conference*, pp. 115–118, Nov. 2007.

[35] K. Onizuka, H. Kawaguchi, M. Takamiya, T. Kuroda, and T. Sakurai, "Chip-to-chip inductive wireless power transmission system for SiP applications," *2006 IEEE Custom Integrated Circuits Conference*, pp. 575–578, Sep. 2006.

[36] M. Ghovanloo and K. Najafi. (2004, Nov.). Fully integrated wideband high-current rectifiers for inductively powered devices. *IEEE Journal of Solid-State Circuits. 39*(11), pp. 1976–1984.

[37] H. Nakamoto, D. Yamazaki, T. Yamamoto, H. Kurata, S. Yamada, K. Mukaida, T. Ninomiya, T. Ohkawa, S. Masui, and K. Gotoh, "A passive UHF RFID tag LSI with 36.6% efficiency CMOS-only rectifier and current-mode demodulator in 0.35μm FeRAM technology," *2006 IEEE International Solid State Circuits Conference*, pp. 310–311, Feb. 2006.

[38] Y. Yuan, Y. Yoshida, N. Yamagishi, and T. Kuroda. (2008, Apr.). Chip-to-chip power delivery by inductive coupling with ripple canceling scheme. *Japanese Journal of Applied Physics*. *47*(4S), pp. 2797–2800. DOI https://doi.org/10.1143/JJAP.47.2797.

[39] Y. Yuan, A. Radecki, N. Miura, I. Aikawa, Y. Take, H. Ishikuro, and T. Kuroda, "Simultaneous 6Gb/s data and 10mW power transmission using nested clover coils for non-contact memory card," *2010 Symposium on VLSI Circuits*, pp. 199–200, Jun. 2010.

[40] Y. Yuan, N. Miura, S. Imai, H. Ochi, and T. Kuroda, "Digital rosetta stone: a sealed permanent memory with inductive-coupling power and data link," *2009 Symposium on VLSI Circuits*, pp. 26–27, Jun. 2009.

[41] D. S. Peck, "Comprehensive model for humidity testing correlation," *24th International Reliability Physics Symposium*, pp. 44–49, Apr. 1986.

[42] D. Ditzel, "Low-cost 3D chip stacking with ThruChip wireless connections," *Hot Chips 26*, Aug. 2014.

[43] K. Shiba, M. Hamada, and T. Kuroda, "3D SoC design with TSV-less power supply employing highly doped silicon via," *2019 International Conference on Solid State Devices and Materials*, pp. 515–516, Sep. 2019.

[44] W. L. Yang, C. Y. Cheng, M. S. Tsai, D. G. Liu, and M. S. Shieh. (2000, May). Retardation in the chemical–mechanical polish of the boron-doped polysilicon and silicon. *IEEE Electron Device Letters*. *21*(5), pp. 218–220.

[45] K. Ueyoshi, K. Ando, K. Hirose, S. Takamaeda-Yamazaki, J. Kadomoto, T. Miyata, M. Hamada, T. Kuroda, and M. Motomura, "QUEST: A 7.49TOPS multi-purpose log-quantized DNN inference engine stacked on 96MB 3D SRAM using inductive-coupling technology in 40nm CMOS," *2018 IEEE International Solid-State Circuits Conference*, pp. 216–217, 2018.

[46] K. Shiba, M. Hamada, and T. Kuroda. (2020, Feb.). 3D system-on-a-chip design with through-silicon-via-less power supply using highly doped silicon via. *Japanese Journal of Applied Physics*. *59*(SGGL04), pp. 1–5.

[47] DISCO Corporation. Introduction of the Gettering DP Wheel. [Online]. Retrieved on Feb. 3, 2020. Available: www.disco.co.jp/eg/solution/apexp/polisher/gettering.html.

[48] Y. S. Kim, S. Kodama, Y. Mizushima, N. Maeda, H. Kitada, K. Fujimoto, T. Nakamura, D. Suzuki, A. Kawai, K. Arai, and T. Ohba, "Ultra thinning down to 4-μm using 300-mm wafer proven by 40-nm node 2Gb DRAM for 3D multi-stack WOW applications," *2014 Symposium on VLSI Technology*, pp. 1–2. Jun. 2014.

[49] J. H. Cho, J. Kim, W. Y. Lee, D. U. Lee, T. K. Kim, H. B. Park, C. Jeong, M. J. Park, S. G. Back, S. Choi, B. K. Yoon, Y. J. Choi, K. Y. Lee, D. Shim, J. Oh, J. Kim, and S. H. Lee, "A 1.2V 64Gb 341GB/s HBM2 stacked DRAM with spiral point-to-point TSV structure and improved bank group data control," pp. 208–209, *2018 IEEE International Solid-State Circuits Conference*, Feb. 2018.

[50] Y. Sugimori, Y. Kohama, M. Saito, Y. Yoshida, N. Miura, H. Ishikuro, T. Sakurai, and T. Kuroda, "A 2Gb/s 15pJ/b/chip inductive-coupling programmable bus for NAND flash memory stacking," *2009 IEEE International Solid-State Circuits Conference*, pp. 244–245, Feb. 2009.

[51] M. Saito, N. Miura, and T. Kuroda, "A 2Gb/s 1.8pJ/b/chip inductive-coupling through-chip bus for 128-die NAND-flash memory stacking," *2010 IEEE International Solid-State Circuits Conference*, pp. 440–441, Feb. 2010.

[52] T. Kuroda, "半導体集積回路装置 [Semiconductor Integrated Circuit Device]," Japanese patent application P2009–197669, Aug. 28, 2009.

[53] N. Miura, Y. Take, M. Saito, Y. Yoshida, and T. Kuroda, "A 2.7Gb/s/mm^2 0.9pJ/b/chip 1coil/channel ThruChip Interface with coupled-resonator-based CDR for NAND flash memory stacking," *2011 IEEE International Solid-State Circuits Conference*, pp. 490–492, Feb. 2011.

[54] A. Sheibanyrad, F. Petrot, and A. Jantsch, Eds. *3D integration for NoC-based SoC architectures*. New York: Springer-Verlag, 2010.

[55] Y. Take, H. Matsutani, D. Sasaki, M. Koibuchi, T. Kuroda, and H. Amano. (2014, Mar.). 3D NoC with inductive-coupling links for building-block SiPs. *IEEE Transactions on Computers. 63*(3), pp. 748–763.

[56] P. Abad, V. Puente, P. Prieto, and J. A. Gregorio, "Rotary router: an efficient architecture for CMP interconnection networks," *34th International Symposium on Computer Architecture*, pp. 116–125, Jun. 2007.

[57] V. Puente, R. Beivide, J. A. Gregorio, J. M. Prellezo, J. Duato, and C. Izu, "Adaptive bubble router: a design to improve performance in torus networks," *1999 International Conference on Parallel Processing*, pp. 58–67, Sep. 1999.

[58] N. Miura, Y. Koizumi, Y. Take, H. Matsutani, T. Kuroda, H. Amano, R. Sakamoto, M. Namiki, K. Usami, M. Kondo, and H. Nakamura. (2013, Nov.). A scalable 3D heterogeneous multicore with an inductive ThruChip Interface. *IEEE Micro. 33*(6), pp. 6–15.

[59] H. Jin, M. Frumkin, and J. Yan (1999, Oct.). The OpenMP implementation of NAS parallel benchmarks and its performance. *NAS Technical Report NAS-99-011*. [Online]. Available: www.nas.nasa.gov/assets/pdf/techreports/1999/nas-99-011.pdf.

[60] A. Nomura, Y. Matsushita, J. Kadomoto, H. Matsutani, T. Kuroda, and H. Amano. (2018). Escalator network for a 3D chip stack with inductive coupling ThruChip Interface. *International Journal of Networking and Computing. 8*(1), pp. 124–139.

[61] R. Tsuchiya, M. Horiuchi, S. Kimura, M. Yamaoka, T. Kawahara, S. Maegawa, T. Ipposhi, Y. Ohji, and H. Matsuoka, "Silicon on thin BOX: a new paradigm of the CMOSFET for low-power high-performance application featuring wide-range back-bias control," *2004 IEEE International Electron Devices Meeting*, pp. 631–634, Dec. 2004.

[62] B. J. Fletcher, S. Das, and T. Mak, "A high-speed design methodology for inductive coupling links in 3D-ICs," *2018 Design, Automation & Test in Europe Conference & Exhibition*, pp. 497–502, Mar. 2018.

[63] B. J. Fletcher, S. Das, and T. Mak. (2019, Mar.). Design and optimization of inductive-coupling links for 3-D-ICs. *IEEE Transactions on Very Large Scale Integration (VLSI) Systems. 27*(3), pp. 711–723.

[64] J. R. Long and M. A. Copeland. (1997, Mar.). The modeling, characterization, and design of monolithic inductors for silicon RF IC's. *IEEE Journal of Solid-State Circuits. 32*(3), pp. 357–369.

[65] B. J. Fletcher, S. Das, C. S. Poon, and T. Mak, "Low-power 3D integration using inductive coupling links for neurotechnology applications," *2018 Design, Automation & Test in Europe Conference & Exhibition*, pp. 1211–1216, Mar. 2018.

[66] S. Stanslaski, P. Afshar, P. Cong, J. Giftakis, P. Stypulkowski, D. Carlson, D. Linde, D. Ullestad, A. T. Avestruz, and T. Denison. (2012, Jan.). Design and validation of a fully implantable, chronic, closed-loop neuromodulation device with concurrent sensing and stimulation. *IEEE Transactions on Neural Systems and Rehabilitation Engineering. 20*(4), pp. 410–421.

[67] I. A. Papistas and V. F. Pavlidis, "Contactless inter-tier communication for heterogeneous 3-D ICs," *2017 IEEE International Symposium on Circuits and Systems*, pp. 2585–2588, May 2017.

[68] T. Kuroda. (2011, Jun.). 自然界の集団同期現象をエレクトロニクスに応用 [Applying nature's collective synchronization phenomena to electronics]. In Japanese. *Nikkei Electronics* (2011.6.13). pp. 85–94.

[69] Spontaneous Synchronization. [Online]. Retrieved on Feb. 3, 2020. Available: www.youtube.com/watch?v=RMVxVbCIPjg.

[70] N. Wiener. *Cybernetics,* 2nd ed. Cambridge: MIT Press, 1961.

[71] C. Peskin. *Mathematica Aspects of Heart Physiology*. New York: Courant Institute of Mathematical Sciences, 1975, pp. 268–278.

[72] A. Winfree. *The Geometry of Biological Time*. New York: Springer-Verlag, 1980.

[73] Y. Kuramoto, "Self-entrainment of a population of coupled nonlinear oscillators," *International Symposium on Mathematical Problems in Theoretical Physics*, pp. 420–422, 1975.

[74] R. E. Mirollo and S. H. Strogatz. (1990, Dec.). Synchronization of pulse-coupled biological oscillators. *SIAM Journal on Applied Mathematics. 50*(6), pp. 1645–1662.

[75] D. Watts and S. H. Strogatz. (1998, Jun. 4). Collective dynamics of "small-world" networks. *Nature. 393*, pp. 440–442.

[76] M. Buchanan. *NEXUS: Small Worlds and the Ground-Breaking Science of Networks*. New York: W. W. Norton & Co., 2002.

[77] R. Adler, "A study of locking phenomena in oscillators," *Proceedings of the IEEE*, pp. 1380–1385, Oct. 1973.

[78] B. Razavi, "Mutual injection pulling between oscillators," *2006 IEEE Custom Integrated Circuits Conference*, pp. 675–678, Sep. 2006.

[79] T. Shibasaki, H. Tamura, K. Kanda, H. Yamaguchi, J. Ogawa, and T. Kuroda. (2007). 18-GHz clock distribution using a coupled VCO array. *IEICE Transactions on Electronics. E90-C*(4), pp. 811–822.

3 Transmission Line Coupler

A Wireless Module Connector

As discussed in Section 1.4.3, TLC is an electromagnetic near-field coupling interconnect technology that addresses the challenges faced by conventional connectors. In particular, it eliminates physical contact and exhibits well-controlled characteristic impedance and crosstalk that enables simple transceiver design, resulting in a low-power and high-bandwidth connector solution with physical robustness. Furthermore, since the contactless connection alleviates the need for a mechanically stable housing, and the coupler is implemented using PCB manufacturing process, TLC offers a better form-factor scaling path than conventional mechanical connectors. In this chapter, we introduce the standard design and electrical characteristics of the interconnect, discuss parameter trade-offs, and present ideas for optimization and adaptation for various applications.

3.1 TLC Fundamentals

3.1.1 Basic Structure and Operating Principle

Figure 3.1 depicts the basic structure of a TLC. It consists of two pairs of differential transmission lines, with each pair implemented on the surface layer of a different PCB, and with the two pairs coupled across a spacer. Each pair of differential transmission lines is laid out in the shape of a horseshoe. The horseshoe shape enables strong interpair coupling across the spacer by increasing intrapair spacing to (1) reduce intrapair coupling and (2) accommodate wider traces for better interpair coupling. Since a signal propagates along a transmission line pair through alternating electric and magnetic fields, a portion of the signal energy in the transmitting pair is transferred to the receiving pair through both capacitive and inductive coupling. Specifically, the changing voltage of the signal induces a current through capacitive coupling, while its changing current induces a voltage through inductive coupling in the receiving transmission line pair. The sum of the two produces the received signal.

Although the electromagnetic coupling in TLC is intentional and utilized to channel energy from the transmitter to the receiver, such coupling in general transmission line applications is often unintended and represents noise in the form of crosstalk. As a result, transmission line crosstalk analysis can be utilized to better understand the operating principle and properties of TLC. For simplicity, single-ended

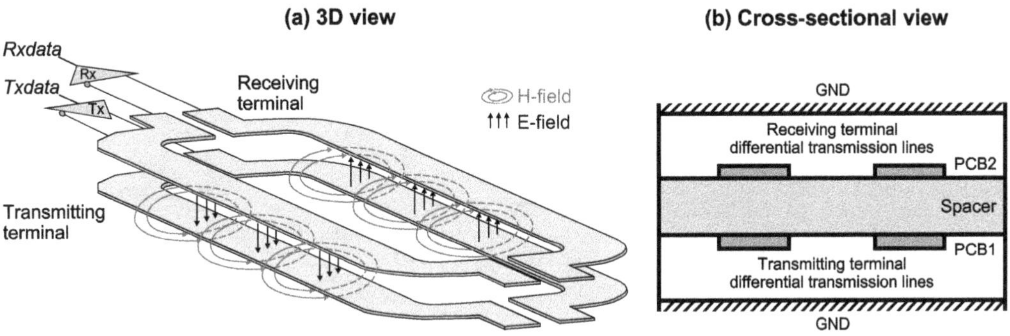

Figure 3.1 Basic structure of TLC. © 2016 Atsutake Kosuge [1]

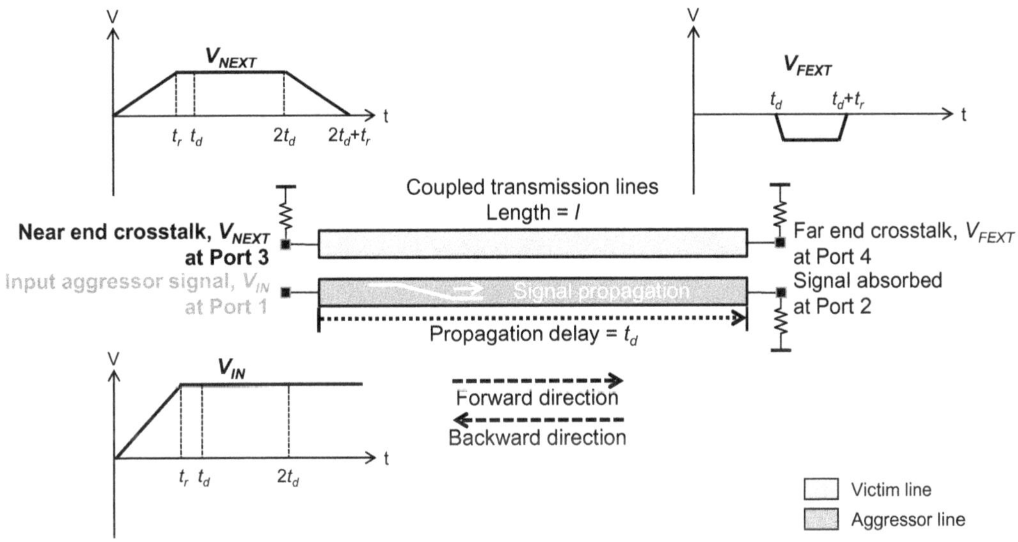

Figure 3.2 Crosstalk in coupled transmission lines.

transmission lines are used for the analysis, although differential transmission lines behave in a similar way.

Figure 3.2 depicts the crosstalk in two coupled, identical transmission lines when a step is injected into Port 1 as the aggressor signal that propagates along the bottom transmission line (the aggressor line) to Port 2. Forward direction is defined as the direction of aggressor signal propagation and is hence from left to right, while backward direction is from right to left. The transmission lines are terminated in their characteristic impedance at Ports 2–4 to eliminate any signal reflection. Through electromagnetic coupling, crosstalk appears in the top transmission line (the victim line) at both Ports 3 and 4. The end of the transmission lines where the input signal originates is known as near end, while the opposite end is far end. Hence crosstalk appearing at Port 3 is near end crosstalk while that at Port 4 is far end crosstalk. Near

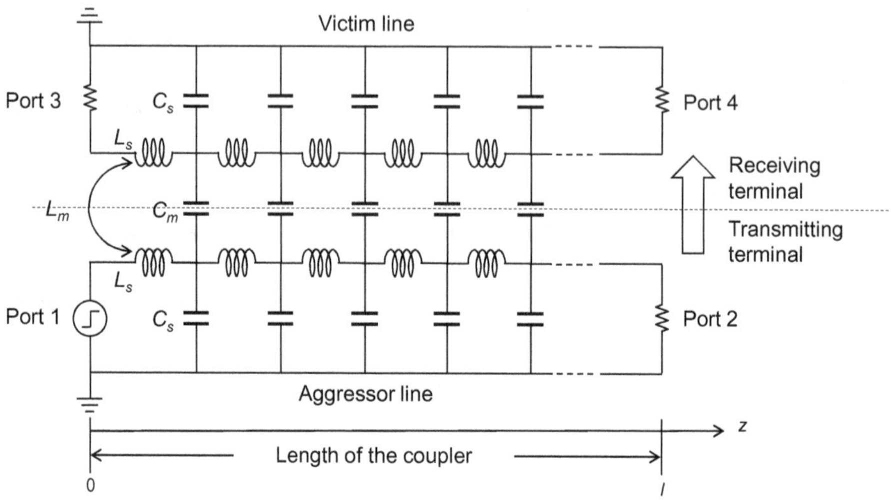

Figure 3.3 Equivalent circuit model of two coupled transmission lines.

end crosstalk V_{NEXT} is generally a trapezoidal pulse with the same polarity as the aggressor signal, while far end crosstalk V_{FEXT} is also a trapezoidal pulse but of opposite polarity. Both pulse shapes depend on the propagation delay of the transmission lines t_d and edge rate t_r of the aggressor signal, but in different ways. The implicit assumption in Figure 3.2 is that the transmission line propagation delay is longer than the edge rate of the aggressor signal ($t_d > t_r$), as illustrated by the input step being entirely inside the aggressor line. The case where $t_d \leq t_r$ will be discussed in the next section.

The crosstalk pulse shapes can be derived using the equivalent circuit model depicted in Figure 3.3, where L_s = self inductance and C_s = self capacitance relative to ground of individual transmission lines, and L_m = mutual inductance and C_m = mutual capacitance between the two transmission lines. All circuit parameters are per unit length distributed parameters. The energy transfer mechanism is modeled by mutual inductance L_m for magnetic coupling and mutual capacitor C_m for electric coupling respectively.

Figure 3.4 illustrates the mechanism for electric/capacitive coupling. The changing voltage dV/dt in the aggressor signal induces a capacitive current I_C in the victim line where

$$I_C = C_m \frac{dV}{dt}. \tag{3.1}$$

This current in the victim line sees equal impedances in both the forward and backward directions. As a result, it splits into two equal components of identical polarity flowing in the two directions. Consequently, the near end and far end crosstalk due to capacitive coupling, $V_{NEXT,C}$ and $V_{FEXT,C}$, have the same magnitude and polarity. As the aggressor signal propagates in the forward direction, it takes time

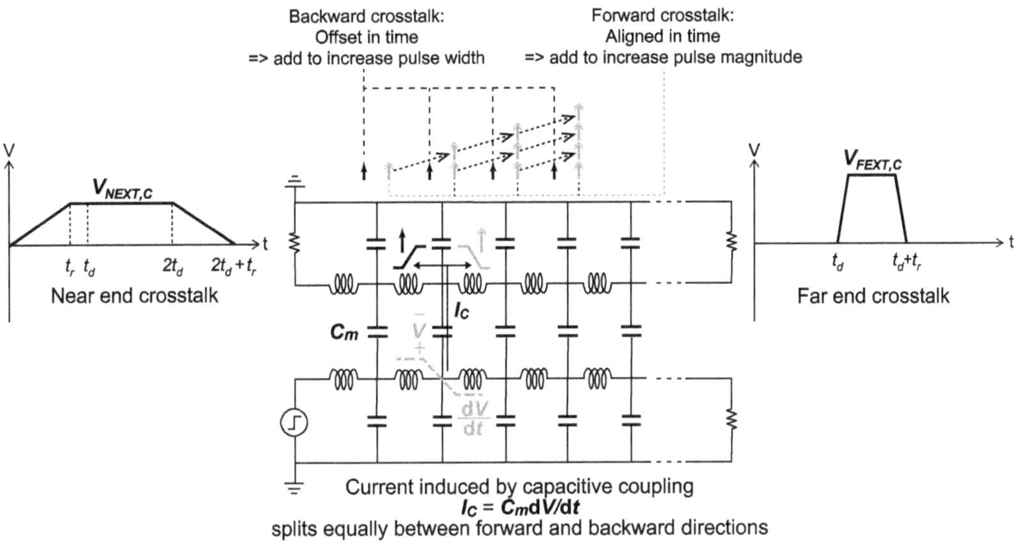

Figure 3.4 Mechanism of electric/capacitive coupling.

for it to reach the next mutual capacitor and for crosstalk through the next mutual capacitor to propagate backward to the previous mutual capacitor. As a result, crosstalk flowing in the backward direction from successive mutual capacitors are offset in time and add to increase the width of $V_{NEXT,C}$. Furthermore, crosstalk from two mutual capacitors sufficiently far apart do not overlap in time and hence do not add to increase magnitude of $V_{NEXT,C}$. In other words, the width of $V_{NEXT,C}$ is proportional to the length l and hence propagation delay t_d of the transmission lines, but the magnitude is not when the length is sufficiently long.

On the other hand, forward crosstalk from the previous mutual capacitor propagates in lock step with the aggressor signal since the aggressor and victim lines are identical and hence have the same propagation delay. Therefore, it arrives at the next mutual capacitor at the same time as the aggressor signal. As a result, forward crosstalk from successive mutual capacitors are aligned in time and add to increase the magnitude but not the width of the crosstalk pulse. In other words, the magnitude of $V_{FEXT,C}$ is proportional to the length l and hence propagation delay t_d of the transmission lines, but the width is not.

Figure 3.5 illustrates the mechanism for magnetic/inductive coupling. The changing current dI/dt in the aggressor signal induces a voltage V_L in the victim line where

$$V_L = L_m \frac{dI}{dt}. \tag{3.2}$$

This voltage is induced to drive a current in the backward direction in order to counter the change in current in the aggressor line. Because this current flows into the impedance to the left, it creates a positive voltage in the backward direction.

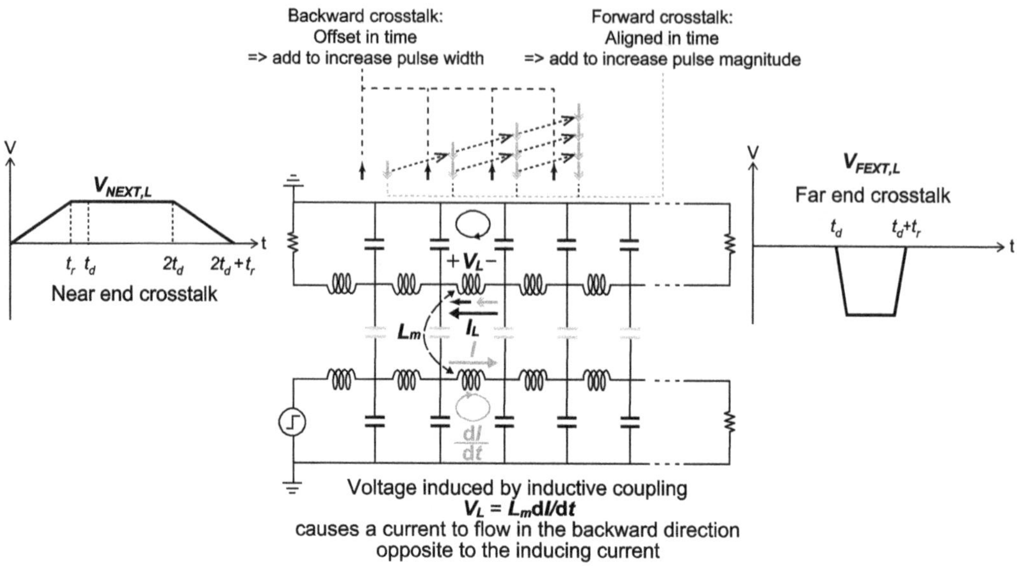

Figure 3.5 Mechanism of magnetic/inductive coupling.

However, since this same current flows out of the impedance to the right, it generates a negative voltage in the forward direction. Consequently, while near end crosstalk due to inductive coupling $V_{NEXT,L}$ has the same polarity as the aggressor signal, far end crosstalk $V_{FEXT,L}$ has the opposite polarity. Furthermore, same as capacitive crosstalk, the width of $V_{NEXT,L}$ is proportional to l and t_d of the transmission lines, but the magnitude is not when the length is sufficiently long. Meanwhile, the magnitude of $V_{FEXT,L}$ is proportional to l and t_d of the transmission lines, but the width is not.

The total near end and far end crosstalk are derived by summing the capacitive and inductive components as illustrated in Figure 3.6. Since inductive crosstalk is generally stronger than capacitive crosstalk, V_{FEXT} is a negative pulse in the general case.

In TLC, the coupled energy in the form of near end and far end crosstalk, which is normally considered noise, is intentionally used for data transmission. In typical usage, $V_{FEXT,C}$ and $V_{FEXT,L}$ are matched in magnitude to reduce V_{FEXT} to 0 (within practical limits), resulting in unidirectional coupling from Port 1 to Port 3 only. In other word, TLC uses near end crosstalk for data transmission.

Based on the circuit model in Figure 3.3, the voltage and current of a signal at any particular point z along the transmission line and at any particular time t can be represented by the following set of equations:

$$\frac{\partial}{\partial z}\begin{bmatrix} V_1 \\ V_2 \end{bmatrix} = -\frac{\partial}{\partial t}\begin{bmatrix} L_{11} & L_{12} \\ L_{21} & L_{22} \end{bmatrix}\begin{bmatrix} I_1 \\ I_2 \end{bmatrix} \tag{3.3}$$

$$\frac{\partial}{\partial z}\begin{bmatrix} I_1 \\ I_2 \end{bmatrix} = -\frac{\partial}{\partial t}\begin{bmatrix} C_{11} & C_{12} \\ C_{21} & C_{22} \end{bmatrix}\begin{bmatrix} V_1 \\ V_2 \end{bmatrix}, \tag{3.4}$$

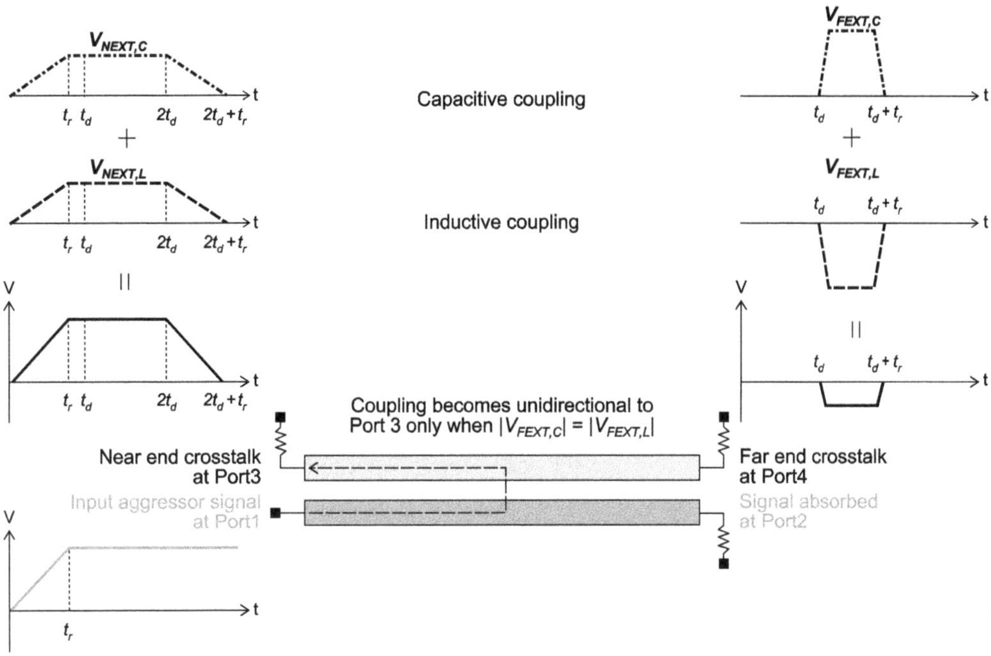

Figure 3.6 Total near end and far end crosstalk.

where $V_1(z,t)$ and $I_1(z,t)$ are the voltage and current for the transmitting terminal, $V_2(z,t)$ and $I_2(z,t)$ the voltage and current for the receiving terminal, $L_{11},L_{22} = L_s$ are the self-inductances of the individual transmission lines, $C_{11},C_{22} = C_s + C_m$ are the total capacitances of the individual transmission lines, $L_{12},L_{21} = L_m$, and $C_{12},C_{21} = -C_m$. (Note that the definition of the inductance matrix is not completely analogous to that of the capacitance matrix in that $L_{11},L_{22} \neq L_s + L_m$.) The transmitter voltage at Port 1 is $V_{TX}(t) = V_1(0,t)$, and the received voltage at Port 3 is $V_{RX}(t) = V_2(0,t)$. By using *Txdata* to modulate the transmitter voltage $V_{TX}(t)$, which in turn induces a voltage $V_{RX}(t)$ in the receiving terminal through the LC network of TLC, the transmitted data can be extracted from $V_{RX}(t)$ to generate *Rxdata* to achieve communication.

3.1.2 Electrical Characteristics and Design Parameters

Since signal transmission in TLC is achieved through capacitive and inductive coupling, its effectiveness is quantified by the capacitive and inductive coupling coefficients, K_C and K_L, given by

$$K_L = \left|\frac{L_{21}}{\sqrt{L_{11}L_{22}}}\right| \tag{3.5}$$

$$K_C = \left|\frac{C_{21}}{\sqrt{C_{11}C_{22}}}\right|. \tag{3.6}$$

In the special case where the transmission line pairs are in a homogeneous medium and all dielectric materials have the same dielectric constant, capacitive and inductive coupling become equal strength, resulting in $K_C = K_L = K$ [2–3]. As a result, the transfer function from transmitter output (Port 1) to receiver input (Port 3), $S_{31} = V_3/V_1$, and that from transmitter output (Port 1) to the far end of the transmitter side of the coupler (Port 2), $S_{21} = V_2/V_1$, can be computed by

$$S_{31} = \frac{jK \sin\left(\frac{\omega l}{v_p}\right)}{\sqrt{1-K^2}\cos\left(\frac{\omega l}{v_p}\right) + j\sin\left(\frac{\omega l}{v_p}\right)} \tag{3.7}$$

$$S_{21} = \frac{\sqrt{1-K^2}}{\sqrt{1-K^2}\cos\left(\frac{\omega l}{v_p}\right) + j\sin\left(\frac{\omega l}{v_p}\right)}, \tag{3.8}$$

where υ_p is the phase velocity of the homogeneous dielectric medium [4]. Therefore, S_{31} and S_{21} are both functions of K and l. Furthermore, at the following center frequency

$$f_c = \frac{v_p}{4l} \tag{3.9}$$

S_{31} reaches its maximum value and S_{21} its minimum value defined by the following equations:

$$|S_{31}|_{\max} = K \tag{3.10}$$

$$|S_{21}|_{\min} = \sqrt{1-K^2}. \tag{3.11}$$

Figure 3.7 depicts the transfer functions S_{31}, S_{21}, and S_{11} (reflection at Port 1) for one design example. It illustrates the fact that TLC characterized by S_{31} is a band pass filter with both low and high cutoff frequencies, and a relatively flat gain in between. The center frequency of the pass band f_c is the peak frequency where coupling (S_{31}) is maximum.

The center frequency can also be derived by considering the coupling of a sine wave as shown in Figure 3.8a. Consider two coupled waveforms a_1 and a_2 that appear at location z, where $0 < z < l/2$ – a_1 is a coupled waveform induced by the transmitter signal at location z, while a_2 is a coupled waveform induced by the same transmitter signal at location $z + l/2$ that has traveled back to location z. The delay between a_1 and a_2 is the time it takes the transmitter signal to travel from z to $z + l/2$, plus the time it takes for a_2 to travel back from $z + l/2$ to z. Since the phase velocity of the transmission lines is υ_p, the delay between a_1 and a_2 is $t_{delay} = l/\upsilon_p$. When the frequency of the sine wave is f_0 such that its wavelength λ_0 is given by

$$l = \frac{\lambda_0}{2}. \tag{3.12}$$

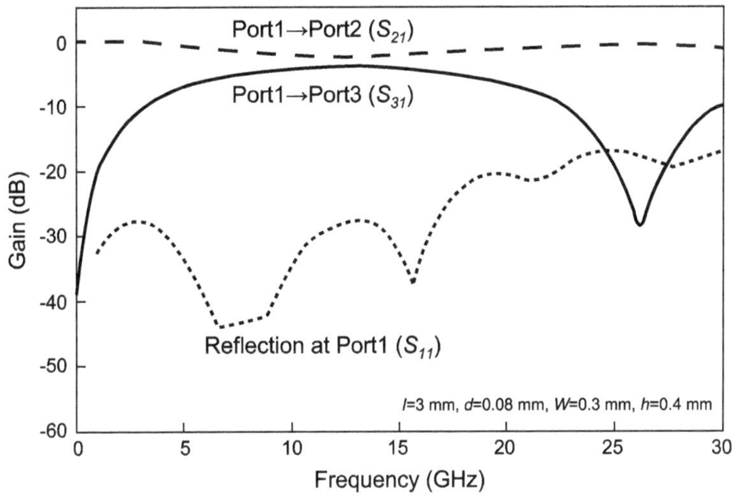

Figure 3.7 Transfer function of a TLC interconnect coupler. © 2015 IEEE. Reprinted, with permission, from [5]

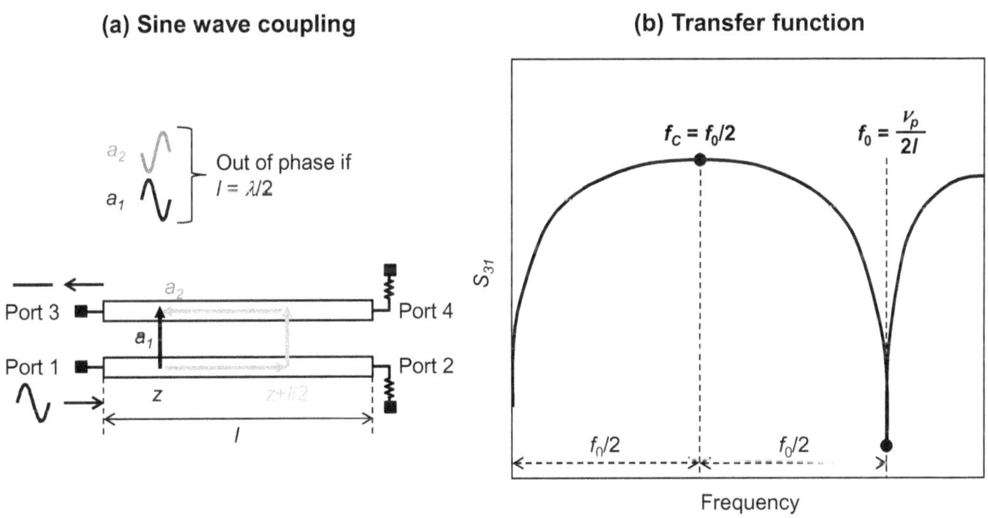

Figure 3.8 Derivation of TLC center frequency.

the delay becomes

$$t_{delay} = \frac{\lambda_0}{2v_p} = \frac{1}{2f_0} = \frac{T_0}{2}, \tag{3.13}$$

where T_0 is the cycle time of the sine wave. In other words, a_2 is half the cycle time or 180 degrees delayed relative to a_1, resulting in complete cancelation and zero net coupling. Consequently, $S_{31} = 0$ at $f = f_0$, where

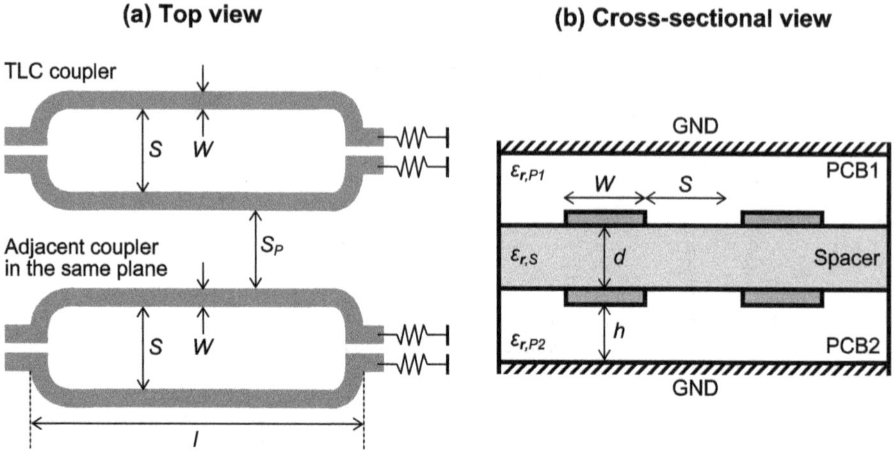

Figure 3.9 Physical design parameters of a standard TLC.

$$f_0 = \frac{v_p}{\lambda_0} = \frac{v_p}{2l}. \tag{3.14}$$

Since TLC transfers energy by changing current and voltage, there is no coupling at DC as well. Therefore, S_{31} has band pass characteristics and the center of the pass band is given by

$$f_C = \frac{f_0}{2} = \frac{v_p}{4l}. \tag{3.15}$$

Figure 3.9 defines the physical design parameters of the standard TLC. The assumption of homogeneous dielectric medium means that both PCBs and the spacer have the same dielectric constant, in other words, $\varepsilon_{r,P1} = \varepsilon_{r,P2} = \varepsilon_{r,S} = \varepsilon_r$.

Since $K_C = K_L = K$, the coupling coefficient K can be computed by considering electric field distribution only, by modeling the coupled traces as parallel-plate capacitors plus fringe effects. Specifically, for identical transmission lines where $C_{11} = C_{22}$, K is given by

$$K = \frac{1}{1 + \frac{C_s}{C_m}} = \frac{1}{1 + \frac{\varepsilon_0 \varepsilon_r \frac{W}{h} + 2C_{s,fr}}{\varepsilon_0 \varepsilon_r \frac{W}{d} + C_{m,fr}}}, \tag{3.16}$$

where $C_{s,fr}$ and $C_{m,fr}$ are the fringe self and mutual capacitances respectively. In other words, K is a function of (W/h) and (W/d). In practice, since h is determined by the thickness of standard PCB laminate layers, it has limited adjustability. As a result, the TLC coupling coefficient is mainly controlled by the spacer thickness d and trace width W.

Meanwhile, dependence of characteristic impedance Z_0 of each of the coupled transmission lines on magnetic field distribution can be accounted for with phase velocity of the medium, v_p, such that Z_0 can also be expressed in terms of capacitances only, by

$$Z_0 = \sqrt{Z_{0e}Z_{0o}} = \frac{1}{v_p\sqrt{C_s(C_s + 2C_m)}}, \tag{3.17}$$

where Z_{0e} and Z_{0o} are the even and odd mode impedances respectively of the coupled transmission lines. (Two coupled transmission lines operate in even mode when the signals propagating in the transmission lines are identical and in phase, and in odd mode when the signals are identical but out of phase.) Therefore, while K is a function of only the ratio of C_m/C_s, Z_0 is a function of both C_m and C_s. As a result, K and Z_0 can be independently optimized. In practice, d is normally used to provide coarse adjustment of K, while W is used for fine adjustment of K as well as tuning of characteristic impedance Z_0. Since 50 Ω is the most common transmission line characteristic impedance, TLC transmission lines are generally designed for a single-ended impedance of 50 Ω, or a differential impedance of 100 Ω.

As previously noted, one motivation for the horseshoe-shaped layout of the differential transmission lines is to reduce intrapair coupling so as to increase interpair coupling. The effect is quantified in Figure 3.10a, where S_{31} at 12 GHz is plotted as a function of S/W. As intrapair spacing increases, S_{31} increases and eventually levels off. The data show that $S = 3W$ provides enough intrapair spacing for S_{31} to approach its maximum value.

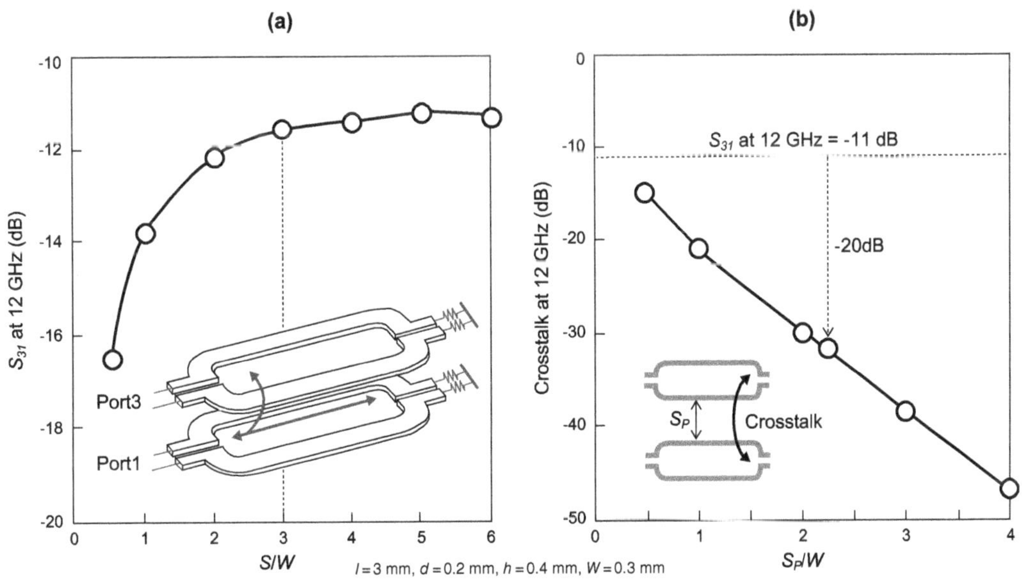

Figure 3.10 Coupling coefficient and crosstalk as a function of spacing. © 2015 IEEE. Reprinted, with permission, from [5]

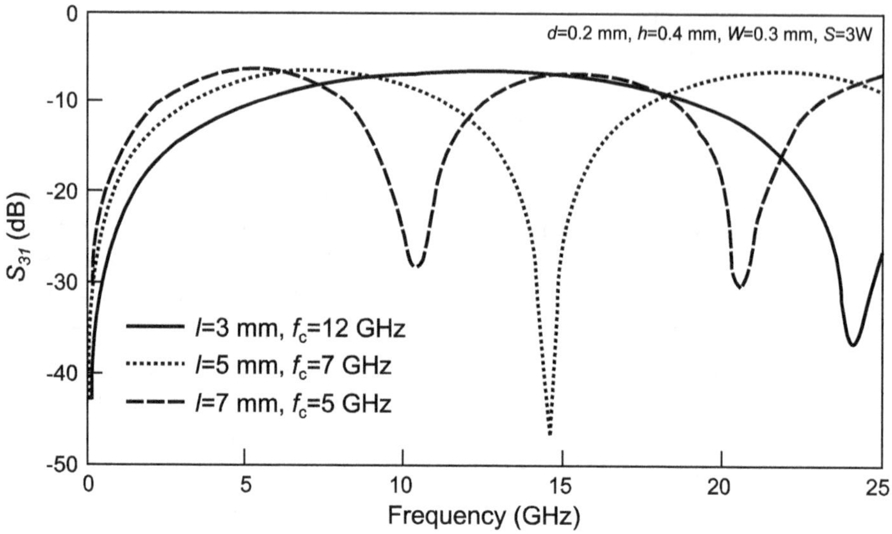

Figure 3.11 TLC interconnect pass band as a function of transmission line length. © 2015 IEEE. Reprinted, with permission, from [5]

Figure 3.10b plots crosstalk from an adjacent differential pair in the same plane as a function of S_P, the spacing in between. By choosing $S_P = 2.3W$, crosstalk noise can be reduced to 20 dB below signal.

Figure 3.8b reveals the band pass property of TLC. Furthermore, (3.14) and (3.15) show that the pass band and center frequency are both functions of the length l of the coupler. The quantitative relationship is depicted in Figure 3.11. As l is increased, the frequency response of S_{31} is compressed toward lower frequency. Consequently, both center frequency f_c and high cutoff frequency decrease. Meanwhile, low cutoff frequency also decreases. Therefore, a shorter coupler is better for passing faster edges with energy in higher frequencies, while a longer coupler is better for passing slower edges with energy in lower frequencies.

To better understand the relationship between coupler length and signal edge rate, analytical equations are developed by using a Sigmoid function to approximate a step output from the transmitter as

$$V_{TX}(t) = \frac{V_{TX,P}}{1 + e^{-\frac{4.4t}{t_r}}}, \tag{3.18}$$

where $V_{TX,P}$ is the step height, and t_r is the 10–90% rise time. From (3.3) and (3.4), the received voltage can be expressed as

$$V_{RX}(t) = \frac{K_C + K_L}{2v_p} \int_0^l \frac{dV_{TX}\left(t - \frac{2z}{v_p}\right)}{dt} dz. \tag{3.19}$$

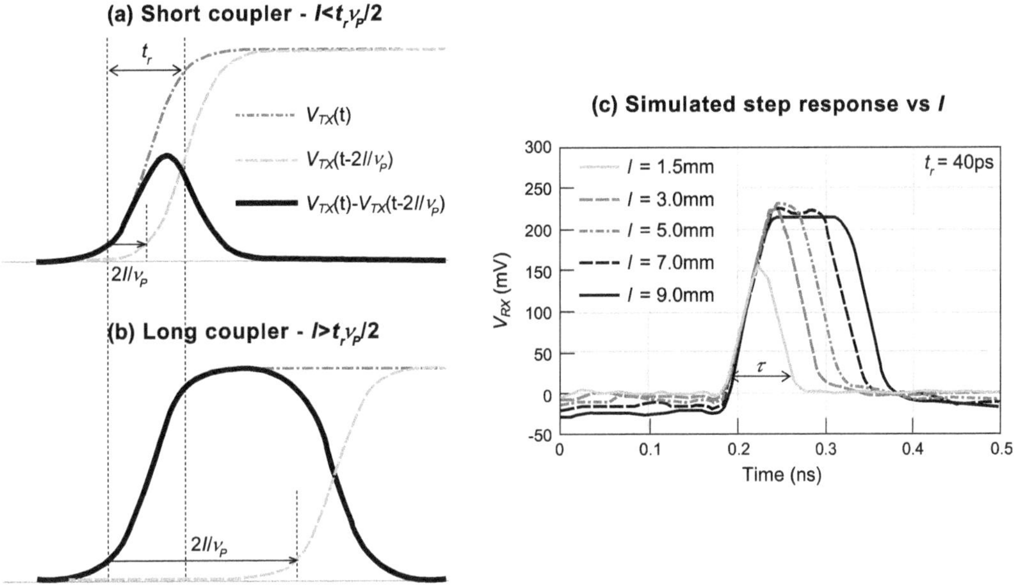

Figure 3.12 Received pulse shape as a function of transmission line length. (c) © 2015 IEEE. Reprinted, with permission, from [5]

Substituting (3.18) into (3.19),

$$V_{RX}(t) = \frac{K_C + K_L}{4}\left\{V_{TX}(t) - V_{TX}\left(t - \frac{2l}{v_p}\right)\right\}. \tag{3.20}$$

Therefore, the received voltage is proportional to a pulse that results from the difference between the transmitter step and a replica of it delayed by a timing offset of $2l/v_p$. Note that $2l/v_p$ is the round trip delay of the transmission lines. Figures 3.12a, b depict the resultant waveforms for $l < t_r v_p/2$ and $l > t_r v_p/2$ respectively. By interpolation, it can be seen that as the length of the coupler is increased, the magnitude and width of the received pulse increase, until the magnitude reaches saturation. This is illustrated in Figure 3.12c, which plots the step response of an example coupler as a function of l.

Specifically, the received pulse magnitude $V_{RX,P}$ is given by

$$V_{RX,P} = \frac{(K_C + K_L)l}{2t_r v_p} V_{TX,P} \; for \; l < \frac{t_r v_p}{2} \tag{3.21}$$

$$V_{RX,P} = \frac{(K_C + K_L)}{4} V_{TX,P} \; for \; l \geq \frac{t_r v_p}{2} \tag{3.22}$$

and the pulse width τ by

$$\tau = t_r + \frac{2l}{v_p}. \tag{3.23}$$

Beyond the saturation point, further increase in the coupler length increases only the pulse width but not its magnitude. Saturation occurs at a length of

$$l = \frac{t_r v_p}{2}. \tag{3.24}$$

$t_r v_p$ is known as the electrical length of the edge since it represents the length that yields a propagation delay equal to the edge rate. The factor of 2 is due to the fact that it takes the round trip delay for the transmitted signal to reach the far end of the coupler and then for the last portion of the coupled energy to travel back to the receiver at the near end of the coupler.

To achieve maximum coupling while avoiding ISI, choose l based on (3.24) and keep received pulse width $\tau < 1$ UI (UI stands for unit interval, which is equal to bit time). Under this condition,

$$\tau = 2t_r < 1\,\mathrm{UI}. \tag{3.25}$$

Note that a faster signal edge has a smaller t_r and hence requires a shorter coupler length to achieve saturation.

When the dielectric medium is homogeneous, $K_C = K_L = K$, and the received pulse magnitude in (3.22) is reduced to

$$V_{RX,P} = \frac{K}{2} V_{TX,P}. \tag{3.26}$$

Table 3.1 summarizes the key physical and electrical parameters of TLC, as well as their design targets. To reiterate, the equations are based on the assumptions of identical transmission line designs in both the transmitting and receiving terminals, as well as homogeneous dielectric medium to illustrate key ideas without loss of generality. In practice, the dielectric constant of PCB laminate materials does not have large variations, and the spacer dielectric constant can be chosen to match them. Therefore, the assumption of homogeneous medium can be met by design within manufacturing tolerances.

3.1.3 Transceiver Design

Unlike commonly used wireless technologies such as cellular and Wi-Fi that transmit data by modulating amplitude, phase, or frequency of a carrier signal, TLC, same as TCI, uses simple pulse modulation of a baseband signal. As a result, the same circuit structure used in conventional transceivers for wired communication can be used for TLC. The challenge is in the front-end design to drive the coupled, differential transmission lines. Because of the low frequency cutoff characteristics of TLC, the digital signal from the transmitter undergoes first-order differentiation when passed through the coupler, generating a pulse from each transition. Data can be recovered by passing the received pulse through a continuous time linear equalizer (CTLE) block for waveform shaping and amplification, followed by a hysteresis comparator for pattern extraction.

Table 3.1. TLC physical and electrical design parameters and targets.

Physical design parameter	Symbol	Electrical parameter	Physical to electrical parameter conversion	Design target
Trace width	W	Coupler transfer function	$S_{31} = \frac{jK\sin\left(\frac{\omega l}{v_p}\right)}{\sqrt{1-K^2}\cos\left(\frac{\omega l}{v_p}\right) + j\sin\left(\frac{\omega l}{v_p}\right)}$	$l = t_r v_p/2$ to maximize received voltage
Intrapair spacing	S	Coupling coefficient	$K = \frac{1}{1+\frac{C_s}{C_m}} = \frac{1}{1+\frac{\varepsilon_0\varepsilon_r\frac{W}{h}+2C_{s,fr}}{\varepsilon_0\varepsilon_r\frac{W}{d}+C_{m,fr}}}$	Maximize K while keeping area overhead small; use d for coarse, W for fine adjustment
PCB dielectric thickness (distance to ground)	h	Characteristic impedance	$Z_0 = \frac{1}{v_p\sqrt{C_s(C_s+2C_m)}}$	$Z_0 = 100\ \Omega$ differential; use W for tuning
Spacer thickness	d	Intrapair coupling		$S > 3W$ to minimize effect on K
Spacing between adjacent pairs	S_P	Same plane crosstalk		$S_P > 2.3\text{W}$ to minimize crosstalk
Coupler length	l	Transmitter rise time for saturation	$t_r = \frac{2l}{v_p}$	$2t_r < 1$ UI to avoid ISI

3.1.3.1 Transmitter Design

The key design targets for the driver stage are rise time, fall time, and power consumption. As noted in the previous section, a faster rise time relative to the length of the coupler results in a received pulse with larger amplitude (until saturation) as well as less ISI. Meanwhile, the driver contributes to more than 60% of the total interface power, making it important to minimize driver power.

Table 3.2 enumerates three options for the driver stage of the transmitter, namely, current mode logic (CML), source series terminated (SSL), and pulse. The incoming data pattern is transmitted as is in the case of both CML and SST. By contrast, the pulse driver first converts each data transition into a short positive or negative pulse, hence the name. Therefore, with a pulse driver, current flows only during data transitions, resulting in significant power reduction. Furthermore, power consumption scales with data rate. Hence, it is well suited for applications such as the MIPI standard used in mobile applications, which demands low power and requires the ability to lower power through speed downgrade. On the other hand, since the received pulse is generated from second-order differentiation of the original data pattern, each transition

Table 3.2. TLC driver options comparison.

Driver type		CML	SST	Pulse
Circuit schematics		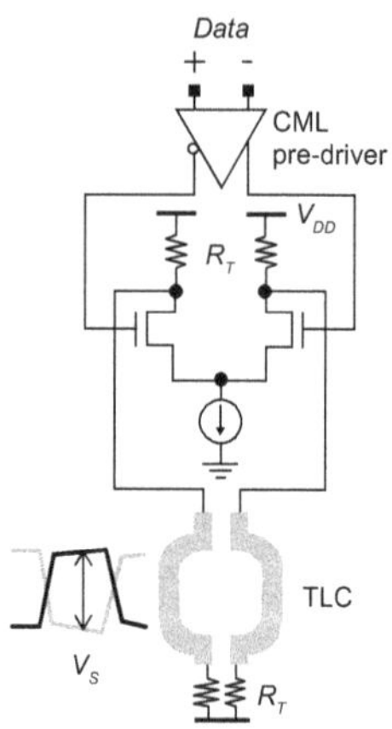	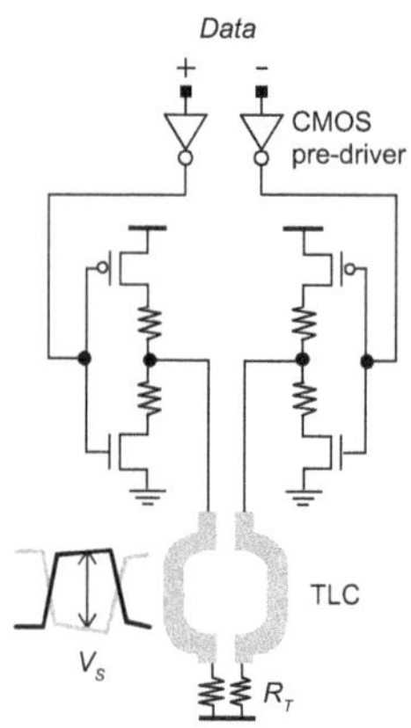	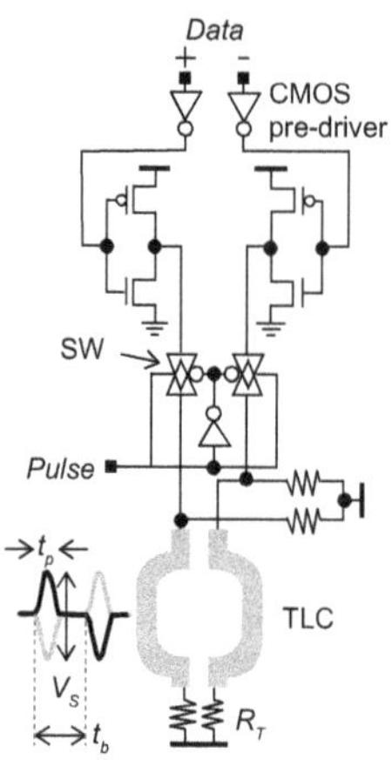
Power		$\frac{2V_S}{R_T} \times V_{DD}$	$\frac{V_S}{2R_T} \times V_{DD}$	$\frac{2}{3} \times \frac{V_S}{R_T} \times \frac{t_p}{t_b} \times V_{DD}$
Relative comparison	Data rate (DR)	++	+	–
	Power	–	+	++
	Power scaling with DR	–	–	+
	Switching noise	+	–	–

results in a double pulse. Therefore, the total width of the received pulse is about twice that of CML and SST drivers [6]. As a result, the maximum achievable data rate is about halved. Consequently, the pulse driver is not suited for high-speed applications (typically >1 Gb/s/link) such as a CPU memory interface. This is the same as the pulse transmitter used for TCI.

The difference between the CML and SST drivers lies in operating speed and power. CML was initially the driver of choice for TLC due to its ability to deliver fast rise and fall times, making it suited for high-speed applications [7]. Furthermore, due to the constant drive current, it has minimal switching current and hence minimal supply switching noise. The downside is its large power consumption. Because the 50 Ω termination at the far end of the coupler is in parallel with the driver termination, the equivalent load is 25 Ω, resulting in a large static current and hence power consumption. By contrast, in SST, the 50 Ω termination of the coupler is in series with the driver termination, resulting in an equivalent load of 100 Ω. Consequently, for the same signal amplitude, SST requires only ¼ the current of CML. Since the predriver consists of CMOS circuits as well, SST driver + predriver consumes less than ¼ the power of CML + predriver. The downside of SST is its speed and supply noise. Its speed was previously limited by the slower PMOS transistors. But the problem is being alleviated as process scales – link speeds of 6.4, 12.5, and 24 Gb/s have been achieved in 90, 65, and 28 nm CMOS processes respectively [8–10]. If advanced processes are available, SST can be used to significantly reduce power consumption compared to CML while maintaining the same signal amplitude, provided that the added supply noise is acceptable. The relative merits of the three driver options are summarized in Table 3.2.

3.1.3.2 Receiver Design

The standard TLC receiver design is depicted in Figure 3.13. It consists of a four-stage CML buffer for amplification followed by a hysteresis comparator to reverse the first-order differentiation of the received waveform to recreate the transmitted data. A hysteresis comparator is used because (1) the two alternating threshold voltages provide better noise margin and (2) its simple structure without a clock reduces timing constraints [11]. The amplified signal V_{AMP} is compared against the threshold voltage (+ or $-V_{TH}$). If it exceeds the threshold, the output is switched and latched. The polarity of the threshold voltage is set based on the state of the last data transition. The hysteresis characteristics are realized through feedback of the cross-coupled NMOS pair with loop gain of more than 1, same as the hysteresis comparator used for TCI.

The threshold voltage V_{TH} must be carefully selected for proper operation. If it is set too low, the receiver can be falsely triggered by noise such as reflected signal resulting in increased BER. If it is set too high, the receiver may miss certain signal transitions. The value is adjusted by varying the ratio of the two current sources I_2/I_1, as shown in Figure 3.13c. In practice, V_{TH} is set to be between the 40 and 90% levels of the input signal amplitude as shown in Figure 3.13b. The CML buffer inserted in front of the hysteresis comparator amplifies the received signal for better V_{TH} margin.

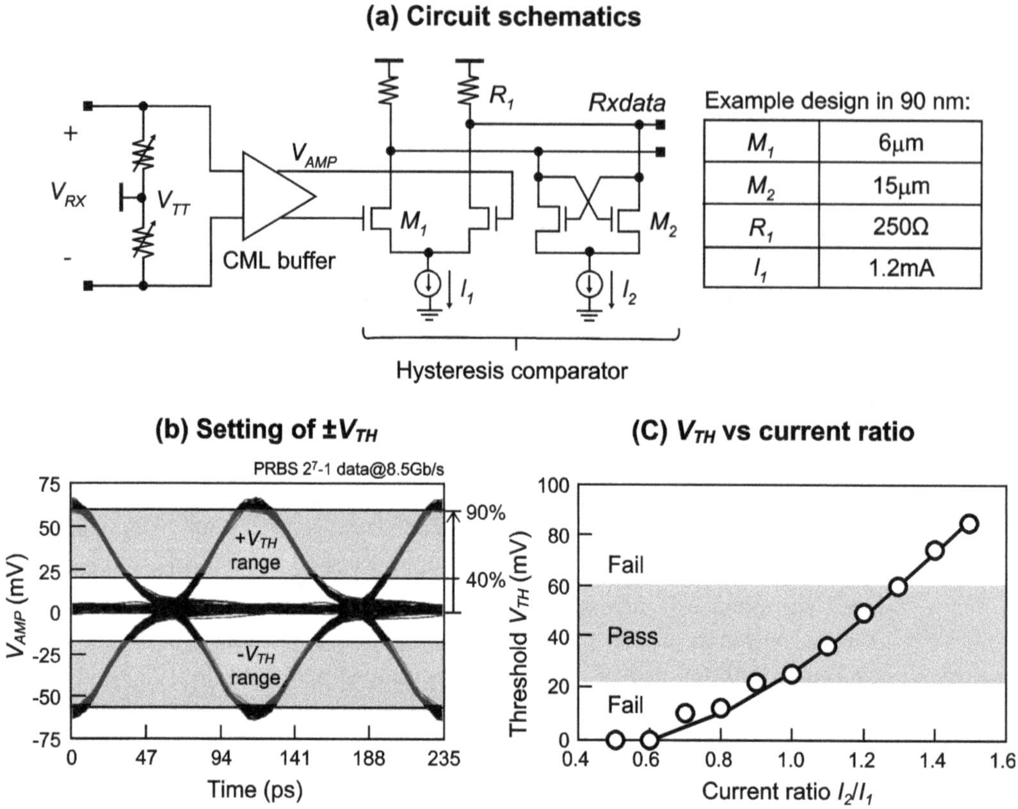

Figure 3.13 TLC receiver design. © 2015 IEEE. Reprinted, with permission, from [5]

3.1.4 Closing Thoughts

This section presents the basic structure of TLC and its operating principle, its key electrical characteristics and how they relate to physical design parameters, as well as the standard transceiver design. As is the case with TCI, various adaptations can be made to the basic design to optimize for various applications. Specific implementation examples will be presented in ensuing sections to better illustrate the characteristics and flexibility of the technology, including implementations for memory module, space, smartphone, automotive, and storage applications.

3.2 TLC Multidrop Bus

To allow for memory capacity expansion or functional extension, a bus is needed to connect multiple modules together. A good example can be found in the connection of a CPU to multiple DRAM modules in a computer memory subsystem. Among other

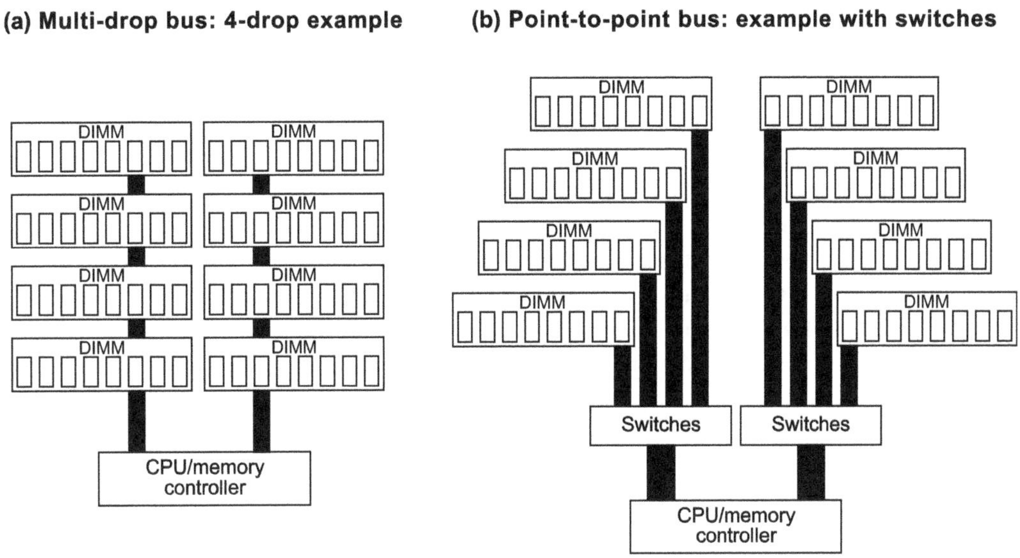

Figure 3.14 Bus topologies for connecting multiple modules together. © 2015 IEEE. Reprinted, with permission, from [5]

benefits, the bus architecture allows the same motherboard design to support a large range of DRAM capacity by adjusting not only the DRAM capacity on a module, but also the number of modules inserted into the bus. Such a system can be implemented using one of two bus topologies – multidrop and point-to-point, as shown in Figure 3.14 [5]. Compared to point-to-point, a multidrop bus has the advantage of having fewer wire connections, the number of which is independent of the number of modules on the bus, resulting in a lower CPU/memory controller package pin count and smaller PCB layout area, and hence a lower cost and more lightweight and compact design. Conversely, a multidrop bus using the same package pin count and PCB layout area can support more modules and hence a larger memory capacity than a point-to-point bus. Unfortunately, each drop (connection point) on the bus adds loading and creates an impedance discontinuity, resulting in signal attenuation, reflection, and distortion. The more connections there are, the larger the total loading and number of reflection points. As a result, the maximum data rate decreases as the number of drops and memory capacity increase. Consequently, the use of multidrop bus is generally limited to lower-speed communication, especially when many modules need to be connected. Therefore, the topology of choice for high-speed communication is a point-to-point bus, which makes a compact design challenging. Furthermore, since the number of wire connections increases with the number of modules, it limits expandability. The use of switches as shown in Figure 3.14b helps improve expandability without increasing upstream wire connections. The trade-off is a higher component count and costs. Furthermore, the number of connections downstream remains a problem.

Other solutions have also been explored, but each suffers limitations in one way or another. Decision feedback equalizer (DFE) can be used to remove reflections generated at the impedance discontinuities [12–14], but a large number of taps are required, especially as the number of drops increases, resulting in complex circuit design and large power consumption. For instance, [13] presented a solution with 12 taps for connecting four modules at a data rate of 2.6 Gb/s/link and energy efficiency of 42.9 pJ/b. Alternatively, reflections can be suppressed using damping resistors and by tuning the characteristic impedance of individual transmission lines as proposed in [15]. But the reduced reflections still limit the data rate to 4.8 Gb/s/link.

Our proposal is to adopt TLC to enable a fast and low-power multidrop bus using simple transceiver design. The implementation reported in [5] achieves a data rate of 8.5 Gb/s/link and energy efficiency of 8.9 pJ/b.

As explained in Section 3.1.2, TLC is designed to have well-controlled characteristic impedance matching that of the rest of the system. Therefore, reflection at module connection points of a TLC multidrop bus is minimal (less than –20 dB over a wide frequency range, as shown in Figure 3.7). Nevertheless, design adjustment must be made to TLC in order to resolve unique challenges presented by a multidrop bus. In this section, design modifications are presented to address challenges in implementing two types of multidrop bus. Sections 3.2.1–3.2.3 discuss respectively the implementation of a master–slave, multidrop bus such as a DRAM memory subsystem, and a multimaster, multidrop bus such as a communication backplane.

3.2.1 EE-TLC for Master-Slave, Multi-Drop Bus

Figure 3.15 illustrates the TLC implementation in [5] of a dual inline memory module (DIMM) multidrop bus that connects a memory controller to multiple DIMMs, each with an interface chip equipped with TLC transceivers that acts as a bridge between the controller and DRAM chips on the module. A signal traveling from the controller down the bus loses part of its energy through TLC to each module it passes. As a result, the farther down the bus it travels, the weaker its strength becomes.

Figure 3.16 depicts the simulated frequency responses and time domain signal waveforms for an implementation of the memory multidrop bus in Figure 3.15 that has eight drops and identical design for all eight TLCs. The received signal at the farthest module (Port3_8) is severely degraded compared to that at the nearest module (Port3_1) even though the coupler gains are identical. This is because the transmitted signal loses energy every time it passes through a coupler. This is illustrated in Figure 3.16b, where the frequency responses at the eight different receivers relative to the output signal of the transmitter at Port1_1 are compared.

Since the spacer and DIMM side PCB dielectric materials are chosen to have a dielectric constant matching that of the PCB, the couplers are embedded in a homogeneous dielectric medium, resulting in zero energy coupled in the forward direction from Port1 to Port4. In other words, $S_{41} = 0$. As a result, from conservation of energy, if one unit of energy enters Port 1, the sum of energy coupled to Port 3 and Port 2 is equal to 1. In other words,

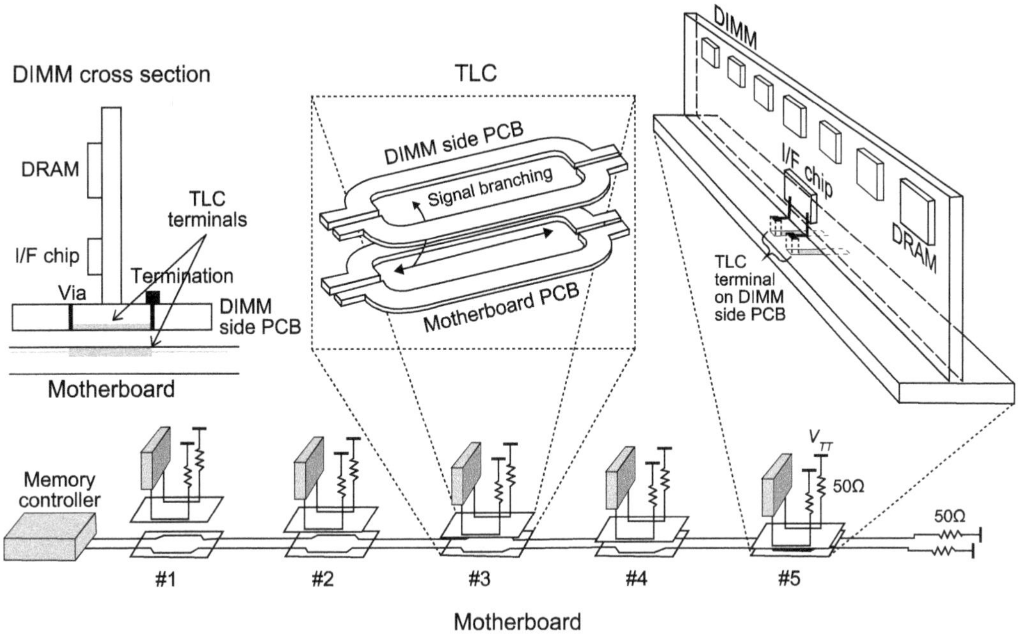

Figure 3.15 TLC implementation of a memory multidrop bus. © 2015 IEEE. Reprinted, with permission, from [5]

$$|S_{31}|^2 + |S_{21}|^2 = 1. \tag{3.27}$$

Therefore, $|S_{21\_1}|^2 = 1 - |S_{31\_1}|^2$, which is less than 1. In other words, the energy entering Port1_2 from Port2_1 is less than the energy entering Port1_1. As a result, the energy entering Port1_i gets progressively smaller as i increases.

Our proposed solution is an energy equipartition TLC (EE-TLC) multidrop bus. The individual couplers are designed the same way as the standard TLC design described in Section 3.1.2. But instead of using identical design across the bus, the individual couplers in an EE-TLC are designed to compensate for the decreasing signal strength to achieve equal net energy delivered to each receiver regardless of its position on the bus, hence the name "energy equipartition." In other words, the couplers are designed to have increasing coupling gain.

Since the couplers are connected in series, the transfer function from the transmitter to the ith receiver, $H_{EE\text{-}TLC,i}(\omega)$, can be expressed as

$$H_{EE-TLC,i}(\omega) = S_{21\_1}(\omega) \times S_{21\_2}(\omega) \ldots \times S_{21\_i-1}(\omega) \times S_{31\_i}(\omega). \tag{3.28}$$

From (3.7), (3.8), (3.27), and (3.28), the condition for equal energy delivery to all receivers can be derived to be

$$|S_{31\_i}|_{peak} = K_i = \sqrt{\frac{1}{N+2-i}}, \tag{3.29}$$

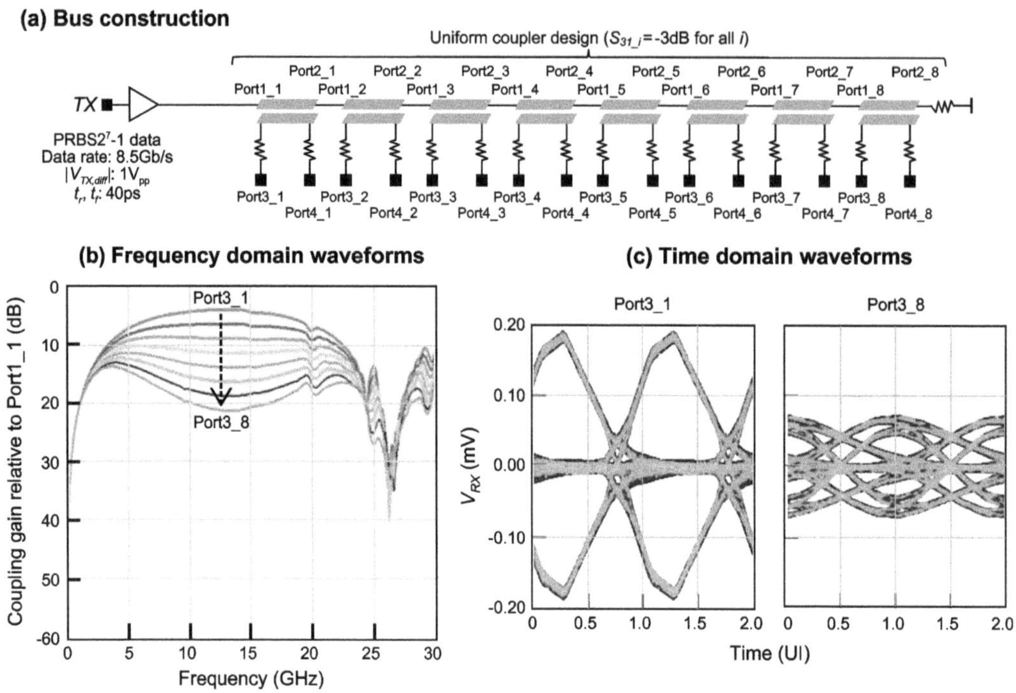

Figure 3.16 Simulated response of a TLC multidrop bus with uniform coupler design. © 2015 IEEE. Reprinted, with permission, from [5]

where N is the number of couplers or drops, and $|S_{31\_i}|_{\text{peak}}$ is the peak coupling gain over frequency of the ith coupler. Under this condition, $K_{EE\text{-}TLC,i}$, the effective coupling coefficient from the transmitter including S_{21} of all preceding couplers becomes the same for all receivers at the peak frequency, and is equal to

$$K_{EE-TLC,i} = K_{EE-TLC} = \sqrt{\frac{1}{N+1}}. \tag{3.30}$$

Figure 3.17 depicts the simulated response of the EE-TLC implementation of the multidrop bus in Figure 3.16. It confirms that by designing for equal energy delivery to each receiver, EE-TLC enables a high-speed multidrop bus with optimized signal quality across the bus, resulting in significant improvement of the received signal toward the end of the bus (Figure 3.17b vs Figure 3.16b).

Based on the design approach discussed in Section. 3.1.2, a larger coupler gain can be achieved by reducing the communication distance d through the use of a thinner spacer. Since varying d changes the electric field distribution and hence capacitance, it has a small effect on the characteristic impedance of the coupler transmission lines. This effect can be ignored on the module side due to the short trace length. Hence the same trace width W_{Module} can be used for all couplers. However, on the motherboard side, since the bus has long total length and includes multiple couplers, fine-tuning of characteristic impedance through tuning of the coupler trace width $W_{Motherboard}$ is

Table 3.3. Design summary of an EE-TLC multidrop memory bus. © 2015 IEEE. Reprinted, with permission, from [5].

	Port1	Port2	Port3	Port4	Port5	Port6	Port7	Port8
d (μm)	395	360	325	290	255	220	150	80
$W_{Motherboard}$ (μm)	340	330	320	310	300	290	250	210
W_{Module} (μm)	300	300	300	300	300	300	300	300
l (mm)	3	3	3	3	3	3	3	3
S (μm)	1050	1050	1050	1050	1050	1050	1050	1050
h (mm)	0.6	0.6	0.6	0.6	0.6	0.6	0.6	0.6

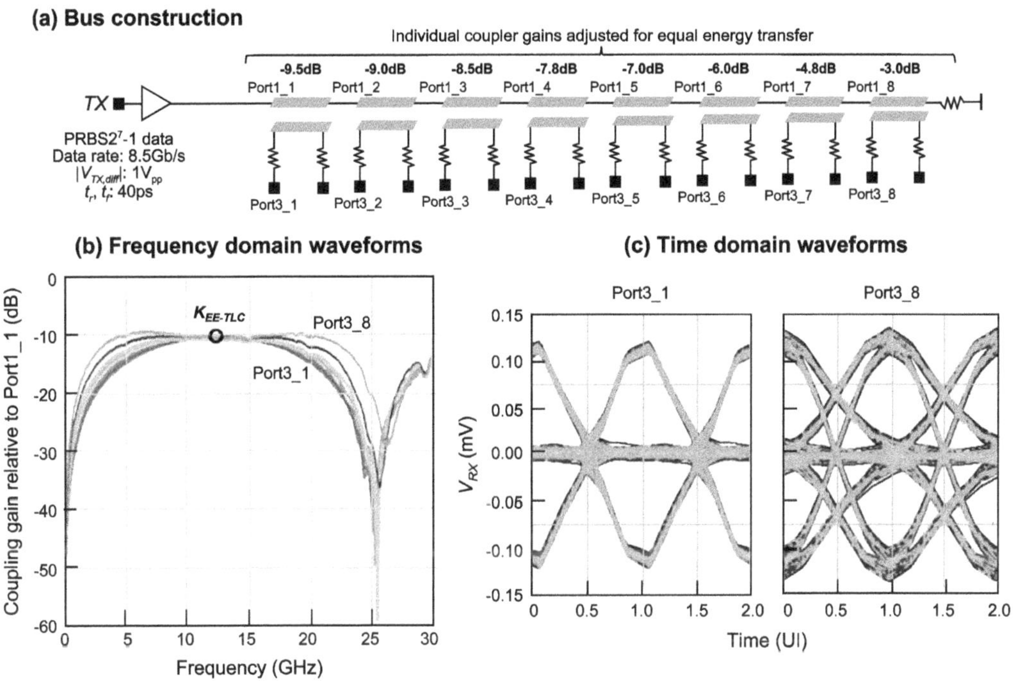

Figure 3.17 Simulated response of an EE-TLC multidrop bus. © 2015 IEEE. Reprinted, with permission, from [5]

recommended to suppress multiple reflections. As an illustration, the design parameters for an example implementation is shown in Table 3.3.

No other design changes are necessary, and the standard transceiver design can be used as is.

3.2.2 Single-Ended to Differential Conversion TLC (SDC-TLC)

Both the standard TLC and EE-TLC are designed for differential signaling that requires a pair of transmission lines for each signal. This doubles the number of signal traces routed on the PCB compared to single-ended signaling. In a system with limited

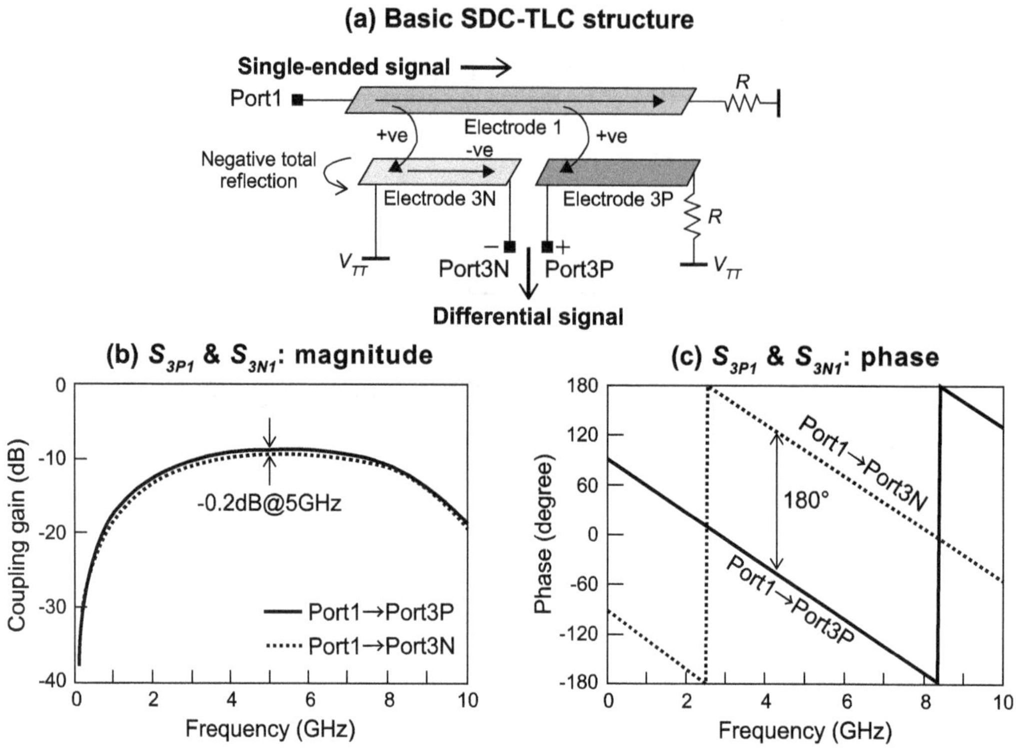

Figure 3.18 SDC-TLC structure and simulated frequency response. © 2015 IEEE. Reprinted, with permission, from [16]

PCB area, it may be challenging to route all the required signals across the board. An example is a test board for DRAM production testing. The cost of DRAM manufacturing is highly sensitive to test time. As a result, a DRAM production test board is built to test as many DRAM chips in parallel as possible. Compared to single-ended signaling, the use of differential signaling significantly reduces the number of DRAM chips that can be mounted on the test board and tested simultaneously for a given board size, increasing per DRAM test time and hence costs. To suppress the increase in test board layout area, one solution is to drive the test board with single-ended signals, then convert them into differential signals through the DRAM sockets. This can be achieved by implementing a single-ended-to-differential conversion TLC (SDC-TLC) that serves as a wireless interface between the test board and DRAM chips while also effecting conversion from single-ended to differential signaling.

Figure 3.18a depicts the basic SDC-TLC structure reported in [16], which consists of a single electrode (Electrode 1) on the single-ended signaling side and a pair of electrodes (Electrodes 3P and 3N) on the differential signaling side that are cascaded lengthwise in the shadow of Electrode 1. A single-ended signal propagating to the right across Electrode 1 induces a signal through electromagnetic coupling in both

Electrodes 3P and 3N simultaneously. The two induced signals are of equal magnitude and sign, and both travel backward to the left. The signal in Electrode 3P exits Port 3P as is. Meanwhile, since the left end of Electrode 3N is connected to V_{TT} ,which is an AC short, it generates a negative total reflection, sending a signal of equal magnitude but opposite sign to the right that exits Port 3N. The signal exiting Port 3N is thus of equal magnitude but opposite sign to that exiting Port 3P, and the combination of the two creates a differential output signal. Conversely, when a differential signal enters Ports 3P and 3N, the positive (+) signal traveling to the right in Electrode 3P is absorbed by the terminating resistor, while simultaneously inducing a single-ended signal in Electrode 1 that travels to the left. Meanwhile, the negative (–) signal traveling across Electrode 3N to the left goes through a negative total reflection, resulting in a positive signal of equal magnitude traveling to the right. This in turn induces a single-ended signal in Electrode 1 traveling to the left, which adds in phase to the signal induced by the (+) signal. As a result, a single-ended signal of double strength is coupled to Port 1.

Figures 3.18b and c depict the simulated magnitude and phase respectively of the frequency response from Port 1 to Ports 3P and 3N. It confirms that the coupling to Ports 3P and 3N are of the same strength (with a difference of 0.2 dB at 5 GHz) and 180 degrees out of phase across a wide frequency band.

The cascading of Electrodes 3P and 3N results in doubling of the coupler length. However, the coupler width is only $1W$, compared to $W + S + W = 5W$ that is required for the standard differential pair coupler. Consequently, there is a significant reduction in the bus width, making SDC-TLC attractive for PCBs with congested routing.

3.2.3 BD-TLC for Multimaster, Multidrop Bus

In the case of a DIMM memory bus, the modules are all slaves and communicate only with the controller in a single master configuration. Consequently, the TLC implementation needs to support energy transfer in one direction only from any particular driver (Figure 3.19a). As a result, the standard coupler design, which is a unidirectional coupler that couples only in the backward direction, is adequate for such applications. In the standard TLC design depicted in Figure 3.20a, the coupling gain in the forward direction, S_{41}, is about 10 dB less than in the backward direction, S_{31}, within the pass band. (Forward direction coupling is not perfectly 0 because in practice it is difficult to fully match the dielectric constants between two PCBs and a spacer, especially given manufacturing variations.) By contrast, in backplane or automotive network applications, it is necessary for a module in the middle to communicate with modules on both sides of the bus in a multimaster configuration (Figure 3.19b). Therefore, the TLC implementation must be able to transfer energy in both the forward and backward directions. A bidirectional TLC (BD-TLC) implementation has been developed to address such applications. Figure 3.20b illustrates that indeed such a coupler is capable of delivering equal coupling gains in both directions, with forward coupling gain S_{41} matching backward coupling gain S_{31}.

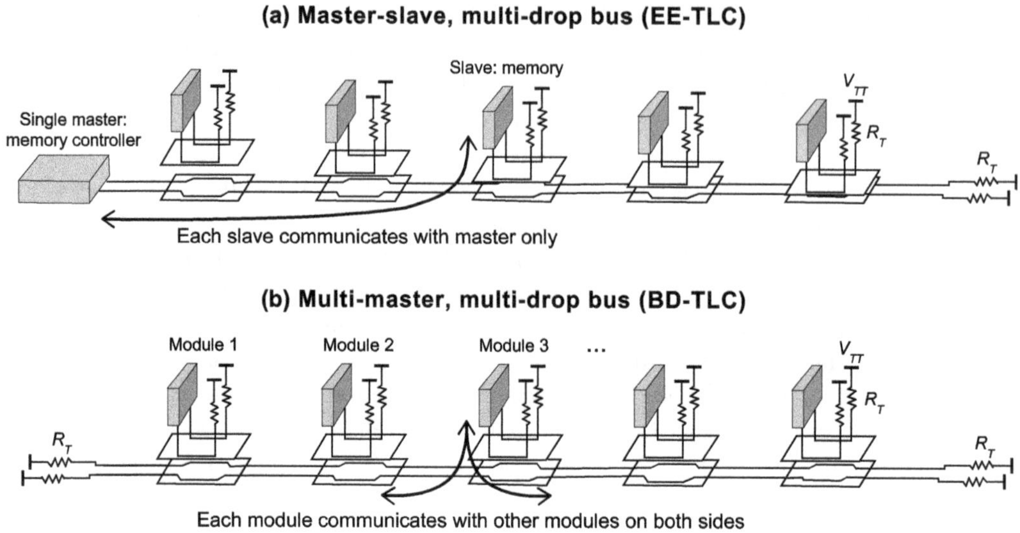

Figure 3.19 Multidrop bus comparison. (a) © 2015 IEEE. Reprinted, with permission, from [5]

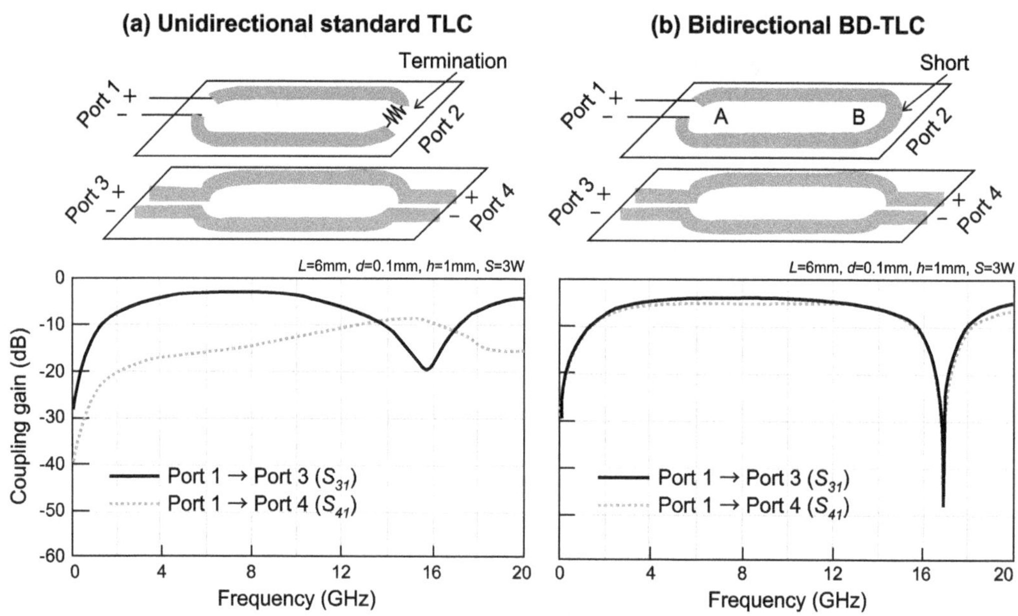

Figure 3.20 Coupler design directionality. © 2016 Atsutake Kosuge [1]

3.2.3.1 BD-TLC Coupler Design

The BD-TLC coupler design is a simple modification of the standard TLC. It makes use of the total reflection that occurs when a signal propagating down a transmission line encounters a short circuit. In both the standard TLC and BD-TLC designs, a signal injected into Port 1 creates a coupled signal in Port 3. When the transmitted signal propagates down the transmission line from Port 1 to Port 2, it is absorbed by the matching termination in the standard TLC design. By contrast, in the BD-TLC design, the transmitted signal undergoes negative total reflection at Port 2 due to the short circuit, preserving its magnitude while reversing its polarity. This reflected signal then propagates back toward Port 1 and is finally absorbed by a matching termination in the transmitter. But as this signal travels from Port 2 to Port 1, it creates a coupled signal in the direction of Port 4. In the ideal case of lossless transmission lines, this second coupled signal traveling in the opposite direction is identical to the first coupled signal except for a reversal in polarity. By adding a preamble to identify the relative location of the transmitter – whether it is to the left or right of the receiver – the polarity of the coupled signal can be anticipated and hence the signal correctly interpreted by the receiver.

Since BD-TLC is created by changing only the termination of the standard TLC design, it shares the same fundamental electrical characteristics and uses the same design methodology as the standard design. Nevertheless, this assumes an ideal short circuit termination with zero delay. In practice, the short is implemented by closing the gap between the differential transmission line pair with a trace connection, which has physical length and hence nonzero delay. The shorting trace length is determined by the intrapair spacing S. To minimize the effect of its delay, the general rule of thumb is to make its length shorter than $1/10\times$ the wavelength of the signal. For instance, since the wavelength of a 10 GHz signal in a PCB dielectric medium with $\varepsilon_r = 4.3$ is about 14.5 mm, the effect of a trace length shorter than 1.45 mm can be considered negligible for a 10 GHz signal. Figures 3.21b and c illustrate that, indeed, for $W = 0.4$ mm and $S = 3W = 1.2$ mm, coupling in both directions have about the same gains S_{31} and S_{41} over a wide frequency range, while $S = 10W = 4$ mm results in divergent gains at high frequencies.

3.2.3.2 BD-TLC Transceiver Design

Reexamination of Figure 3.19 reveals another important difference between a master–slave, multidrop bus and a multimaster, multidrop bus, namely the difference in the number of times a signal passes through a coupler before reaching the receiver. The former involves a single pass (one-pass TLC bus), while the latter two passes (two-pass TLC bus). Since each pass results in a first-order differentiation of the signal, passing through the TLC twice results in effectively a second-order differentiation of the transmitted signal. Consequently, while a one-pass TLC bus has a frequency response that exhibits a 20 dB/dec (dec = decade) roll-off toward low frequencies and converts a step into a pulse, a two-pass TLC bus has a 40 dB/dec low-frequency roll-off and converts a step into a double pulse with roughly twice the total pulse width (Figure 3.22).

The consequences are twofold. First, the faster roll-off of coupling gain results in additional attenuation at low frequencies. Second, the double pulse results in approximately half the maximum achievable data rate. To compensate for the additional

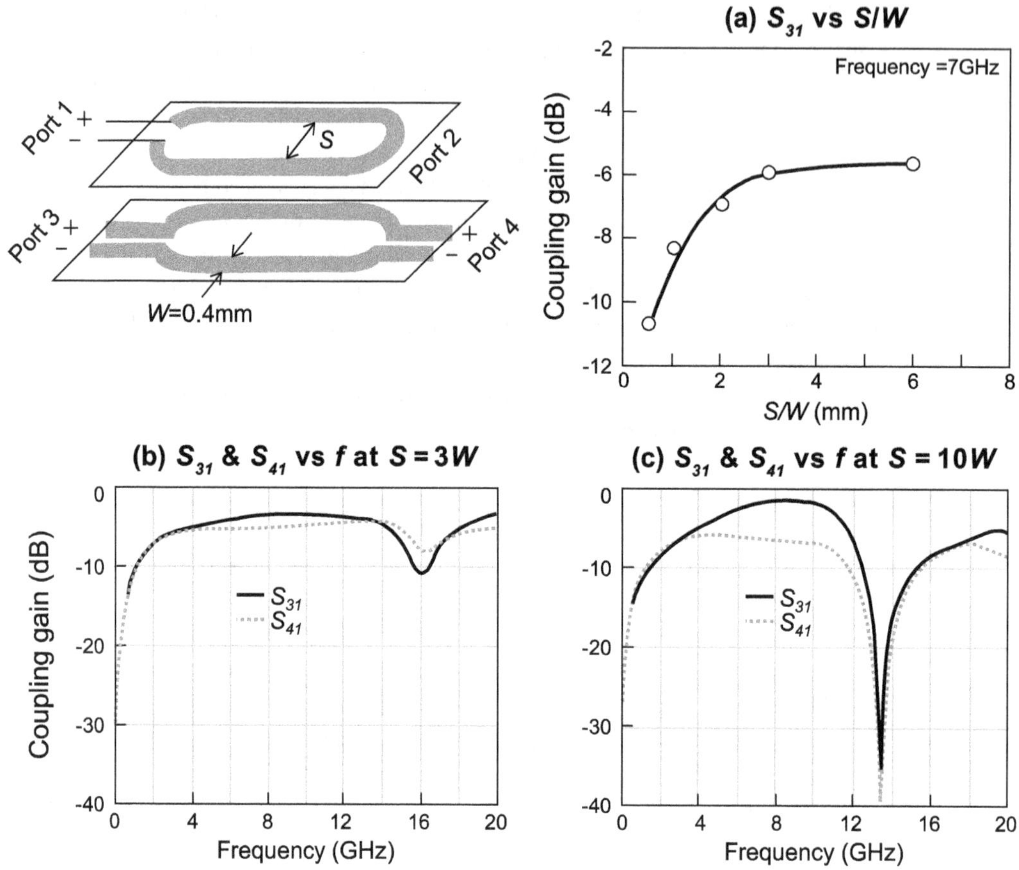

Figure 3.21 Simulated BD-TLC gain as a function of intrapair spacing *S*. © 2016 Atsutake Kosuge [1]

attenuation, a low-frequency compensation equalizer (LFC-EQ) is inserted in front of the hysteresis comparator in the BD-TLC receiver to boost the low-frequency gain such that it rolls off at the same rate of 20 dB/dec as the one-pass TLC bus, so as to suppress ISI and keep the signal eye open in the time domain (Figure 3.23).

Figure 3.24 depicts the schematics of the LFC-EQ circuit. The circuit consists of two paths for the received signal – an upper path intended for low-frequency components with gain of 20 dB/dec, and a lower path intended for high-frequency components that includes a peaking amplifier to ensure that high-frequency components are not attenuated. The processed signals from the two paths are then added together before being sent to the hysteresis comparator for data recovery. The amount of low-frequency compensation is determined by the weights α and (1-α), which are adjusted using current sources within the adder.

The NMOS cross-coupled pair has a negative conductance of $-g_{m,EQ}$. When its magnitude is set to $1/R_1$, where R_1 is the loading resistor of the preceding current source buffer, the effects of the two cancel each other. The remaining loading of the

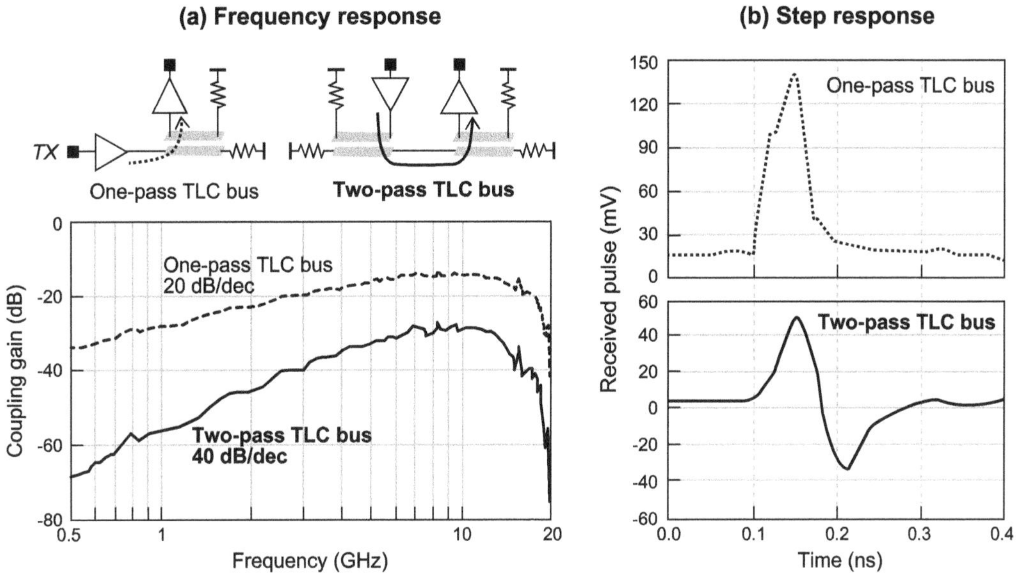

Figure 3.22 Simulated response comparison between one- and two-pass TLC bus. © 2016 Atsutake Kosuge [1]

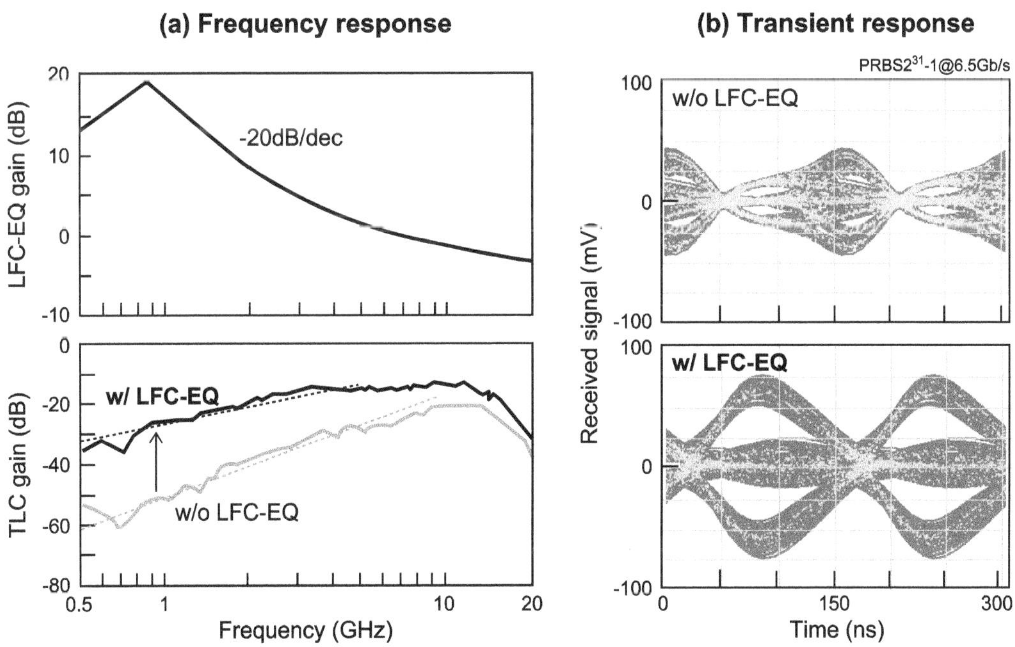

Figure 3.23 Simulated response and enhancement of LFC-EQ. © 2015 IEEE. Reprinted, with permission, from [16]

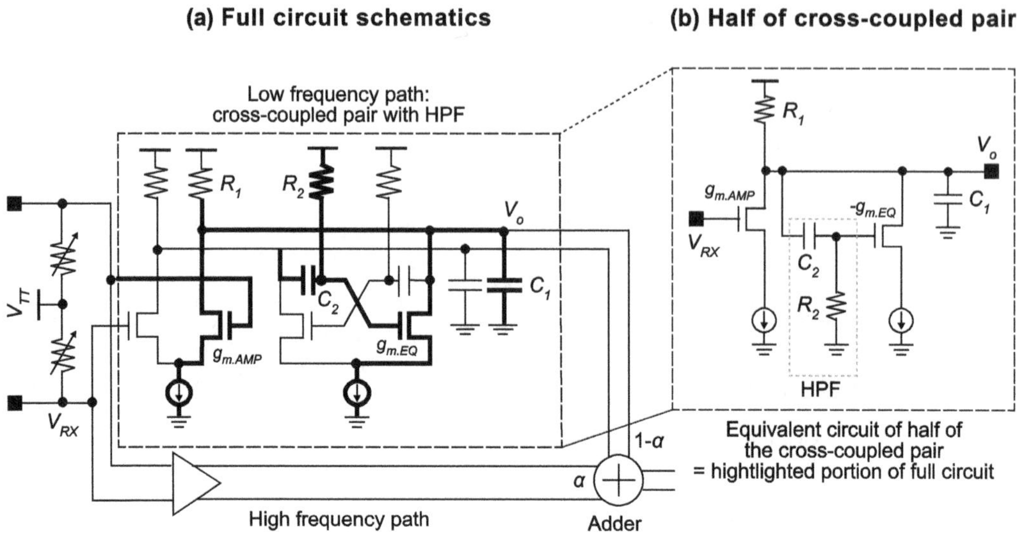

Figure 3.24 LFC-EQ circuit schematics. © 2015 IEEE. Reprinted, with permission, from [16]

half cross-coupled pair output is that of the parasitic capacitance C_1 of the final stage. Ignoring effect of the high pass filter (HPF) consisting of R_2 and C_2 for the moment, the transfer function of the amplifier block becomes the following (see Figure 3.25a):

$$H(\omega) = \frac{g_{m,amp}}{j\omega C_1}. \tag{3.31}$$

In other words, the amplifier has the property of an integrator, and hence amplifies low-frequency components with a gain of –20 dB/dec. Meanwhile, since its DC gain is infinite, any DC offset between the differential pair, however small, will cause the output to get stuck at 1 or 0, preventing the hysteresis comparator from correct data recovery. Since DC offset between the differential pair can result from manufacturing variations, it is impossible to completely eliminate it. To suppress the effect of DC offset, the R_2-C_2 HPF is inserted in the feedback loop of the cross-coupled pair, resulting in the following revised transfer function (see Figure 3.25b):

$$H(\omega) = \frac{g_{m,amp}}{j\omega C_1 + \dfrac{g_{m,EQ}}{1 + j\omega R_2 C_2}}. \tag{3.32}$$

The DC gain is thus reduced to a finite $g_{m,amp}/g_{m,EQ}$. Furthermore, at frequencies where the following condition is satisfied

$$j\omega C_1 \gg \frac{g_{m,EQ}}{1 + j\omega R_2 C_2}, \tag{3.33}$$

the transfer function becomes approximately equal to (3.31), achieving low frequency compensation.

For the transmitter, the standard design can be used without modification.

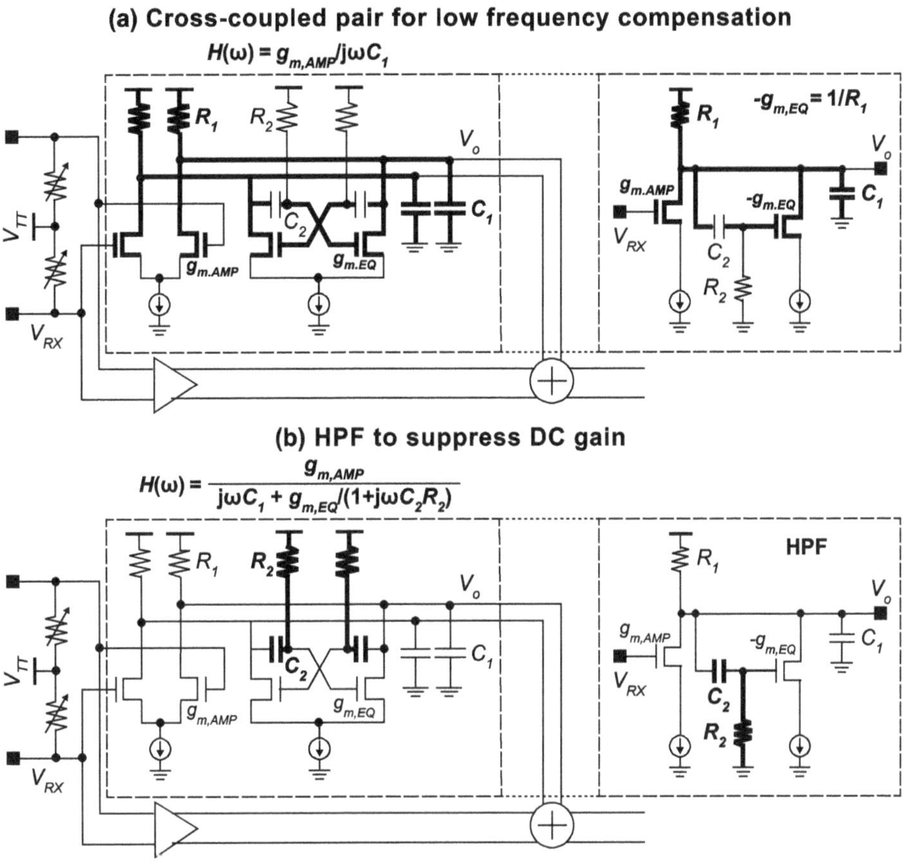

Figure 3.25 Functional analysis of LFC-EQ. © 2016 Atsutake Kosuge [1]

3.2.4 Case Study: TLC in Satellite Processor Systems

Figure 3.26a depicts the block diagram of a conventional satellite processor system that consists of processing, sensing, and memory modules connected by a backplane. Since the cost of launching a satellite is highly dependent on the size and weight of the payload, being small and lightweight is a key requirement for the processor system. Meanwhile, high data communication speed reaching 20 Gb/s and large storage capacity reaching 1 TB are required. To achieve the high data rate, point-to-point backplane connection is the conventional solution, resulting in a large number of signal connections and hence large connectors. On the other hand, the large storage capacity requires using multiple memory modules. Furthermore, since launching a satellite is extremely expensive, the system must have very low probability of failure. A common solution is to have redundancy to enable postlaunch replacement. In the example system, the full set of hardware is duplicated, resulting in twice as many modules on the backplane as is needed for normal operation. As a result, additional

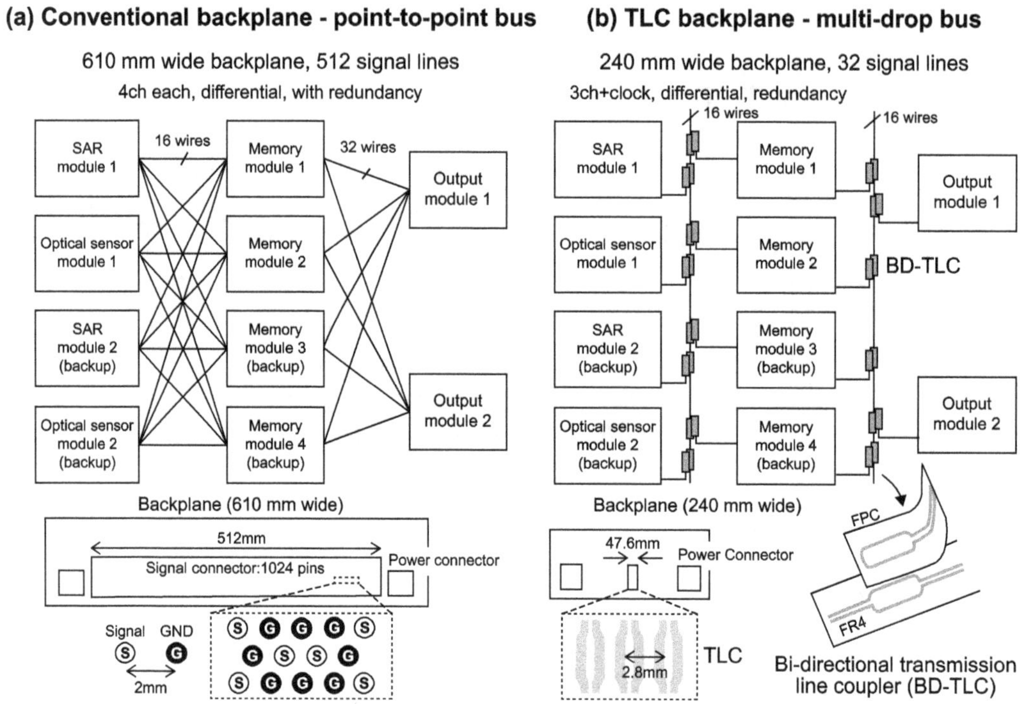

Figure 3.26 A satellite processor system. © 2015 IEEE. Reprinted, with permission, from [16]

modules must be connected to the backplane, further increasing the number of signal connections and connector size. The example system consisting of 10 modules connected by point-to-point buses on the backplane requires 512 signal lines, a connector measuring 512 mm in width with 1,024 signal and ground pins, and a backplane that is 610 mm wide.

In [16], the TLC derivatives BD-TLC and SDC-TLC are adopted to reduce the backplane size of the satellite processor system, resulting in both small form factor and light weight, while maintaining high Gb/s data rates (Figure 3.26b).

BD-TLC is used to replace point-to-point intermodule connections with multimaster, multidrop buses. The use of the BD-TLC transceiver design from Section 3.2.3.2 that includes LFC-EQ in the receiver enables the bus to run at 6.5 Gb/s, while drastically reducing the number of signal lines by a factor of 1/16 from 512 to 32. The BD-TLC connector prototype is depicted in Figure 3.27. The coupler is formed between an FR4 PCB backplane and a flexible printed circuit (FPC) attached to the module board, where coupler length l is 5 mm and width W is 0.4 mm. Using a coupler pitch of 2.8 mm, the connector form factor can be reduced from $512 \times 8 \times 8$ to $47.6 \times 7 \times 0.4$ mm^3, achieving a factor of 1/246 reduction in volume.

Meanwhile, SDC-TLC is applied to create a multidrop bus to connect each group of five DRAM chips to the controller chip on each module as depicted in Figure 3.28a, thereby reducing the module width. It converts single-ended 3.2 Gb/s DDR4 signals at

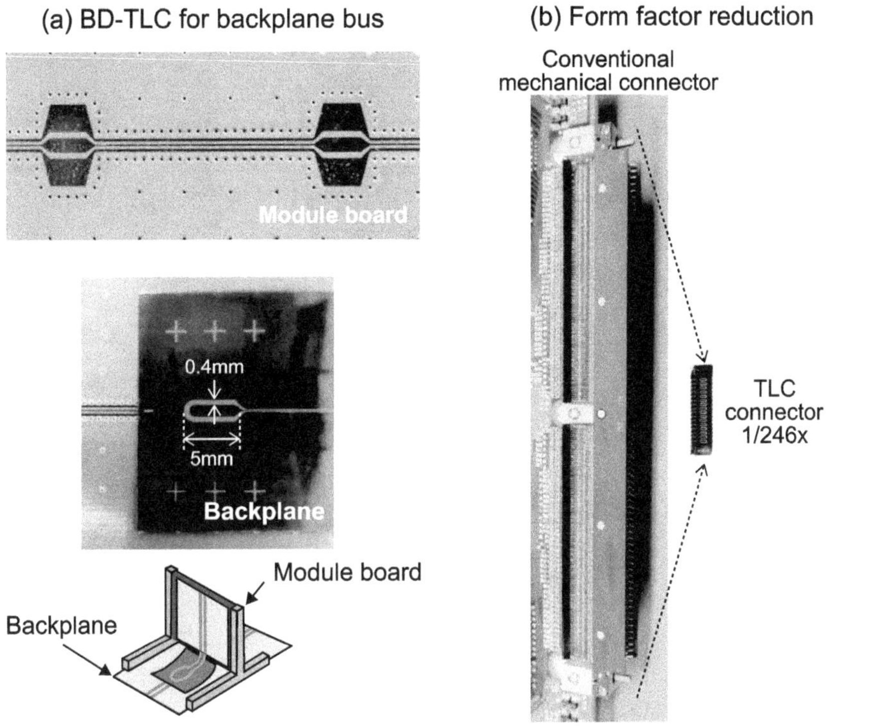

Figure 3.27 BD-TLC prototype and comparison with conventional connector. (a) © 2015 IEEE. Reprinted, with permission, from [16]

each branching point of the bus into differential signals. The number of signals is reduced by 80% compared to the use of point-to-point buses, resulting in a memory module that is 240 mm wide.

The SDC-TLC prototype is built with a communication distance of 1.6 mm and a trace width of 0.6 mm and a length of 7 mm for Electrode 1. Electrodes 3P and 3N are therefore each about 3.5 mm long (Figure 3.28b).

A six-drop bus is constructed using BD-TLC and SDC-TLC respectively, and their BER timing margins are measured and presented in Figure 3.29. Figure 3.29a confirms that the LFC-EQ receiver opens up the data eye to achieve a healthy margin of 0.31 UI at BER $= 10^{-12}$ and a data rate of 6.5 Gb/s even when driving the farthest receiver. Meanwhile, on the SDC-TLC bus, even though the timing margin at the farthest receiver RX5 is smaller than that at the nearest receiver RX1, it is still healthy at 0.67 UI at a BER of 10^{-12} and a data rate of 3.2 Gb/s.

A critical requirement for satellite applications is robust operation under intense vibration during a rocket launch. Using the setup shown in Figure 3.30a, the BD-TLC prototype is tested under the same condition as is experienced during the initial 60 seconds of a small rocket launch, with vibration up to 20.76 G_{rms}. Results from Figure 3.30b shows that the connector is able to sustain data transmission at 100 Mb/s for 60 seconds with no bit errors.

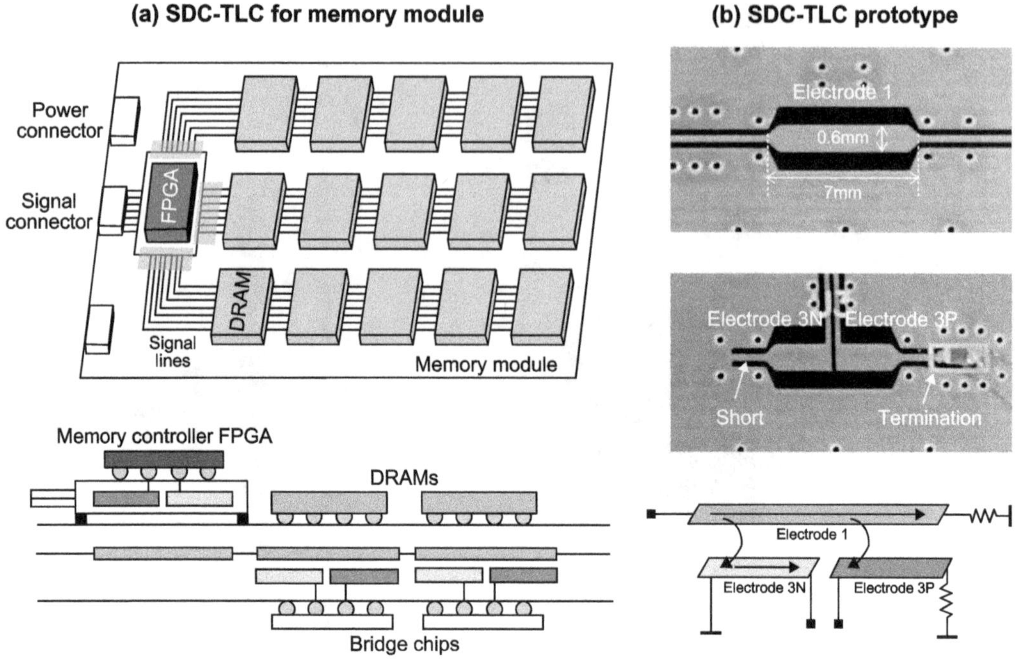

Figure 3.28 SDC-TLC prototype and its application to memory module. © 2015 IEEE. Reprinted, with permission, from [16]

3.2.5 Closing Thoughts

Table 3.4 summarizes the three TLC derivatives for implementing a multidrop bus and compares them against the standard TLC. Using the same principle of electromagnetic coupling between parallel transmissions lines, the standard TLC structure can be modified by varying communication distance d, dielectric constant ε_r of the materials (mostly the spacer since commonly used PCB materials do not vary much in ε_r), and termination (short vs. matched termination) to create derivatives to address a wide range of system requirements, supporting different bus topologies including multidrop, master–slave, and multimaster configurations, for different applications including memory system and backplane, as well as satellite applications, which demand robustness and light weight.

3.3 TLC for Smartphone

Because of the merits of a TLC connector – low power, high bandwidth, physical robustness, elimination of mechanical parts, and headroom for downsizing – it has the potential to deliver a lot of value to mobile applications when used to replace conventional connector solutions. In particular, by concealing its metal terminals,

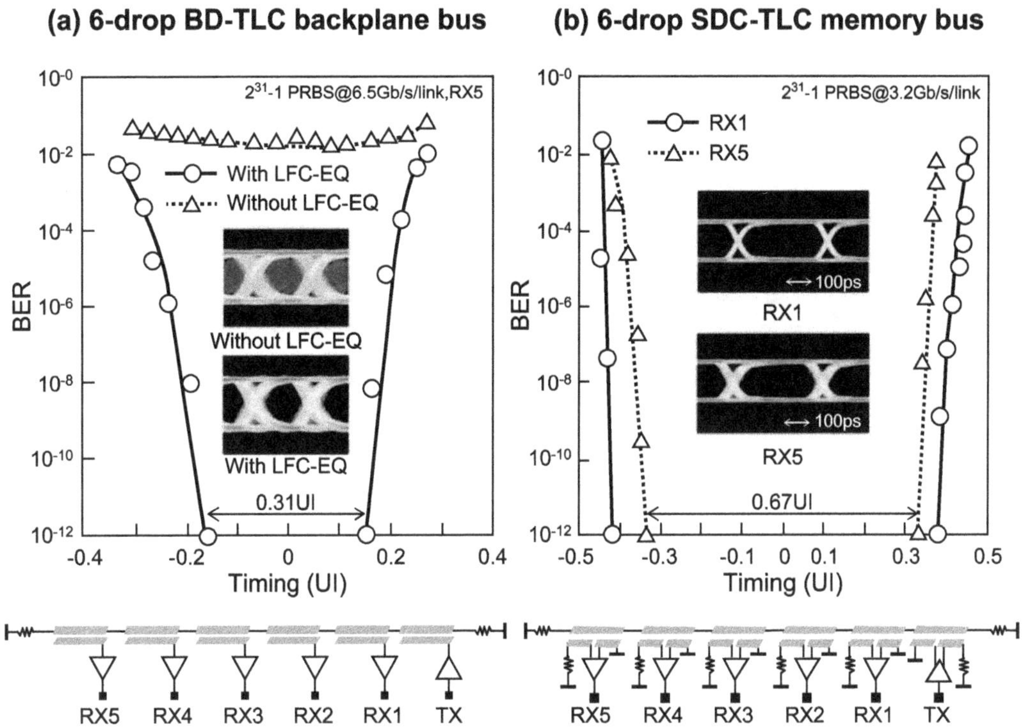

Figure 3.29 Measured BER timing margin of BD-TLC and SDC-TLC buses. © 2015 IEEE. Reprinted, with permission, from [16]

TLC connection is robust against short by water particles or open due to contaminants. Furthermore, it eliminates the high profile and large mounting area required by a mechanical connector that includes a spring mechanism-based receptacle and needs mechanical stability (Figure 3.31). As a result, TLC can enable significant connector miniaturization within the smartphone.

Nevertheless, there are several challenging system-level requirements for the adoption of TLC in a smartphone. The first challenge is footprint. The general interface in a mobile application includes not only high-speed data, but also clock and control signals. It is therefore necessary to implement at least three couplers, one for each class of signal, if the standard TLC design is employed. To minimize the board real estate required for TLC, two small footprint derivatives have been developed: vertical directional TLC (VD-TLC) and twofold TLC (T-TLC). Meanwhile, the use of embedded clock together with CDR can eliminate the coupler for clock to further reduce footprint.

The second challenge is EMC. With the myriad of wireless technologies adopted in the small form factor of a smartphone, including LTE and Wi-Fi that emit signals of up to 30 dBm in magnitude, and GPS, whose receiver has sensitivity down to –130

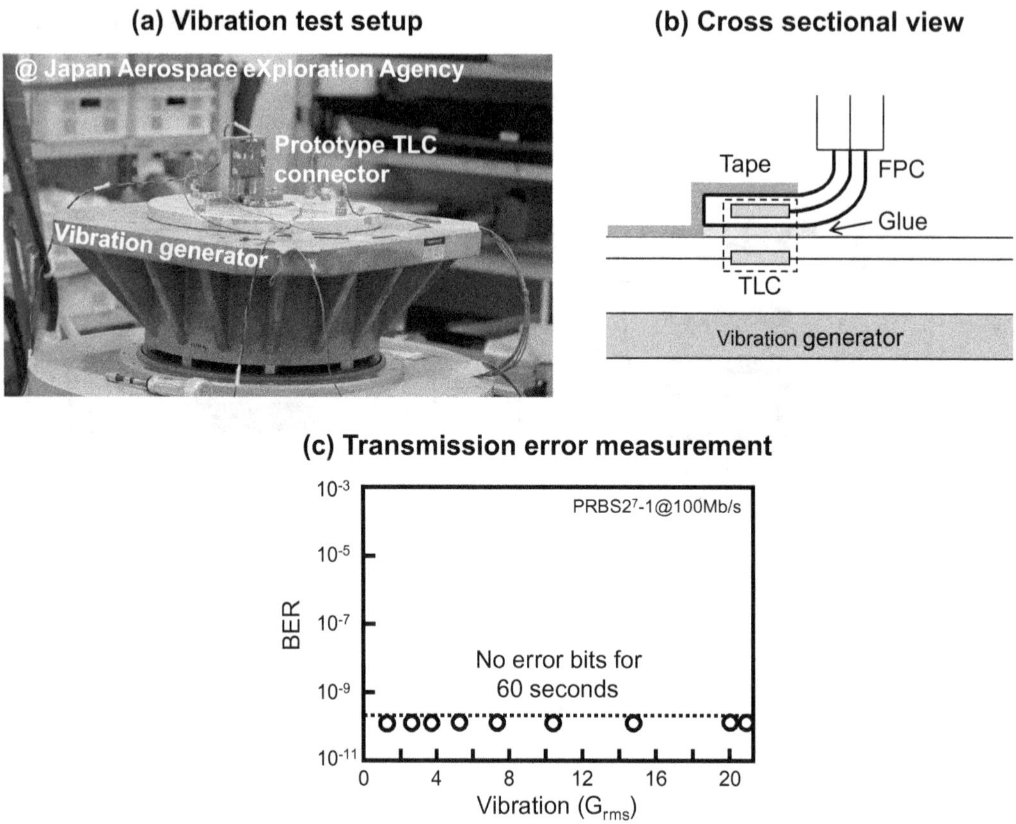

Figure 3.30 Rocket launch vibration test of TLC. (b) and (c) © 2015 IEEE. Reprinted, with permission, from [16]

dBm, any high-performance interface must have both low electromagnetic susceptibility (EMS) and electromagnetic interference (EMI). This is especially challenging for TLC, being a wireless interface. The small signal it uses compared to wired interfaces means it is more susceptible to noise. Furthermore, its coupler is a good antenna both for radiating energy and for picking up radiated noise.

The third challenge is achieving both high data rate and high energy efficiency. The MIPI M-PHY standard that is widely adopted in smartphones requires a data rate of more than 6 Gb/s. At the same time, since battery technology continues to be outpaced by increase in smartphone functionality, there is a constant need for better and better interface energy efficiency.

Taken altogether, smartphone applications require a connector technology to be small footprint, of high EMC immunity (low EMI and EMS), and both high speed and highly energy efficient. In this section, we discuss technologies to optimize the standard TLC design to meet all these requirements.

Table 3.4. Summary of TLC derivatives for multidrop bus.

TLC derivative	Design goal	Design feature	Coupler design	Transceiver design: Transmitter	Transceiver design: Receiver	Interconnect structure	Application
Standard TLC	Contactless connector	Electromagnetically coupled pair of differential transmission lines	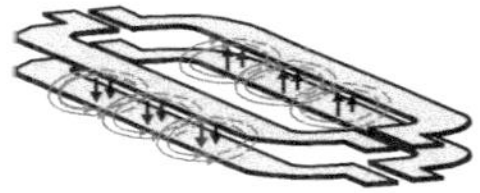	Standard (CML, SST, or pulse)	Standard (hysteresis comparator) 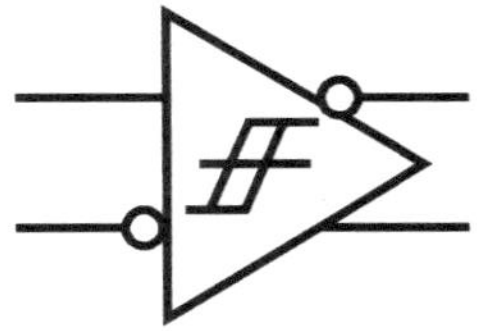		
EE-TLC	Equal energy to all receivers on multidrop bus	Increasing coupling gain down a multidrop bus through decreasing *d*		Standard	Standard 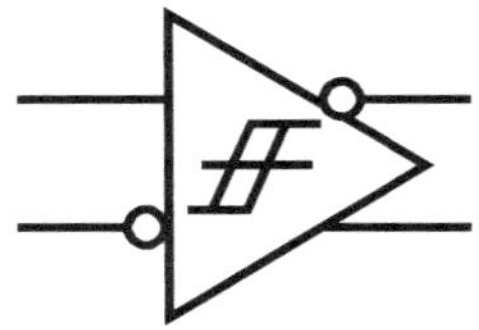	Master–slave, multidrop bus	DIMM memory bus
SDC-TLC	Enabling single-ended signal routing	Single-ended to differential conversion coupler with short circuit termination	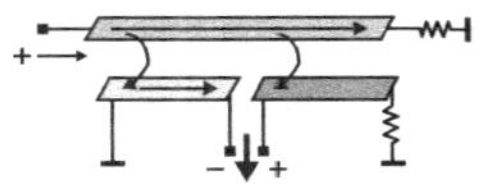	Standard	Standard 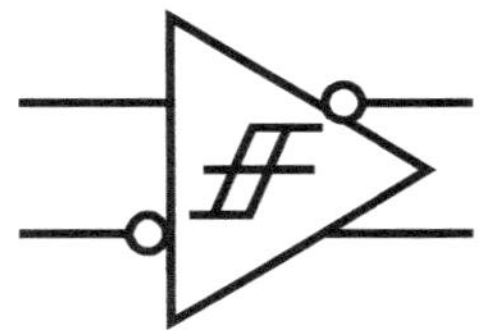	Master–slave, multidrop bus with limited PCB area	DIMM memory bus, DRAM test board
BD-TLC	Bidirectional coupling	Standard TLC with short circuit termination on module	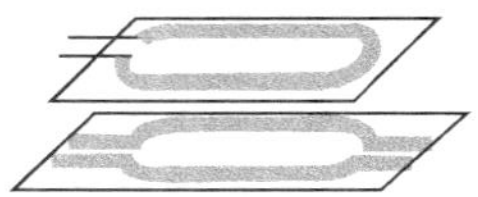	Standard	LFC-EQ + standard 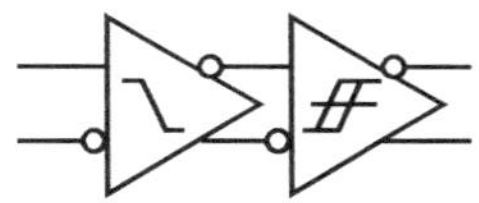	Multimaster, multidrop bus, two-pass TLC	Backplane, automotive network

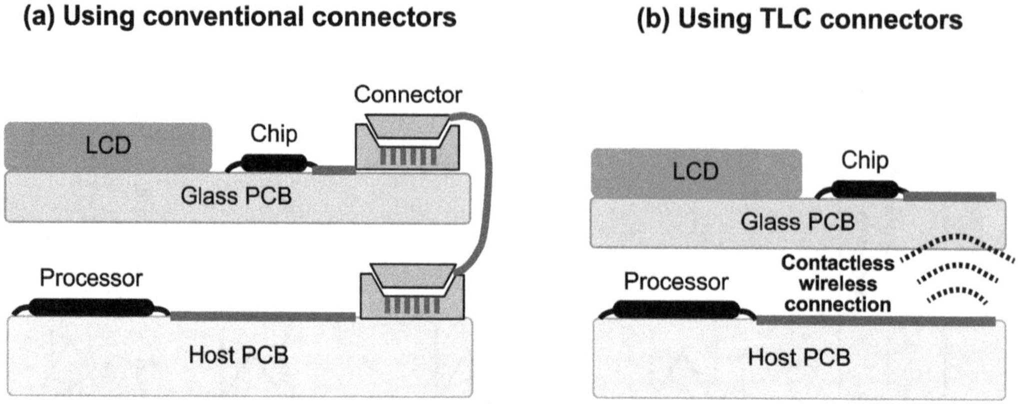

Figure 3.31 TLC vs. mechanical connection in a smartphone.

3.3.1 Small Footprint TLC Derivatives for Smartphone

3.3.1.1 VD-TLC for Simultaneous Dual Channel Transmission

VD-TLC is a coupler that can be used to transmit two signals simultaneously, each in a different direction. In other words, the same physical coupler is used to implement two separate channels with opposite directions of transmission. An example is illustrated in Figure 3.32, where two channels (*Data1* and *Data2*, *Data3* and *Clock*) share one VD-TLC for simultaneous communication in a smartphone LCD interface. Consequently, using VD-TLC, only half the number of couplers is required compared to the standard TLC, resulting in 50% reduction in PCB footprint.

The VD-TLC design uses the same construction as the standard TLC design, with additional design constraints to ensure that unintended cross-channel coupling is negligible (Figure 3.33a). The design constraints can be derived by using the analysis model in Figure 3.33b, where $e^{j\beta z}$ and $e^{-j\beta z}$ represent waves traveling toward Port 3 and Port 4 respectively when a voltage V_{IN} is applied to Port 1. Each wave can be further broken down into odd and even mode components characterized by the odd and even mode impedances, Z_{0o} and Z_{0e}, and phase constants, β_{0o} and β_{0e}, of the VD-TLC. From [18], the transfer function from Port 1 to Port 4, $H_{41}(\omega)$, can then be expressed by

$$H_{41}(\omega) = \frac{Z_L Z_{0e}}{{Z_L}^2 - {Z_{0e}}^2} \times \frac{e^{j\beta_{0e} l}}{\frac{Z_L + Z_{0e}}{Z_L - Z_{0e}} e^{2j\beta_{0e} l} - \frac{Z_L - Z_{0e}}{Z_L + Z_{0e}}} - \frac{Z_L Z_{0o}}{{Z_L}^2 - {Z_{0o}}^2} \times \frac{e^{j\beta_{0o} l}}{\frac{Z_L + Z_{0o}}{Z_L - Z_{0o}} e^{2j\beta_{0o} l} - \frac{Z_L - Z_{0o}}{Z_L + Z_{0o}}}. \tag{3.34}$$

Based on [19], when the VD-TLC is in a homogeneous dielectric medium and terminated with matching impedance Z_L at all ports, the following applies:

$$j\beta_{0e} = j\beta_{0o} = j\beta \tag{3.35}$$

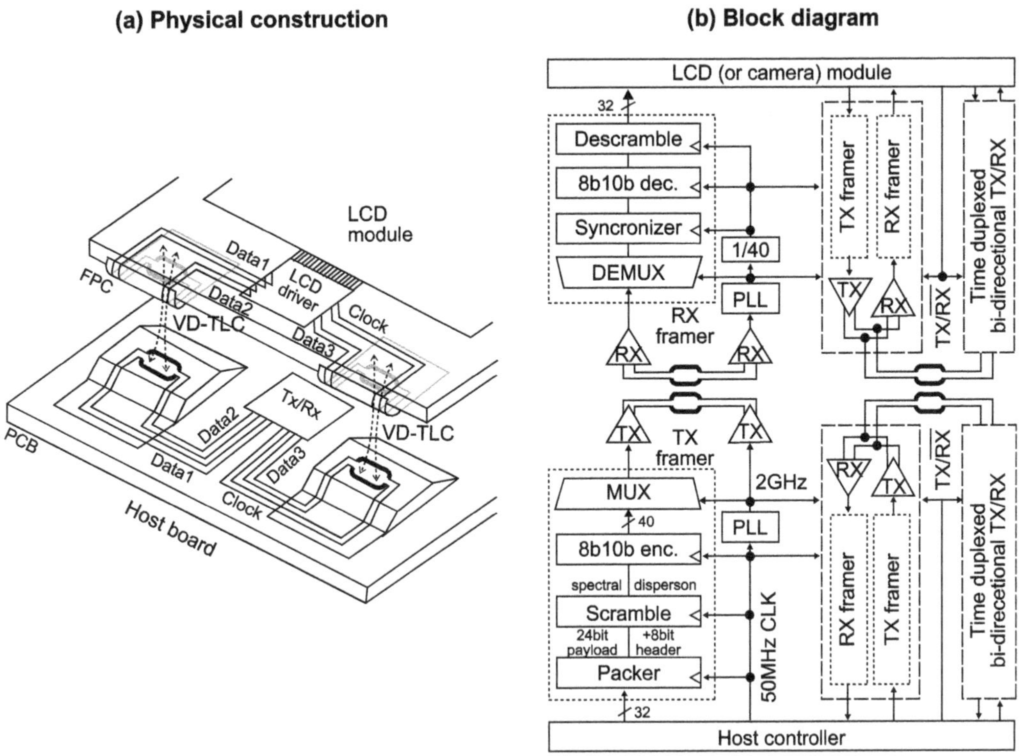

Figure 3.32 Two-channel simultaneous communication using VD-TLC. © 2013 IEEE. Reprinted, with permission, from [17]

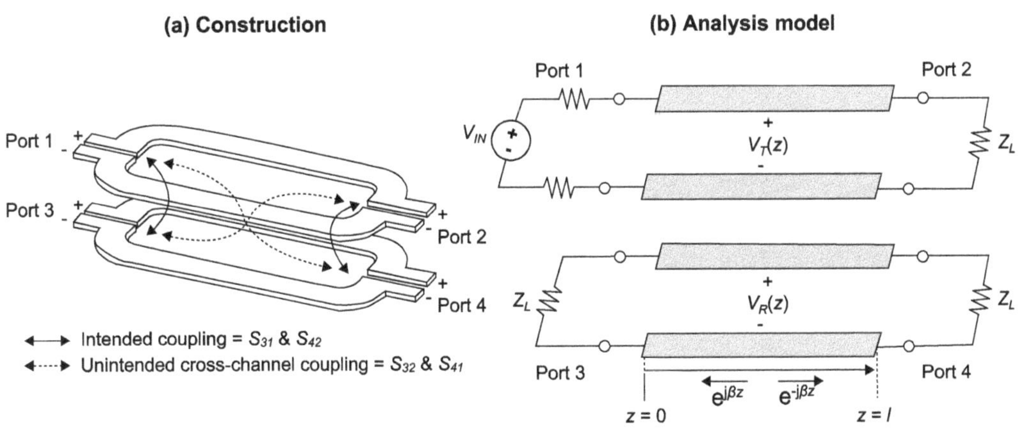

Figure 3.33 VD-TLC construction and analysis model. © 2014 IEEE. Reprinted, with permission, from [18]

$$Z_L = \sqrt{Z_{0e} Z_{0o}}. \tag{3.36}$$

It then follows that

$$\frac{Z_L + Z_{0e}}{Z_L - Z_{0e}} = -\frac{Z_L + Z_{0o}}{Z_L - Z_{0o}} \tag{3.37}$$

$$\frac{Z_L Z_{0e}}{{Z_L}^2 - {Z_{0e}}^2} = -\frac{Z_L Z_{0o}}{{Z_L}^2 - {Z_{0o}}^2}. \tag{3.38}$$

Hence $H_{41}(\omega) = 0$. In other words, when a VD-TLC is in a homogeneous dielectric medium and is impedance matched, there is no cross-channel coupling, making simultaneous dual channel transmission possible. Note that in normal usage the standard TLC design assumes the same conditions. But for VD-TLC, they are strictly enforced, and deviations must be quantified and minimized.

In practice, it is difficult to perfectly match impedances or the dielectric constants of the three dielectric layers in the construction of the VD-TLC. As a result, there is always some residual, unintended cross-channel coupling. If the signal from cross-channel coupling is –20 dB compared to that from intended coupling, then its magnitude is an order of magnitude smaller, and there is sufficient margin in the hysteresis comparator to reject it. Therefore, the design goal of VD-TLC is to suppress cross-channel coupling to –20 dB relative to intended coupling.

Figure 3.34 depicts the simulated cross-channel coupling S_{41} for a VD-TLC design implemented in [18] using FPC. It is more than 20 dB below the intended coupling gain S_{31} for up to 6 GHz.

Figure 3.35 depicts the sensitivity of cross-channel coupling to variations in the spacer (adhesive joining the two FPCs in this case) dielectric constant $\varepsilon_{r.AD}$ and communication distance d, at 2 GHz. Based on Figure 3.34, S_{41}/S_{31} is –29 dB at

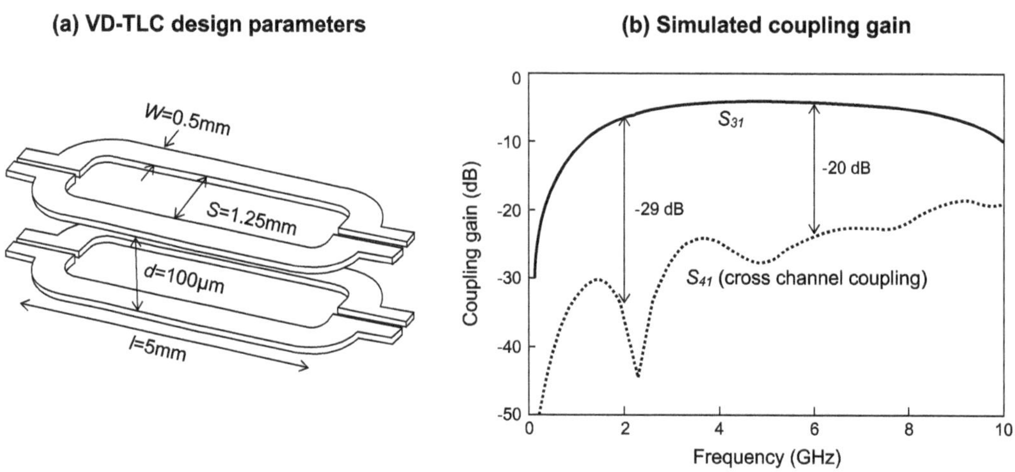

Figure 3.34 Evaluation of VD-TLC cross-channel coupling. © 2014 IEEE. Reprinted, with permission, from [18]

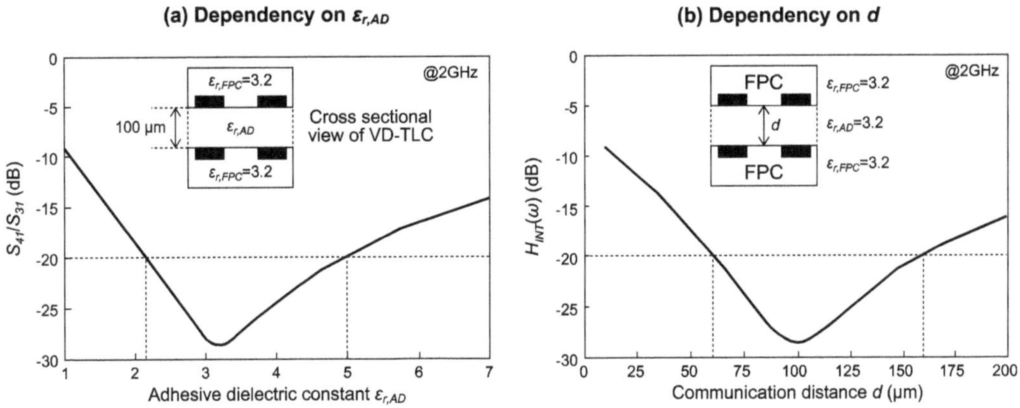

Figure 3.35 Cross-channel coupling sensitivity to design parameter variations. © 2014 IEEE. Reprinted, with permission, from [18]

2 GHz, so there is margin to absorb certain variations in design parameters. Figure 3.35 shows that the ratio stays below –20 dB for $2.2 < \varepsilon_{r,AD} < 4.9$ and $65 < d < 160$ µm, or more than ±30% from their respective nominal values, which should be sufficient to cover normal manufacturing variations.

The VD-TLC transceiver consists of a pulse driver and the standard hysteresis comparator receiver. The pulse driver from Table 3.2 is chosen because VD-TLC is intended for smartphone applications that demand very low power. Furthermore, because a pulse driver consumes power only during a data transition, it has an energy efficiency (pJ/b) that is independent of the data rate, as well as power that scales with the data rate (at fixed supply voltage), as required by the MIPI standard.

A test system built to confirm the VD-TLC design in Figure 3.32 is demonstrated in [18]. The VD-TLC consists of two FPCs joined with adhesive that results in a communication distance $d = 75$ µm and total thickness of 150 µm. The FPCs are made of polyimide with $\varepsilon_r = 3.2$, while the adhesive is made of polyester with $\varepsilon_r = 3.0$, so there is a small mismatch in dielectric constant. Nevertheless, as the measured eye diagrams and BER show, the degradation due to cross-channel coupling is around 3% only at BER = 10^{-12} (Figure 3.36).

3.3.1.2 T-TLC for Extended Communication Distance

Figure 3.37 presents the cross-sectional view of a smartphone where TLC is applied to communicate between different modules across its thickness, including camera, LCD, and other modules. With smartphone thickness being on the order of 10 mm, TLC must support a communication distance on the order of 5 mm.

As explained in Section 3.1, the coupling coefficient of TLC is mainly controlled by the communication distance d and trace width W, or more specifically, the ratio W/d. In other words, as communication distance increases, TLC trace width and hence its footprint must increase accordingly to maintain the coupling strength of the connection. Typically, W/d is chosen to be 0.5, resulting in a coupling gain S_{31} of

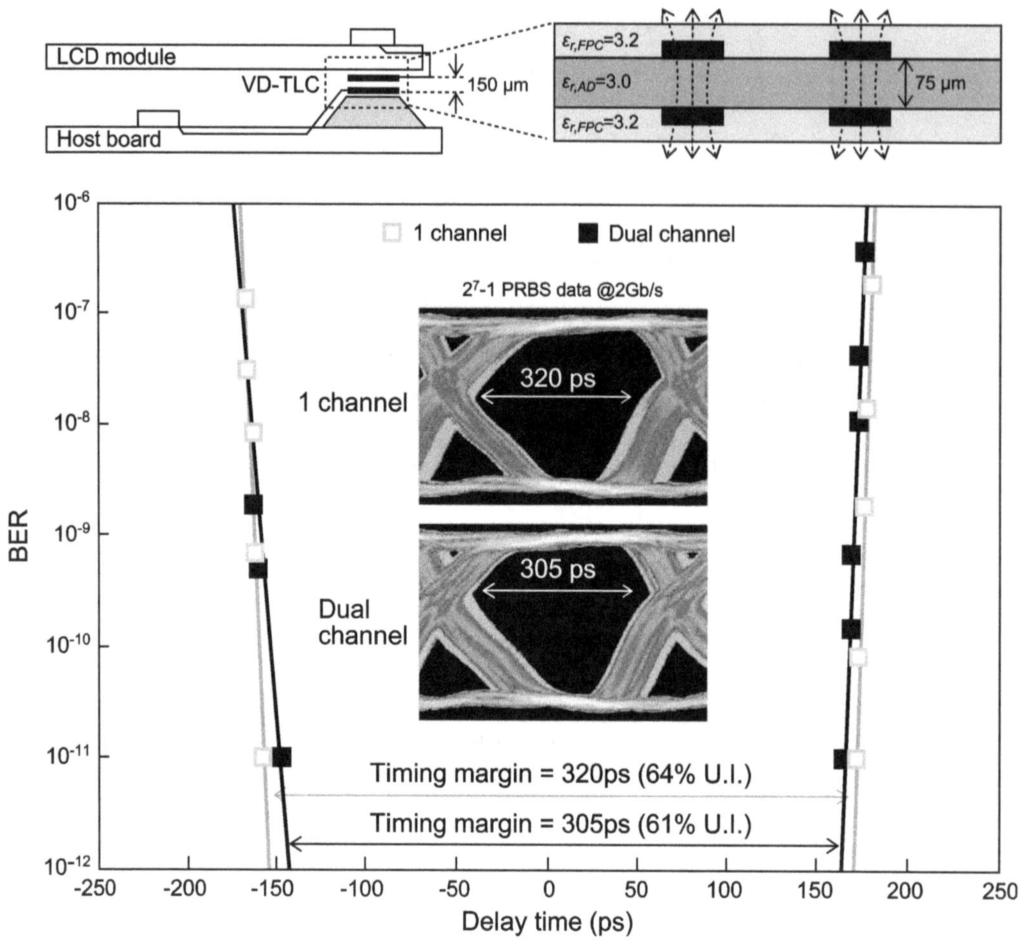

Figure 3.36 Measured eye diagrams and BER of test VD-TLC system. © 2013 IEEE. Reprinted, with permission, from [17]

about –24 dB at 6 GHz (Figure 3.38a). Since the recommended intrapair spacing is $S = 3W$, the total width of a TLC coupler for $d = 5$ mm is $W + 3W + W = 5W = 5*(0.5d) = 12.5$ mm (Figure 3.38b). This is large for smartphone applications, where dimensions of the smallest modules are around 20×20 mm^2.

The TLC footprint can be significantly reduced through two optimizations of the standard structure. First, since there are two electrodes in each coupling terminal of a standard TLC, for P and N polarities respectively of the differential signal, the total terminal width is $5W$. If both P and N polarities of the signal use the same electrode for coupling, the terminal width can be reduced to W. Second, in a standard TLC, in the general case, not all the coupled energy is directed to the receiver. Figure 3.39a illustrates how coupled energy from the transmitter is transferred to the receiver in a standard TLC. Since TLC operates through electromagnetic coupling, the coupled

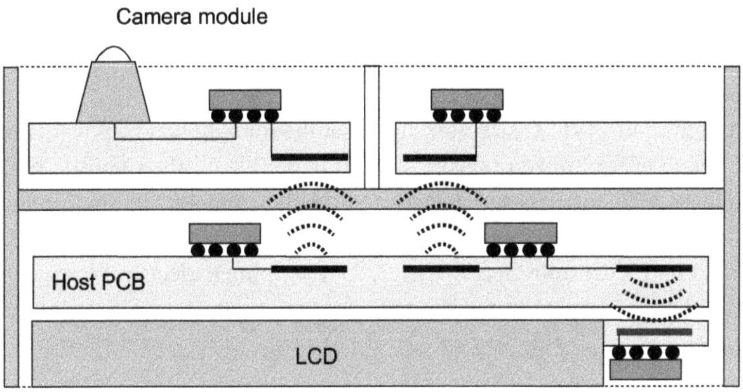

Figure 3.37 Cross-sectional view of TLC in a smartphone. © 2015 IEEE. Reprinted, with permission, from [20]

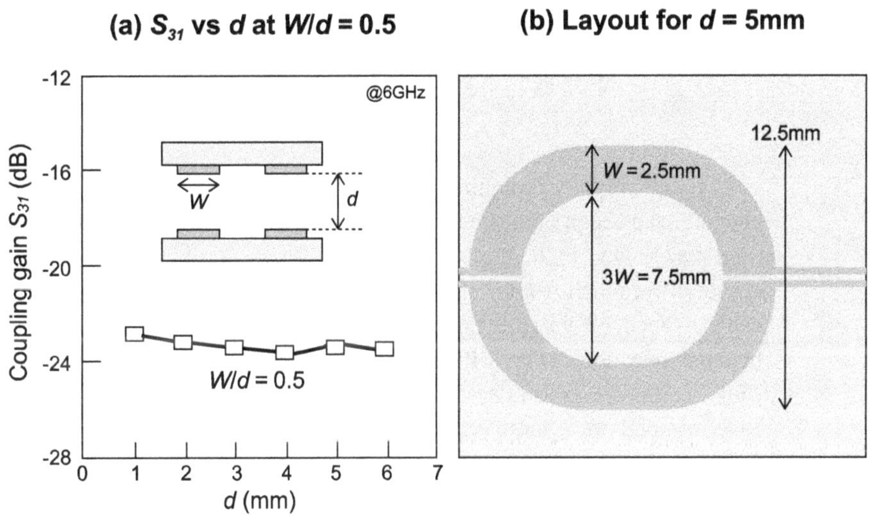

Figure 3.38 Standard TLC for a 5 mm communication distance.

energy travels in both the forward (fwd) or same direction, and backward (bwd) or opposite direction as the transmitter signal. In a standard TLC, only the backward coupled energy is directed to the receiver; any forward coupled energy due to mismatched dielectric constants is absorbed by the far end termination. If the forward coupled energy is also directed to the receiver, the terminal footprint can be reduced for the same S_{31}.

Figure 3.39b depicts the two-fold TLC (T-TLC) structure that implements these two optimizations. Each polarity of the received signal consists of two components: a forward and a backward component, one coupled from the P, the other N polarity of the transmitter signal. Specifically,

$$RXP = bwd\_P + fwd\_N \tag{3.39}$$

$$RXN = fwd\_P + bwd\_N. \tag{3.40}$$

By designing the T-TLC structure such that the forward and backward components of the coupled energy have opposite signs (Figure 3.39c), the opposite polarities of the transmitter differential signal will lead to the two components reinforcing each other, resulting in more energy coupled to the receiver than in a standard TLC.

Based on (3.22) from Section 3.1.2, for the optimal electrical length of $l = \frac{t_r v_p}{2}$, the backward component that is equal to the received pulse magnitude in a standard TLC is given by

$$V_{bwd} = \frac{K_C + K_L}{4} V_{TX}. \tag{3.41}$$

As discussed in Section 3.1, the two components in the forward and backward directions from capacitive coupling are both of the same polarity as the transmitter signal. Meanwhile, the backward component from inductive coupling is of the same polarity while the forward component is of the opposite polarity. Therefore, in contrast to (3.41), the forward component of the coupled energy is given by

$$V_{fwd} = \frac{K_C - K_L}{4} V_{TX}. \tag{3.42}$$

Consequently, by designing the T-TLC to achieve $K_L > K_C$, the forward component will have a sign opposite to that of the backward component. Since the commonly used materials in PCB fabrication are nonferromagnetic, their relative permeabilities do not differ much from each other, making it difficult to adjust K_L. By contrast, K_C can be tuned by varying the dielectric constant of the spacer $\varepsilon_{r,S}$ through proper material selection. This is illustrated in Figure 3.40 for when the PCB is made of FR4. Figure 3.40a shows how the forward waveform flips in sign when $\varepsilon_{r,S}$ is changed from a small to a large value relative to the dielectric constant of FR4, while Figure 3.40b plots the ratio of the peak voltages $V_{fwd,P}/V_{bwd,P}$ as a function of $\varepsilon_{r,S}$. As noted in Section 3.1, $K_C = K_L$ when the spacer has the same dielectric constant as the PCBs, which in this case is 4.4. When $\varepsilon_{r,S}$ falls below 4.4, K_C becomes smaller and V_{fwd} becomes negative.

In the standard TLC, even with a recommended intrapair spacing of $S = 3W$, there is still some intrapair coupling that weakens the interpair coupling. Therefore, the T-TLC structure with a single electrode also has inherently higher coupling due to the lack of intrapair interference.

Figure 3.41a compares S_{31} of T-TLC with that of standard TLC, as a function of frequency, for the same trace width $W = 1.5$ mm, length $l = 4$ mm, and communication distance $d = 5$ mm. At 6 GHz, due to the combined effect of the reinforcing forward coupled energy and elimination of intrapair interference, T-TLC has 9 dB more coupling than standard TLC. To achieve comparable S_{31}, the standard TLC trace width needs to be increased to 2.5 mm, resulting in $W/d = 0.5$ for standard TLC vs. 0.3 for T-TLC (Figure 3.41b). Consequently, the use of T-TLC enables a coupler width that is $1.5/(2.5 * 5) = 1/8.3\times$ that of standard TLC for comparable S_{31}.

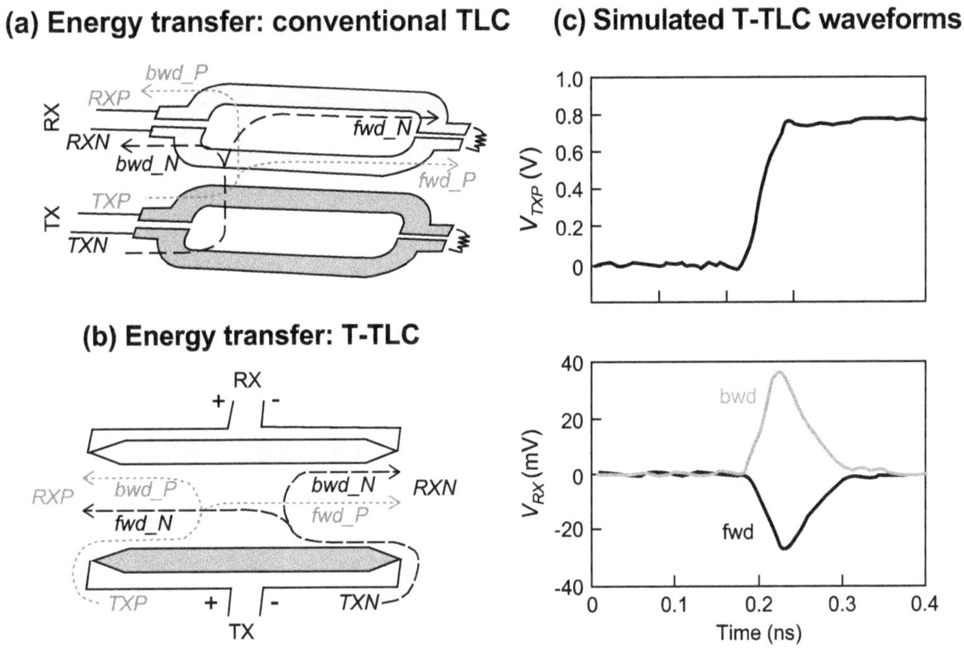

Figure 3.39 Comparison of T-TLC vs. TLC in energy transfer. © 2016 IEEE. Reprinted, with permission, from [21]

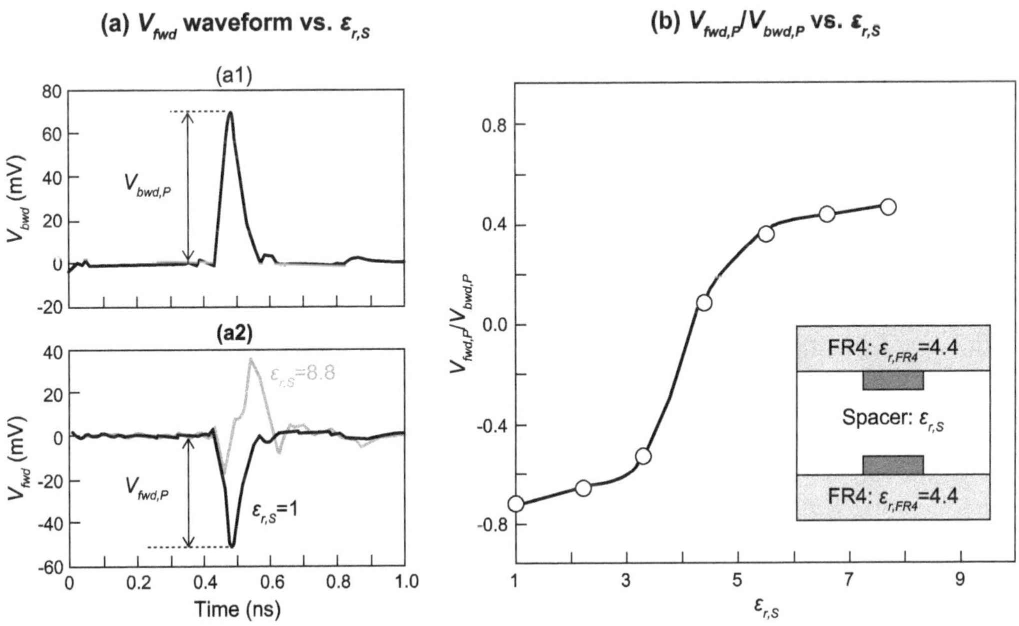

Figure 3.40 Forward coupling vs. dielectric constant of spacer. © 2016 IEEE. Reprinted, with permission, from [21]

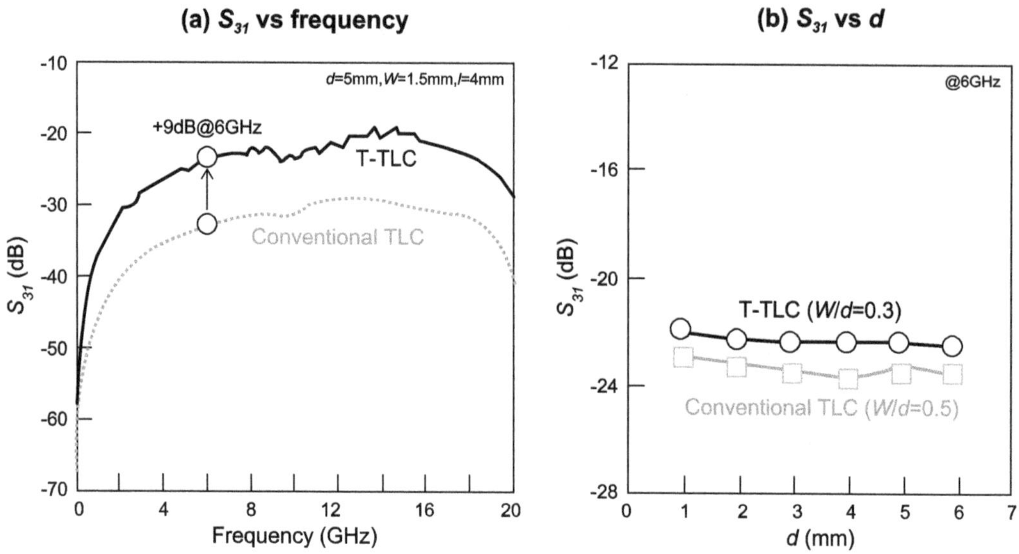

Figure 3.41 Comparison of T-TLC vs. standard TLC in S_{31}. © 2015 IEEE. Reprinted, with permission, from [20]

Furthermore, termination resistors are eliminated, resulting in additional reduction in board layout area. Note that there is no change to the TLC bandwidth since it is determined by the trace length and independent of trace width.

Since T-TLC is a single-ended transmission line, it is more susceptible to crosstalk noise than the differential transmission line used in the standard TLC. Figure 3.42b plots the crosstalk from a single-ended, 50 Ω, coplanar aggressor transmission line as a function of its distance from T-TLC, for both orthogonal and parallel orientation as shown in Figure 3.42a. For an aggressor amplitude of 1 Vpp, to limit the crosstalk to less than 1 mVpp, the required separation is > 2 mm from an orthogonal aggressor and > 4.5 mm from a parallel aggressor. Since $W = 1.5$ mm, this corresponds to a separation of $1.3W$ and $3W$ respectively.

3.3.2 Smartphone TLC Transceiver with High EMC Immunity

Figure 3.43 depicts the block diagram of a TLC transceiver with high EMC immunity that is used in conjunction with a T-TLC prototype for smartphone in [21]. It consists of a biphase pulse transmitter and a synchronous CDR receiver with a noise rejection digital filter. The CDR circuit includes an edge counter and an injection-locked array oscillator (ILAO) to achieve high noise immunity and accurate clock recovery.

The simulated transceiver waveforms for a 65 nm process are plotted in Figure 3.44. In the biphase pulse transmitter, current flows only when clock is high. Therefore, within every clock cycle, when clock is high, a data pulse is generated that is positive if *Txdata* is high, and negative otherwise. This in turn induces a positive

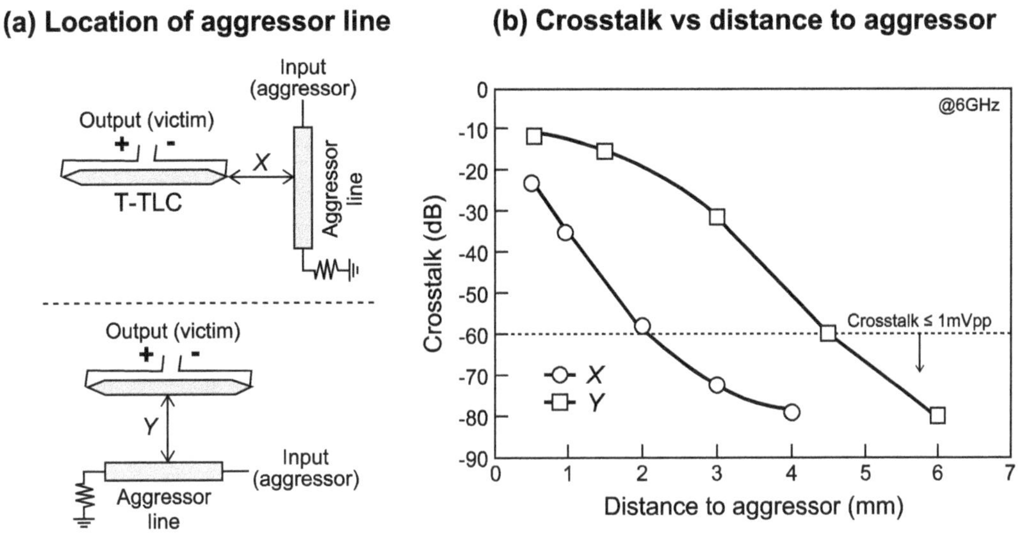

Figure 3.42 Crosstalk from a single-ended, coplanar aggressor line to T-TLC vs. distance. © 2016 IEEE. Reprinted, with permission, from [21]

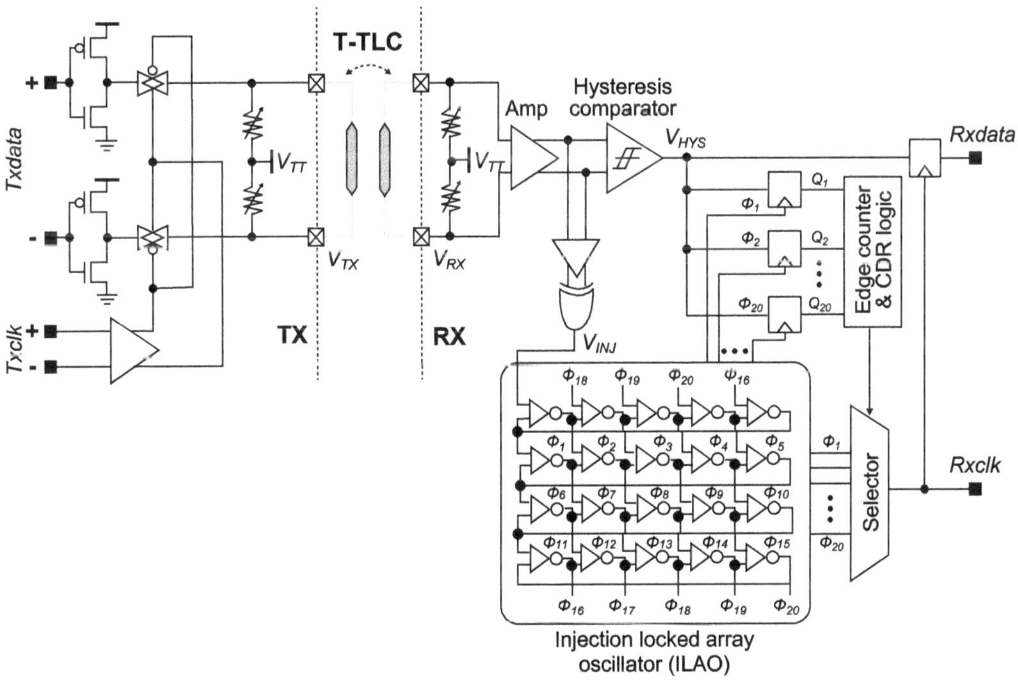

Figure 3.43 TLC transceiver with high EMC immunity. © 2015 IEEE. Reprinted, with permission, from [20]

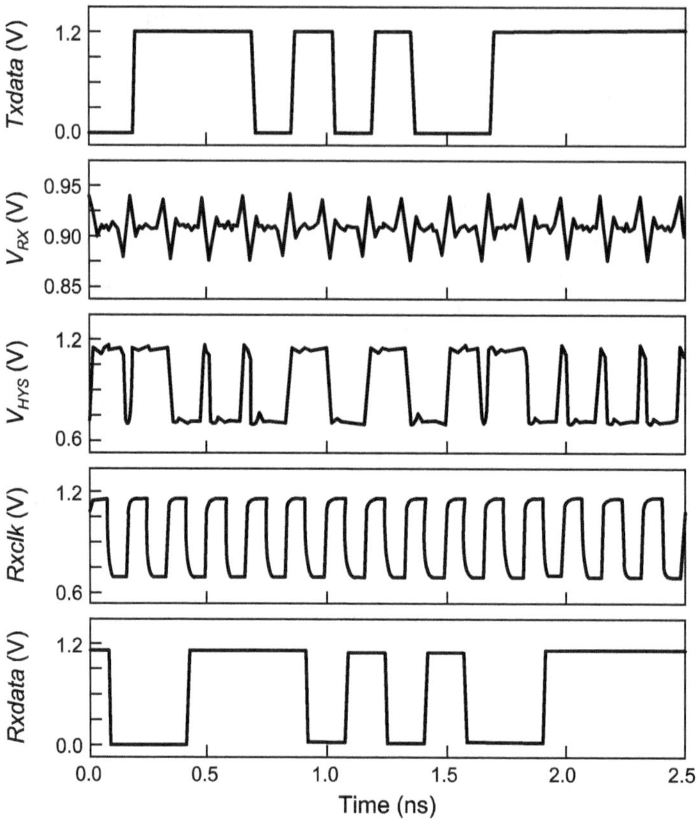

Figure 3.44 Simulated waveforms of biphase transceiver. © 2016 IEEE. Reprinted, with permission, from [21]

pulse followed by a negative pulse in the receiver when *Txdata* is high and a dual pulse of opposite polarity otherwise (see V_{RX} in Figure 3.44). This is unlike the non-return-to-zero (NRZ) signaling used in the standard TLC transmitter, which generates a pulse only when there is a data transition. In other words, the biphase transmitter encodes an NRZ signal into a biphase signal. As a result, the radiated spectrum is twice as wide with reduced magnitude. This is illustrated in Figure 3.45, which compares measured signal power spectrum between an NRZ signal and an equivalent biphase signal of the same signal amplitude of 1 Vpp and at the same data rate of 6 Gb/s. The comparison shows a 5.9 dBm reduction in magnitude at the GPS frequency of 1.575 GHz. The drawback of the biphase transmitter is that its pulse width is twice as wide and so the maximum data rate is half that of NRZ signaling. Nevertheless, with a wide bandwidth of up to 12 Gb/s, a data rate of 6 Gb/s is still achievable.

In addition to lower EMI, the biphase transmitter consumes less power for two reasons. Since there is no current flow during half of the clock cycle when clock is low, the driver stage consumes half the power as the standard CML driver. Moreover, it enables the use of low-power CMOS digital circuits in the predriver, further

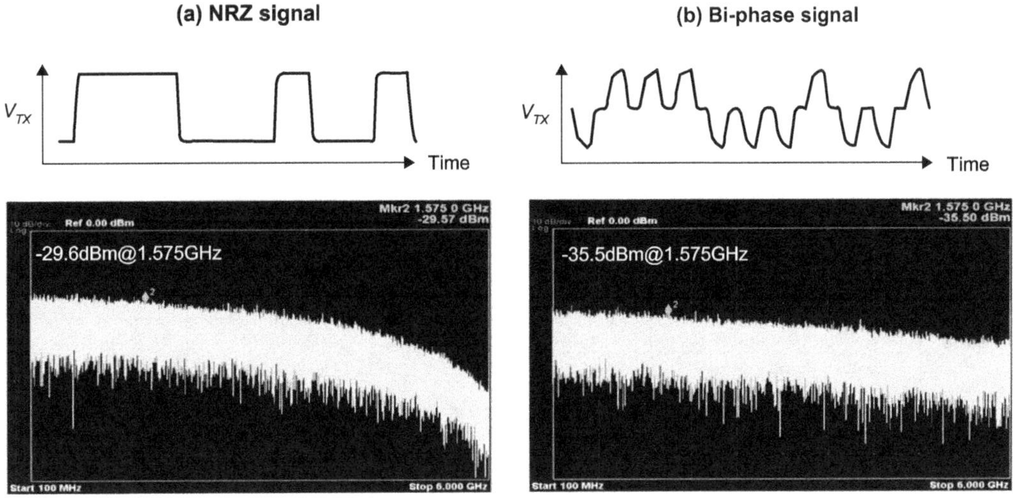

Figure 3.45 Comparison of measured signal power spectrum. © 2016 IEEE. Reprinted, with permission, from [21]

reducing power compared to a CML predriver. Consequently, while the biphase transmitter requires additional circuit elements on the clock path and hence a higher gate count, it reduces power consumption by more than 40% compared to the CML transmitter that consumes constant power.

The synchronous CDR receiver in Figure 3.43 implements multiple measures to improve noise immunity. The amplifier inserted in front of the hysteresis comparator boosts the magnitude of the received signal to improve the comparator threshold margin. In [21], a 6 dB gain is used to enable the comparator to detect very small signals. In addition, since output of the hysteresis comparator, V_{HYS}, is sampled synchronously using the received clock extracted by the CDR circuit, noise that occurs outside of the sampling window can be safely ignored. Furthermore, because pulses are received at every clock cycle, the receiver is not subject to error propagation, unlike NRZ signaling.

3.3.2.1 Edge-Counting CDR for High Noise Immunity

The challenge in using a synchronous CDR receiver is to correctly extract timing information from a noisy signal. The use of CDR helps reduce the footprint of the TLC connector for smartphone applications by eliminating the need for a dedicated coupler for clock. In a conventional CDR solution, an edge detector is used to detect data transitions in the received signal to extract timing information. While the edge detector expects only one data edge in each clock cycle, noise in the received signal can create additional edges, causing an edge detector to lock incorrectly, resulting in timing errors.

To correctly extract timing information, an edge-counting CDR circuit is employed to oversample the noisy data (Figure 3.46). When more than one edge is detected

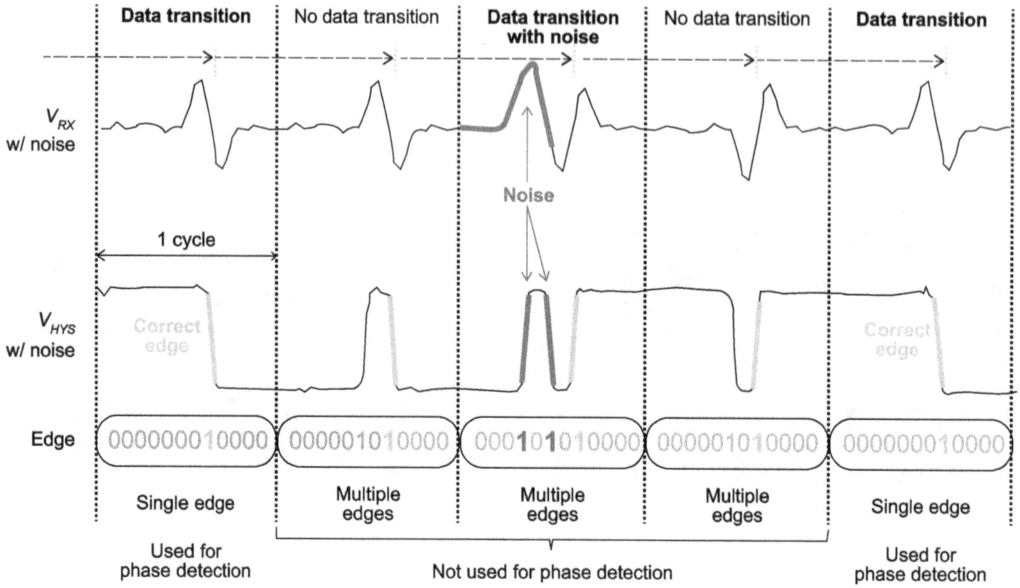

Figure 3.46 Phase detection using an edge counter. © 2016 IEEE. Reprinted, with permission, from [21]

within a clock cycle, there is the possibility of noise and the edges are not used for phase detection. Conversely, when only one edge is detected, the signal is considered clean and the edge is used for phase detection. Within a biphase transmitter, there are two edges within a clock cycle if there is no data transition in the original NRZ data. Therefore, phase detection occurs in a clock cycle only when there is a data transition in the original data, and when there is no noise causing additional edges.

The optimal sampling timing occurs right after a data edge. Hence, the optimal clock phase that follows a data transition is chosen from a multiphase receiving clock circuit to recover the clock for sampling *Rxdata*.

The typical solution for generating a multiphase clock for oversampling utilizes a delay locked loop (DLL) [22]. However, the number of phases that can be generated using such a solution is limited by the gate delay of each delay cell. For instance, given a simulated gate delay of around 30 ps for a 65 nm current-starved delay cell, the maximum number of phases that can be generated at 6 GHz is 5. With such coarse resolution, it may not be possible to detect narrow noise pulses, resulting in locking to the incorrect phase.

Instead of a DLL, the receiver in Figure 3.43 employs an array oscillator for phase generation. It consists of an interconnected two-dimensional array of ring oscillators that oscillate to interpolate phases [23]. As a result, a large number of phases can easily be generated at 6 GHz. As shown in Figure 3.47, each oscillator in the array is implemented as a dual-input, current-starved delay buffer. The array in 65 nm process is shown in simulation to be able to generate 20 phases, resulting in four times better resolution than a DLL-based solution.

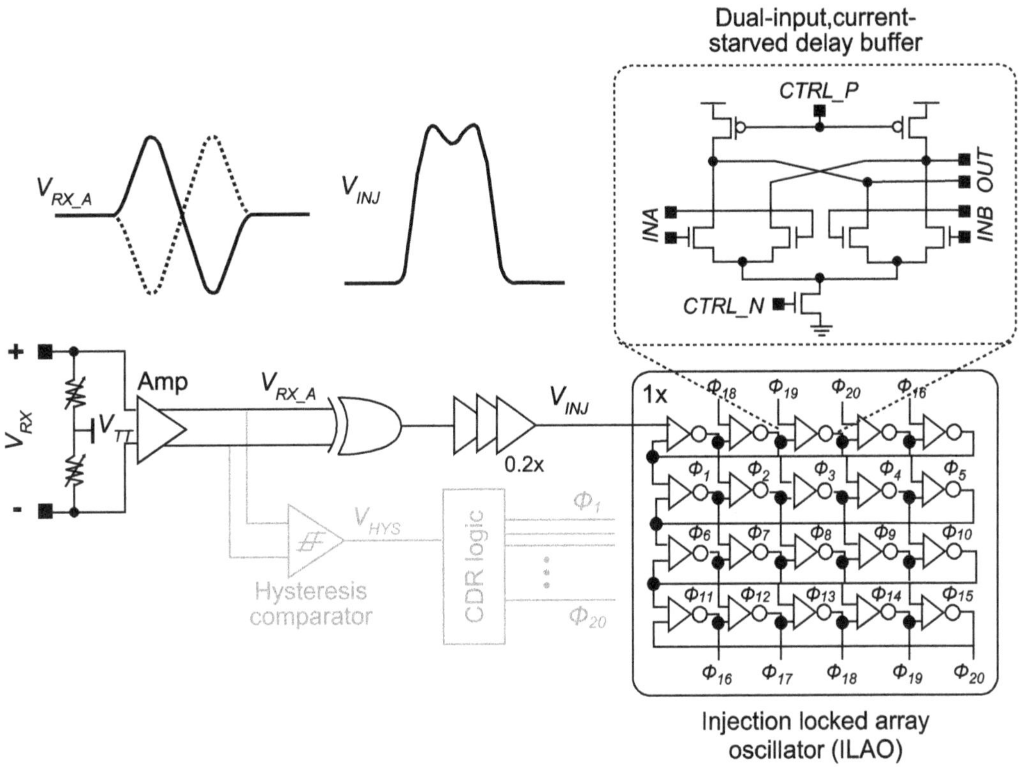

Figure 3.47 Injection-locked array oscillator block diagram. © 2016 IEEE. Reprinted, with permission, from [21]

However, the output frequency from the array oscillator may not be fully matched with that of data. Thus, the implementation in Figure 3.47 utilizes an ILAO that employs injection locking to remove such frequency mismatch. Since the biphase transmitter transmits a pulse in every clock cycle with regularity, the received signal can be used in injection locking to recover the correct frequency. Nevertheless, since the phase of the received biphase pulse can jump by 180 degrees depending on the polarity of the original data, it cannot be directly used for injection. Instead, it is necessary to first pass the received differential signal through an XOR gate to generate a single-phase pulse at each clock cycle that is independent of the polarity of the original data. This single-phase pulse can then be injected into the ILAO to recover a clock with matching frequency.

If there is noise in the received signal, such as noise from Wi-Fi signal at 2.4 GHz, the width of the injected pulse may fluctuate, resulting in high-frequency jitter in the recovered clock. However, since the ILAO has the same low pass characteristics as conventional injection-locked oscillators [24], it can filter out such high-frequency jitter. The bandwidth of the ILAO depends on the injection strength, where a lower injection strength results in a lower cutoff frequency.

Since the lock range of injection locking is narrow, it is necessary to ensure that the free-running frequency of the ILAO is close to the signal frequency. This can be achieved by either calibrating the free-running frequency of the ILAO using an external device, or by placing a reference phase locked loop (PLL) nearby to adjust the frequency automatically [25].

3.3.3 Smartphone TLC Experimental Results

Figure 3.48a depicts the test setup of a T-TLC prototype with a 5 mm communication distance. The T-TLC electrode measures 4 mm in length l and 1.5 mm in width W. It is fabricated on the back side of a four-layer PCB. The reference ground plane is located in the top layer with a separation of 1.5 mm from the electrode. The measured S_{31} shown in Figure 3.48b reveals a wideband frequency response up to and beyond 20 GHz. Furthermore, PRBS2^{31}–1 data can be transmitted with good eye opening at 12 Gb/s as demonstrated in Figure 3.48c.

The high EMC immunity transceiver from Section 3.3.2 is fabricated in a 65 nm process and tested for EMC. Driving the T-TLC prototype at 6 Gb/s and with a supply voltage of 1.1V, power consumption is 24.6 mW for the transmitter, 3.0 mW for the receiver, and 8.5 mW for CDR, for a total energy efficiency of 6 pJ/b. The test setup is depicted in Figure 3.49. The transceiver is connected to the T-TLC prototype using a shielded SMA cable and placed far from the T-TLC to avoid interference with the measurement. The setup is placed in an anechoic chamber together with an EM probe whose placement around the T-TLC is controlled by a robotic arm with spatial resolution of 0.5 mm. EMI measurement is done using an external spectrum analyzer that is connected to the EM probe. On the other hand, EMS measurement is done by driving the EM probe with an external noise source while measuring T-TLC

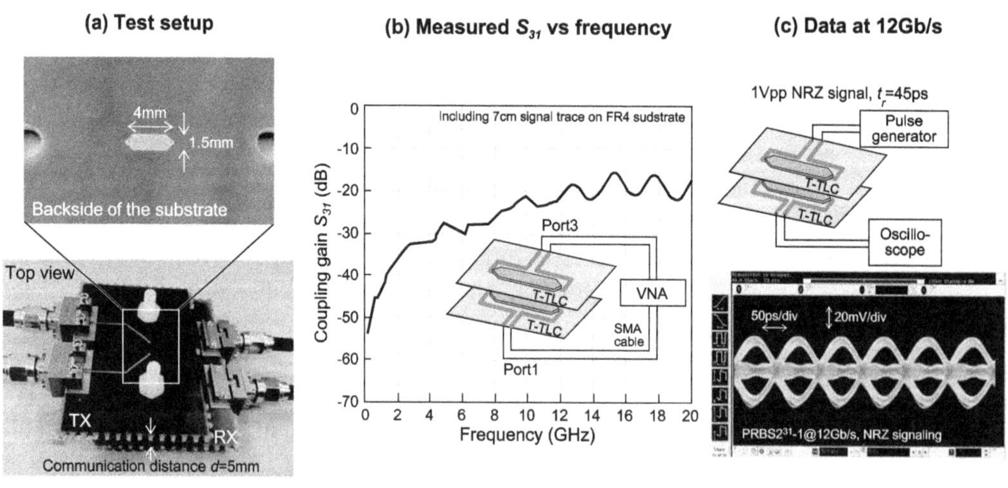

Figure 3.48 T-TLC test setup and measured frequency and time domain responses. © 2016 IEEE. Reprinted, with permission, from [21]

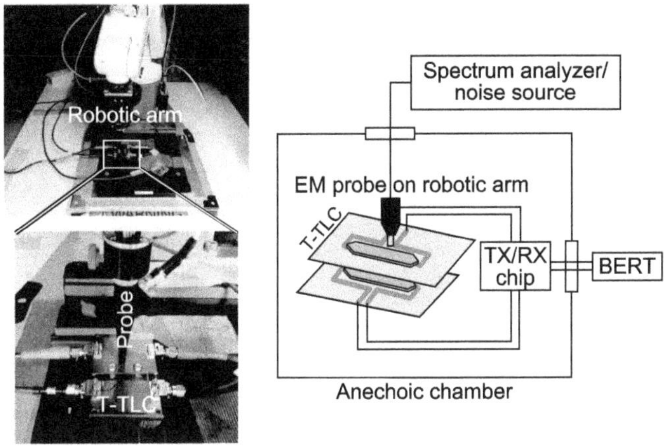

Figure 3.49 EMC test setup for T-TLC driven by proposed transceiver.

transmission BER with an external BERT connected to the test chip, using a PRBS $2^{31}-1$ data pattern.

As noted in [26], a GPS receiver is capable of detecting a –130 dBm signal when the noise level is lower than –111 dBm at the GPS antenna. Therefore, any noise source of less than –121 dBm should cause negligible degradation in the GPS performance. This is in agreement with [27], which notes that when the GPS module is isolated with a –30 dB shield, it is sufficient to suppress radiation noise in the GPS band to below –91 dBm. The measured EMI at the GPS frequency of 1.575 GHZ from the T-TLC prototype in Figure 3.50a shows that at a distance of 10 mm, in *x*-, *y*-, or *z*-direction, radiation in the GPS band from the T-TLC interface drops below –91 dBm and can thus be considered to not interfere with the operation of a shielded GPS module.

Figure 3.50b plots the measured EMS in terms of BER of the T-TLC interface when subject to a 30 dBm noise at 0.7 GHz (LTE frequency) and 2.4 GHz (Wi-Fi frequency) respectively as a function of vertical distance between the noise source and the interface. Horizontally the noise source is placed at $X = 1.5$ mm and $Y = 0$ mm near the edge of the coupler that is farthest from the center point of the differential transmission line pair where the worst differential noise is expected. The results show that at a vertical distance of more than 2 mm, the T-TLC interface can operate at 6 Gb/s with a BER $<10^{-12}$ in the presence of 30 dBm LTE or Wi-Fi noise.

EMS of the transceiver CDR to Wi-Fi is also evaluated by placing a 30 dBm, 2.4 GHz noise source at a vertical distance of 2 mm from the T-TLC interface. BER is measured by varying *Txclk* phase while holding the receiver clock phase constant. The recovered clock and data waveforms in the presence of noise are shown in Figure 3.51a. The peak-to-peak jitter of the recovered clock increases by only 5.9–14.4 ps, while the RMS jitter is 2.4 ps. Furthermore, the lock range of the ILAO is 4.1% around the center frequency of 6 GHz. Meanwhile, timing margin at BER $= 10^{-12}$ degrades only slightly by 0.02–0.2 UI.

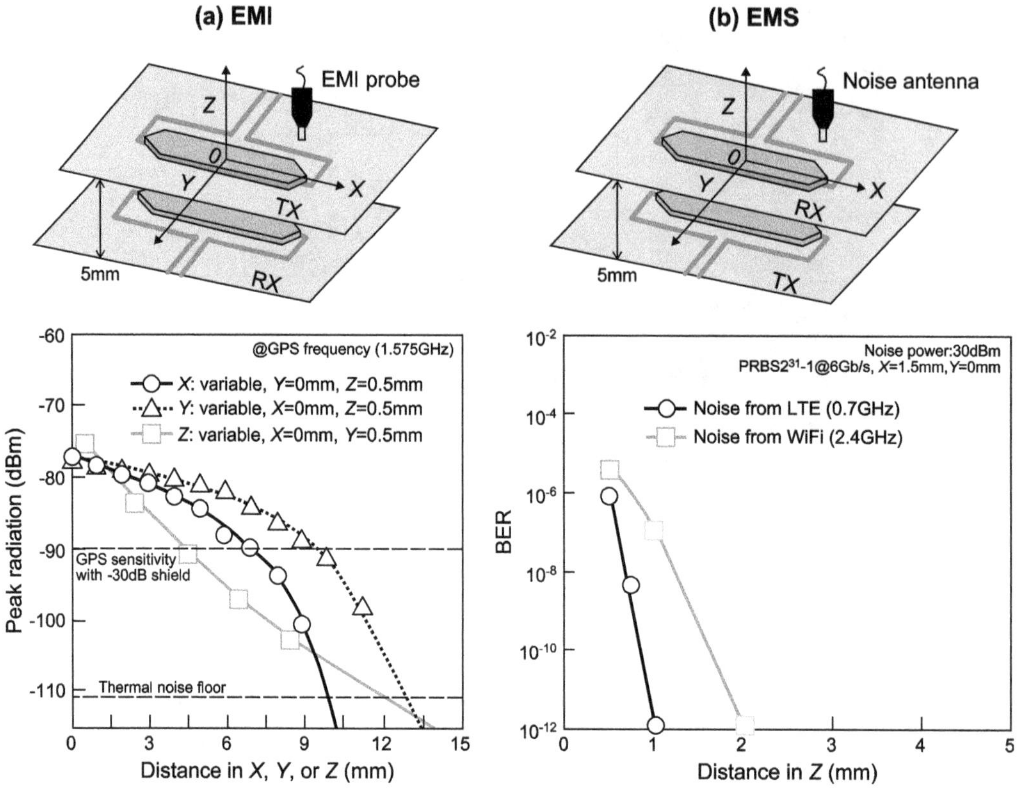

Figure 3.50 EMC measurement results. © 2015 IEEE. Reprinted, with permission, from [20]

3.3.4 Closing Thoughts

TLC holds promise in miniaturizing the connectors in a smartphone while enhancing robustness due to the contactless nature of its interface. To minimize its footprint, the vertical directional TLC (VD-TLC) derivative can be used to reduce the number of couplers by 50% by using the same coupler for simultaneous dual channel transmission. Alternatively, the twofold TLC (T-TLC) derivative can be used to reduce the coupler footprint to 1/8.3× that of the standard TLC while supporting a long communication distance on the order of 5 mm. Comparison of the two derivatives with the standard TLC is summarized in Table 3.5

Meanwhile, a biphase transmitter together with a synchronous hysteresis comparator receiver that uses edge counting and an injection-locked array oscillator for CDR can be used to not only improve EMC but also reduce power. Initial measurements show both low EMI to GPS and low EMS from LTE and Wi-Fi. The use of CDR also enables the clock coupler to be eliminated to further reduce the connector footprint.

The combination of T-TLC and high EMC immunity transceiver results in an attractive wireless interface solution for smartphone applications in terms of footprint,

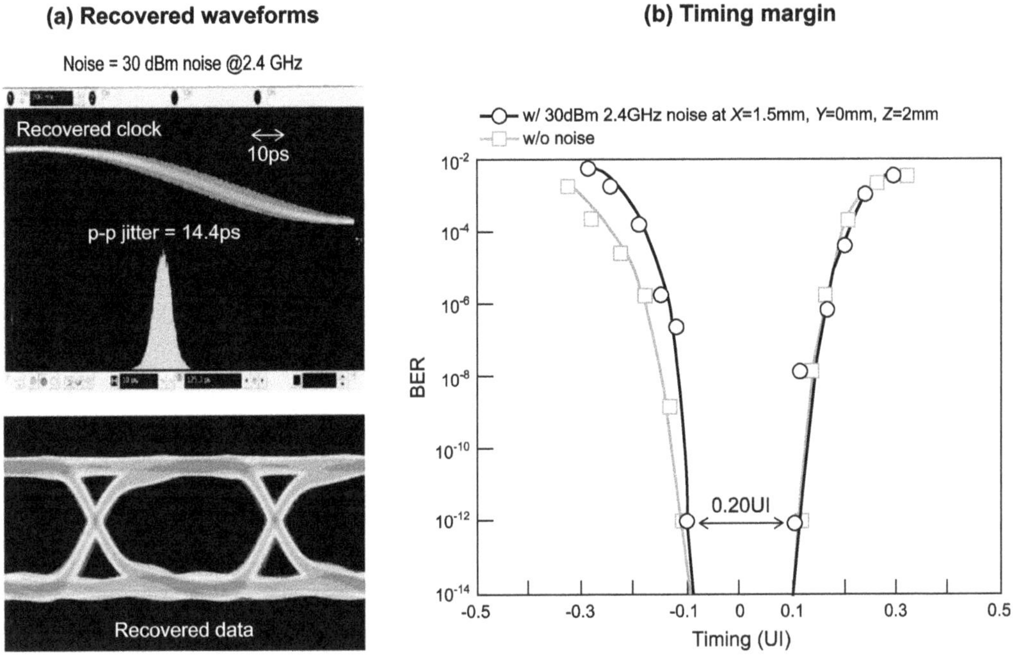

Figure 3.51 Recovered waveforms and timing margin in the presence of Wi-Fi noise. © 2015 IEEE. Reprinted, with permission, from [20]

communication distance, power, and data rate. Figure 3.52 compares this solution from 2016 to six other published wireless solutions from the 2010 to 2013 time frame [28–33]. There was only one other solution capable of communicating across a 5 mm distance, but its coupler size is 24 times larger. Meanwhile, while there are three other solutions capable of running at ≥6 Gb/s, their power consumptions are more than twice as high.

Nevertheless, there are wired solutions with better energy efficiency. As an example, the solution in [34] from 2015 achieves 0.8 pJ/b at a data rate of 8 Gb/s, which is 7.5× better in energy efficiency. The faster process (22 vs. 65 nm) does not fully account for the difference. Energy efficiency of the T-TLC solution can be traded off against other performance metrics. For instance, simulation shows that energy efficiency can be improved by 1.7× by increasing the coupler width from 4 to 6 mm, resulting in a 50% larger footprint. Therefore, the solution can be further optimized based on both the design constraints and performance requirements of the particular target system.

3.4 TLC for Automotive LAN

The rapid development of both the electric vehicle and automated driving has resulted in the proliferation of electronic components in the automotive vehicle. However, the

Table 3.5. Summary of TLC derivatives for smartphone.

TLC derivative	Design goal	Design feature	Coupler design	Transceiver Design: Transmitter	Transceiver Design: Receiver	Interconnect structure
Standard TLC	Contactless connector	Electromagnetically coupled pair of differential transmission lines	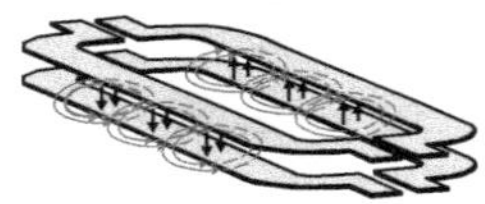	Standard (CML, SST, or pulse)	Standard (hysteresis comparator) 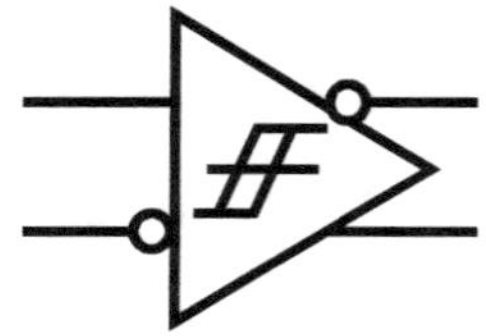	
VD-TLC	Simultaneous dual channel transmission	Matching dielectric constants and impedances	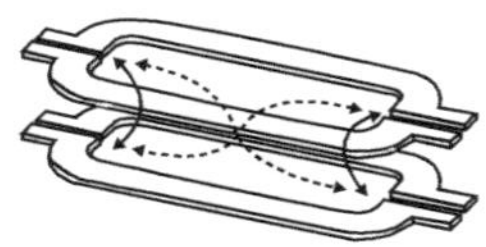	Pulse	Standard 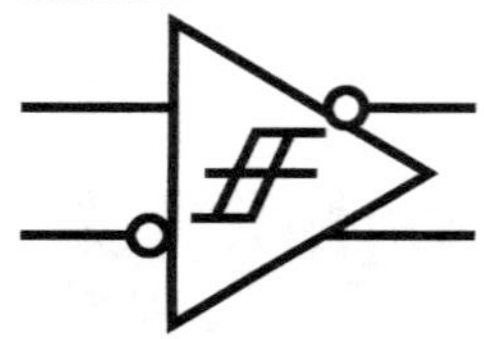	Multilink
T-TLC	Minimum footprint	Single-electrode coupler with $K_C < K_L$ (spacer with small ε_r)	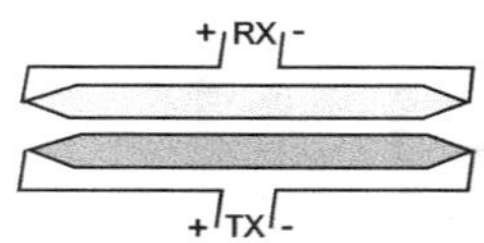	Biphase	Synchronous CDR 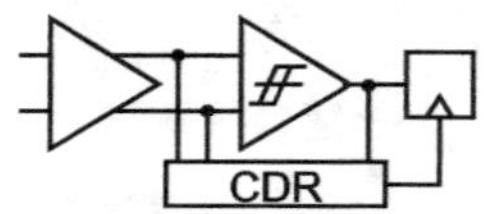	Large d

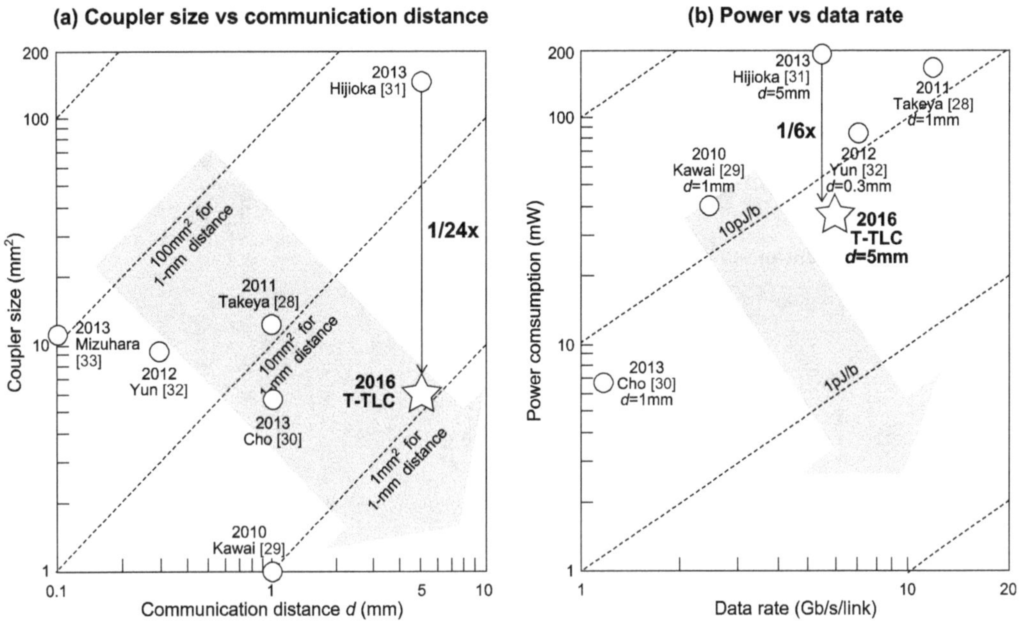

Figure 3.52 Advanced wireless interface solutions from 2010–2016. © 2015 IEEE. Reprinted, with permission, from [20]

automotive LAN that connects these components together is facing two mounting challenges. The first challenge is the increasing weight of wiring, which negatively impacts gas mileage. The demand for ever higher performance and increasing functionality of automotive electronic equipment results in exponential growth in the number of electrical control units (ECUs), sensors, and other electronic components. By 2015, the number of ECUs per vehicle had exceeded 100, and the weight of wiring harness connecting them together exceeded 30 kg [35]. One reason for the increasing weight is the connector. With the conventional mechanical connector, intense vibration within the vehicle can disengage a connector and lead to communication failure. To prevent such disengagement, connectors are heavily protected and firmly attached to the body of the vehicle. The protection results in large and heavy connectors that are aggregated in a junction box to simplify the task of providing mechanical stability. Consequently, even the wiring between two adjacent ECUs or sensors must be routed through the junction box, resulting in long and heavy wiring (Figure 3.53a and b).

The second mounting challenge faced by the conventional automotive LAN is the difficulty in raising the data rate to keep up with the ever-increasing data traffic. While the traditional automotive controller area network (CAN) operates at less than 1 Mb/s, by 2011 the highest data rate demanded of automotive LAN had reached 100 Mb/s [37]. However, since the complex wiring inside the vehicle presents a harsh operating environment to electrical signals, it is challenging to operate at more than a few Mb/s. In order to minimize wiring weight, multidrop is the preferred bus topology [38–39].

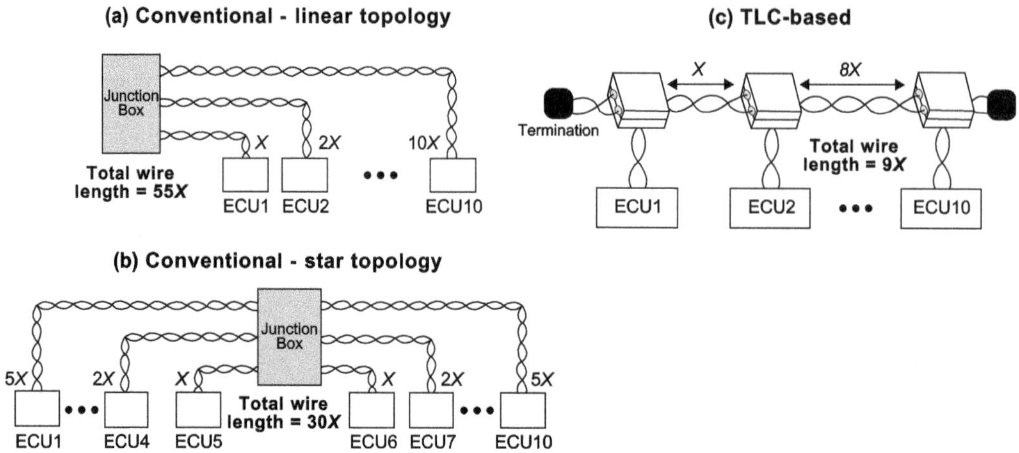

Figure 3.53 Comparison of automotive LAN topologies. © 2016 IEEE. Reprinted, with permission, from [36]

But as noted in Section 3.2, due to impedance discontinuity and hence signal reflection at each branching point, it is difficult to preserve signal quality while operating a multidrop bus at a high data rate, especially as the number of drops (ECUs) increases. Furthermore, transceivers on the ECUs require strong ESD protection, resulting in tens of pF of parasitic capacitance [40–41]. Such large capacitive loading on the transceivers only further exacerbates the signal integrity challenges of the automotive LAN.

Both the weight and data rate challenges can be overcome by replacing the conventional automotive connectors with TLC. Because of the contactless nature of TLC, communication failure due to connector disengagement can be eliminated, even though intense vibration can cause variation in the communication distance. Consequently, protection of the connectors against vibration can be greatly reduced, resulting in lighter and smaller connectors. Furthermore, it becomes unnecessary to route all connections through a junction box. Instead, wiring can run directly between the source and destination components, thereby significantly reducing wiring length and weight. As an illustration, Figure 3.53 compares the total wiring length between two widely used network topologies and a TLC-based topology, for an example network of 10 ECUs with regular spacing of X between neighboring ECUs. Compared to the linear network topology, TLC enables an 83.6% reduction in total wiring length. For the average wiring harness, wiring contributes about 35% of the total weight. Therefore, TLC can shave about 30% off the total weight of the wiring harness. The average midsize vehicle weighs about 1,200 kg, to which wiring harness contributes about 30 kg [42]. Consequently, TLC can reduce vehicle weight by about 9 kg compared to a linear topology, resulting in about 0.5% improvement in gas mileage [43]. Similarly, compared to a star topology, TLC can reduce total wiring harness weight by 25%, resulting in about 0.4% improvement in gas mileage.

On the other hand, as noted in Section 3.2, thanks to properly matched impedance, TLC generates minimal reflection at the branching point of a multidrop bus. As a result, the data rate can be significantly improved. Furthermore, damages from electrostatic discharge can be prevented by sealing the TLC electrodes with an electrically insulating material, thus eliminating the need for ESD protection. Consequently, capacitive loading on TLC drivers is significantly lower than a conventional interface, enabling it to run at higher speed.

Additionally, the band pass characteristics of TLC provide DC isolation of individual components, thus preventing a short circuit at one location from affecting the rest of the network. Moreover, it allows ECUs with different supply voltages and ground levels to connect directly without isolators, reducing component count and costs.

3.4.1 EM-Clip: Twisted Pair TLC Derivative

Since it is necessary for individual ECUs to communicate with each other, an automotive LAN is a multimaster bus. However, because automotive LAN uses twisted pair cables for interconnection instead of PCB traces, the BD-TLC derivative implemented in a layered substrate (PCB and FPC) that is described in Section 3.2.3 for a multimaster, multidrop bus is not applicable. Instead, a variant of the BD-TLC derivative known as an electromagnetic clip connector (EM-Clip, or alternatively C-TLC) has been developed and reported in [36]. It has two semicylindrical electrodes formed on FR4 substrate to enable direct electromagnetic coupling with a twisted pair. Since the EM field can penetrate through the wire covering of the twisted pair, the twisted pair can be used as is without modification, making installation simple. The basic construction and key design parameters are depicted in Figure 3.54. There are five design parameters. In addition to the coupler length l, intrapair spacing S, and communication distance d, there are also the diameter of the semicylindrical electrode D and the electrode central angle θ (Figure 3.54c). Same as for the standard TLC, the coupler length l determines the band pass characteristics of the EM-Clip. A shorter coupler results in a higher center frequency and a wider pass band. As will be discussed in Section 3.4.2, EM-Clip uses a coding scheme that up-converts the signal frequency to 1.4 GHz. Therefore, a coupler length of 20 mm is chosen to achieve a center frequency of around 1.4 GHz.

The coupling strength of the EM-Clip is determined by the communication distance d and electrode central angle θ. Electrically, the coupling coefficient K is determined by the mutual capacitance C_m between a twisted pair wire and the electrode, as well as the capacitance C_s between the electrode and the outside world by the following equation:

$$K = \frac{C_m}{C_s + C_m}. \tag{3.43}$$

As θ increases, C_m goes up, resulting in a larger coupling coefficient. However, based on the simulated data depicted in Figure 3.55, the increase in coupling coefficient slows down around $\theta = 180$ degrees and gradually tapers off. This is because as the

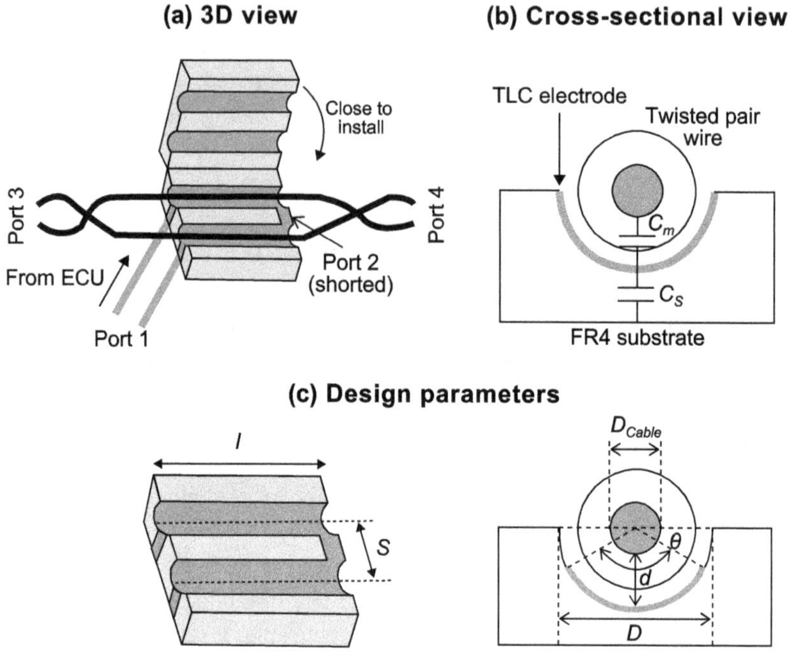

Figure 3.54 EM-Clip TLC for automotive LAN. © 2016 IEEE. Reprinted, with permission, from [36]

electrode area increases, not only C_m but also C_s increases. As θ increases beyond 180 degrees, the TLC electrode will have to extend into the upper half of the clip, making it necessary to electrically connect the upper to lower half. In addition to complicating installation, this makes it necessary to expose the electrode, which will necessitate ESD protection. Therefore, θ is chosen to be 180 degrees to achieve high coupling coefficient, high electrical performance, and simple installation.

The tuning of coupling coefficient is therefore achieved through adjustment of d, which is determined by the twisted pair wire core diameter D_{Cable} and electrode diameter D. As explained in Section 3.2.1, every time a signal passes through a TLC coupler, it loses some of its energy through coupling, hence the signal attenuates as it propagates down the bus. Therefore, while the coupling coefficient needs to be large to couple sufficient energy to drive the receiver, it cannot be excessively large or else not enough energy is transmitted toward the far end of the bus. Because it is necessary to connect up to around 10 ECUs on one LAN [39], a reasonable target coupling coefficient is around 0.1, which is around –10 dB. Based on the simulated coupling coefficient as a function of d depicted in Figure 3.56, a communication distance of 0.5 mm is chosen, resulting in a coupling coefficient of around –10.5 dB. For the AWG 24 cable that is widely used in automotive applications, the wire coating is 0.25 mm thick. Therefore, a 0.25 mm thick tape is inserted in between the electrode and wire to create a 0.5 mm communication distance.

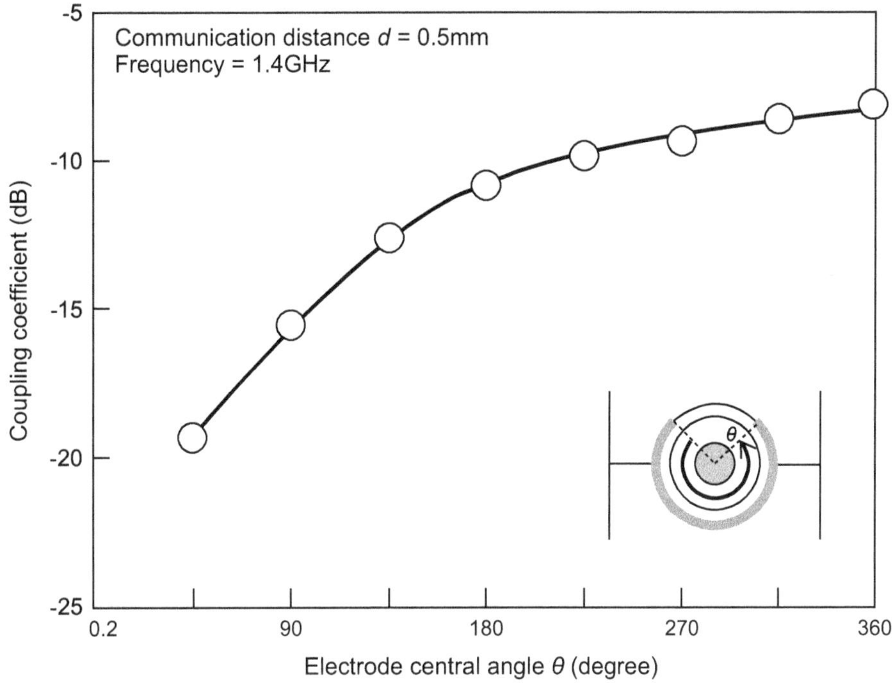

Figure 3.55 EM-Clip coupling coefficient vs. central angle. © 2016 IEEE. Reprinted, with permission, from [36]

Since EM-Clip is a variant of BD-TLC that supports multimaster communication, Port 2 is terminated with a short circuit that is designed the same way as described in Section 3.2.3.1.

3.4.2 Automotive TLC Transceiver with High EMC Immunity

Safety is of the utmost importance in automotive applications. As a result, automotive electronic systems must operate reliably without failure. EMC is therefore a critical design consideration, and automotive electronic systems must meet the strict EMS requirements defined in ISO 11452-4 [44] as well as the EMI requirements in CISPR 25 [45], as depicted in Figure 3.57a. Meanwhile, inverters and motors, as well as current spikes caused by lightning, can generate EMI noise with magnitude on the order of 1 V. On the other hand, the received signal of TLC has magnitude on the order of only 100 mV. Therefore, TLC must have low EMS in order to operate reliably in an automotive environment (Figure 3.57b).

The ISO specification requires error-free communication even in the presence of up to 30 dBm of noise at 400 MHz. However, when such a large noise is imposed on TLC, the noise amplitude at the receiver can be on the same order as the supply voltage as shown in Figure 3.58a. Because of the band pass property of TLC,

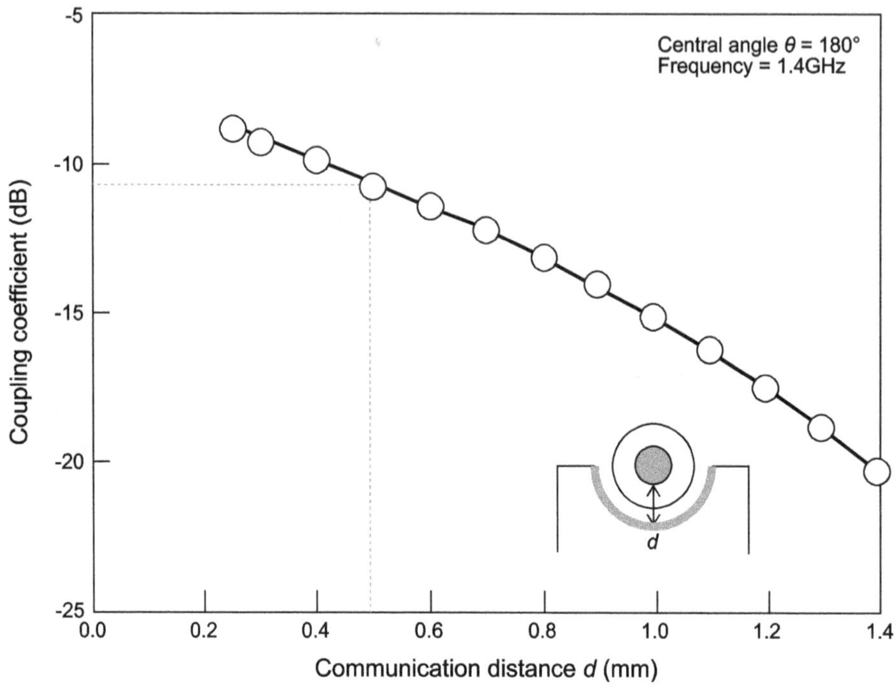

Figure 3.56 EM-Clip coupling coefficient vs. communication distance. © 2016 IEEE. Reprinted, with permission, from [36]

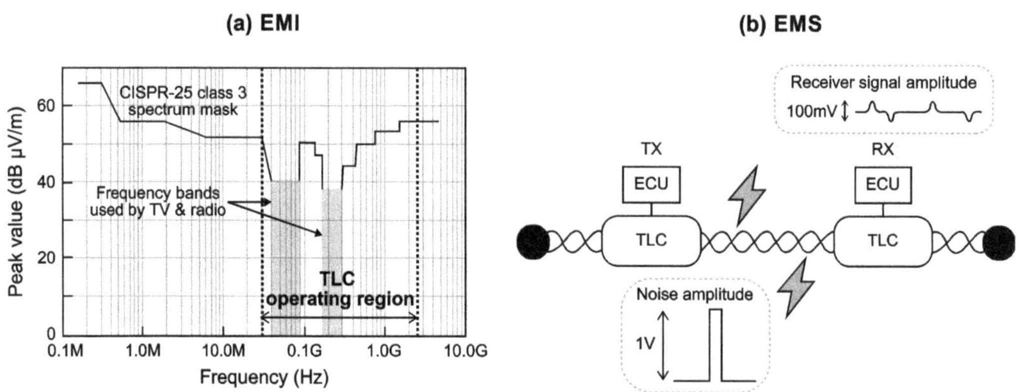

Figure 3.57 EMC requirements for automotive TLC. © 2016 Atsutake Kosuge [1]

low-frequency noise is significantly attenuated. However, noise at 100 MHz and above is able to pass through, resulting in noise amplitude of more than 1 Vpp. Even though the noise is common mode, its large amplitude can saturate the differential amplifier in the front end of the TLC receiver, reducing its differential gain, as depicted in Figure 3.58b. This in turn results in distortion of the amplified waveform

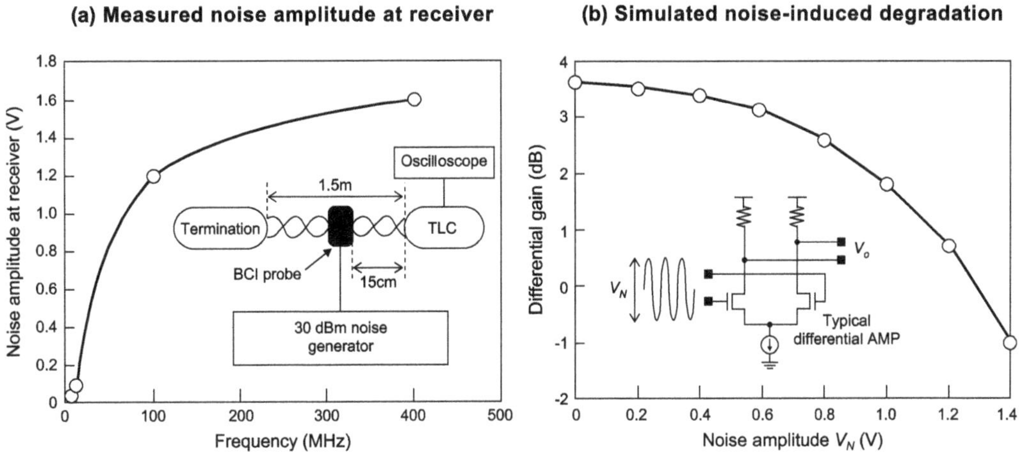

Figure 3.58 EMS evaluation of TLC. (a) © 2016 Atsutake Kosuge [1]. (b) © 2016 IEEE. Reprinted, with permission, from [36]

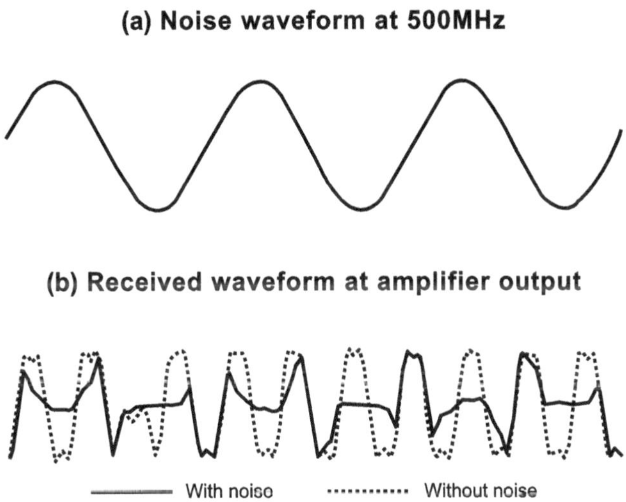

Figure 3.59 Simulated waveform distortion by noise-saturated amplifier. © 2016 Atsutake Kosuge [1]

as illustrated in Figure 3.59. Consequently, it is difficult to ensure error-free operation of the standard TLC in the noisy automotive environment.

3.4.2.1 Oversampling Manchester Coding

To improve TLC noise immunity, an oversampling Manchester coding scheme is adopted in place of the standard NRZ signaling for automotive applications. The transceiver block diagram and simulated waveforms are shown in Figure 3.60. A fast

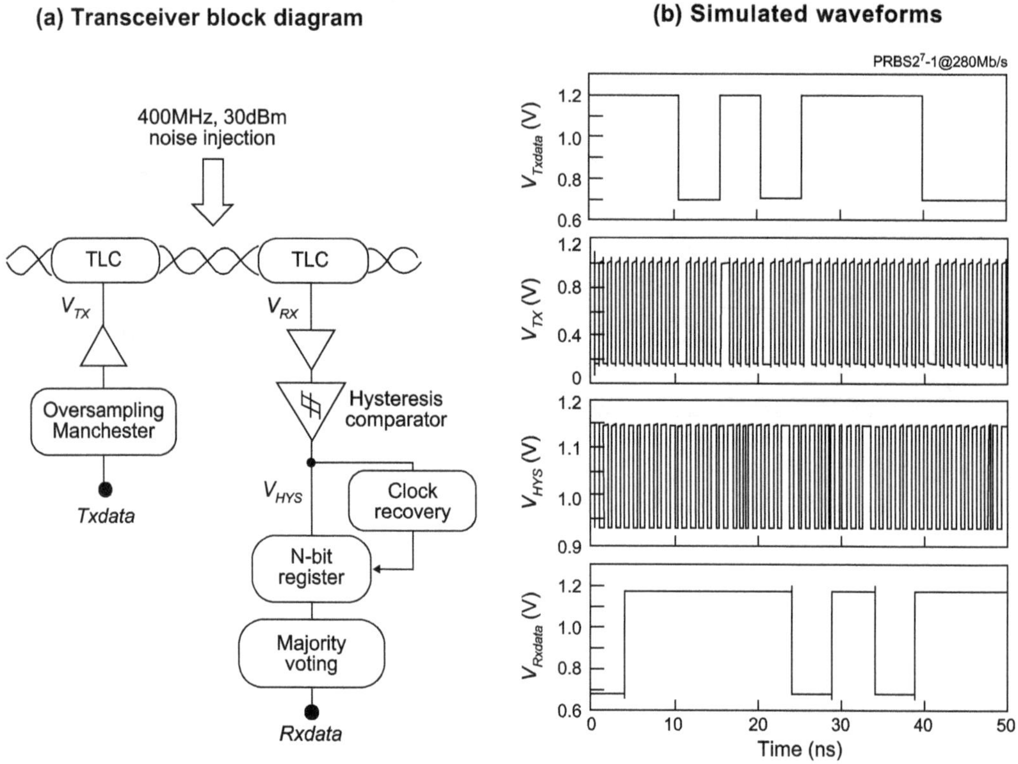

Figure 3.60 Simulated waveforms of oversampling Manchester coding transceiver. © 2016 IEEE. Reprinted, with permission, from [36]

clock with N times the frequency of the original NRZ digital data is used to encode the data, creating N pulses for each bit of data. In other words, the oversampling rate is N. In this way, even if noise creates an error within the N pulses, it can be removed by majority voting at the receiver. When large noise causes saturation of the differential amplifier that results in bit errors, the error rate is bounded by the noise frequency f_{Noise}. Since oversampling Manchester coding uses majority voting to remove error bits, it can successfully remove the errors as long as more than half of the pulses within an NRZ bit are correct. In other words, the oversampling clock frequency f_{CLK} must be more than twice of f_{Noise}:

$$f_{CLK} > 2f_{Noise}. \tag{3.44}$$

Since the highest noise frequency in the ISO specification is 400 MHz, an oversampling frequency larger than 800 MHz is required.

Because oversampling Manchester coding up-converts the signal frequency spectrum, as will be explained next, it can also solve the EMI problem. This is illustrated by the relationship between signal spectrum and the oversampling frequency f_{CLK} and oversampling rate N. Since the result of oversampling Manchester coding is phase modulation of the oversampling clock as a carrier, the time domain encoded signal $s_M(t)$ can be expressed by

$$s_M(t) = \frac{2}{\pi}\sum_{k=1}^{\infty}\frac{\sin\{(2k-1)\omega_{CLK}t + \pi d(t)\}}{2k-1}. \tag{3.45}$$

where $d(t)$ is the data pattern of the original digital data, and ω_{CLK} is the angular frequency of the oversampling clock. By defining time series $m(t)$ as

$$m(t) = \cos\{\pi d(t)\}, \tag{3.46}$$

$s_M(t)$ can be rewritten as

$$s_M(t) = \frac{2}{\pi}\sum\nolimits_{k=1}^{\infty}\frac{m(t)}{2k-1}\left\{e^{j(2k-1)\omega_{CLK}t} + e^{-j(2k-1)\omega_{CLK}t}\right\}. \tag{3.47}$$

The frequency spectrum of the encoded signal can thus be derived by taking the Fourier transform of (3.47), resulting in

$$s_M(\omega) = \frac{2}{\pi}\sum\nolimits_{k=1}^{\infty}\frac{1}{2k-1}M(\omega-(2k-1)\omega_{CLK}), \tag{3.48}$$

where $M(\omega)$ is the Fourier transform of $m(t)$. Since $m(t)$ and $d(t)$ are related as

$$m(t) = \begin{matrix} 1 \ \ when \ \ d(t) = 0 \\ -1 \ \ when \ \ d(t) = 1, \end{matrix} \tag{3.49}$$

$M(\omega)$ is a sinc function. Therefore, ignoring higher harmonics, (3.48) can be simplified to

$$s_M(\omega) = \frac{4}{\pi(\omega-\omega_{CLK})}\sin\left(\frac{N(\omega-\omega_{CLK})}{2f_{CLK}}\right). \tag{3.50}$$

Consequently, the effect of oversampling Manchester coding on the original signal spectrum is to up-convert it to the oversampling clock frequency f_{CLK}, with a spectrum width that is inversely proportional to N.

The CISPR has strict EMI regulations for frequency range from 50 MHz to 1 GHz since the spectrum is used for television, radio, and mobile applications. By contrast, since there is not a lot of automotive equipment operating above 1 GHz, the requirements for the higher-frequency range are more relaxed. As a result, the EMI problem can be alleviated by choosing $f_{CLK} > 1$ GHz. To avoid the GPS frequencies of 1.23 and 1.58 GHz, [36] selects $f_{CLK} = 1.4$ GHz for its prototype implementation. Since the data transmission rate is f_{CLK}/N, the smallest possible N should be used to maximize the data rate. Based on simulated spectrum as a function of N depicted in Figure 3.61, $N = 5$ is chosen to avoid interference with GPS operation at 1.23 and 1.58 GHz, resulting in a data rate of 280 Mb/s.

Figure 3.62 depicts the oversampling Manchester coding transceiver design. The coding is done by combining the differential NRZ data with the oversampling clock using two XOR gates. A voltage mode driver is used to reduce power consumption. The receiver consists of a front-end amplifier, a hysteresis comparator, and a CDR block whose design is to be discussed in the next section. The amplifier consists of a

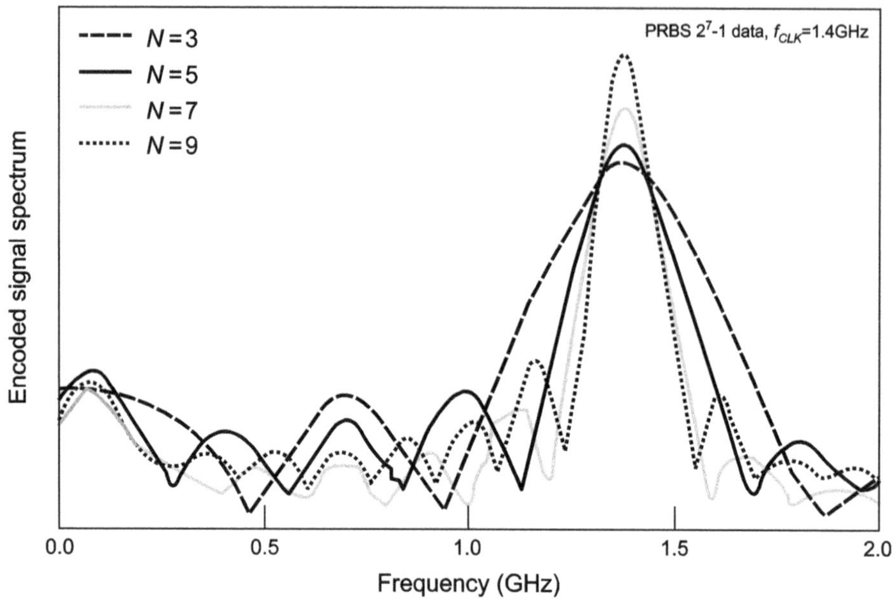

Figure 3.61 Simulated spectrum of oversampling Manchester coding vs *N*. © 2016 IEEE. Reprinted, with permission, from [36]

cross-coupled pair M1 and M2, which serves to deliver large gain for differential signal and small gain for common mode signal. An RC circuit is inserted into the feedback loop to create a high-pass filter to attenuate noise. As a result, while the amplifier achieves a gain of almost 15 dB at 1.4 GHz, it attenuates 400 MHz noise to less than 0 dB (Figure 3.62d).

Compared to the standard transceiver using NRZ signaling, the oversampling Manchester coding transceiver has high EMC immunity. However, it suffers from two drawbacks. First, its bandwidth utilization is reduced by $1/N$. For example, the data rate of the prototype in [36] is reduced from 1.4 Gb/s to 280 Mb/s. Fortunately, the maximum required data rate for automotive LAN is currently at around 100 Mb/s, which is an order of magnitude lower than the operating range of TLC. Furthermore, since TLC suppresses reflections on a multidrop bus and reduces loading capacitance, there is headroom to scale up its data rate. Therefore, the bandwidth utilization drawback should not be a limiting factor in the foreseeable future.

The second drawback is degraded energy efficiency. Compared to NRZ signaling at the same data rate, oversampling Manchester coding transceiver operates at a data rate that is N times higher, resulting in higher energy consumption. However, since current flows constantly in the driver stage that consumes the most energy within the transceiver, a large part of the total energy consumption is independent of the data rate. Likewise, since both the hysteresis comparator and amplifier utilize CML circuits, their power consumption is independent of data rate. Consequently, even though the oversampling Manchester coding transceiver runs at a data rate that is N times higher, the

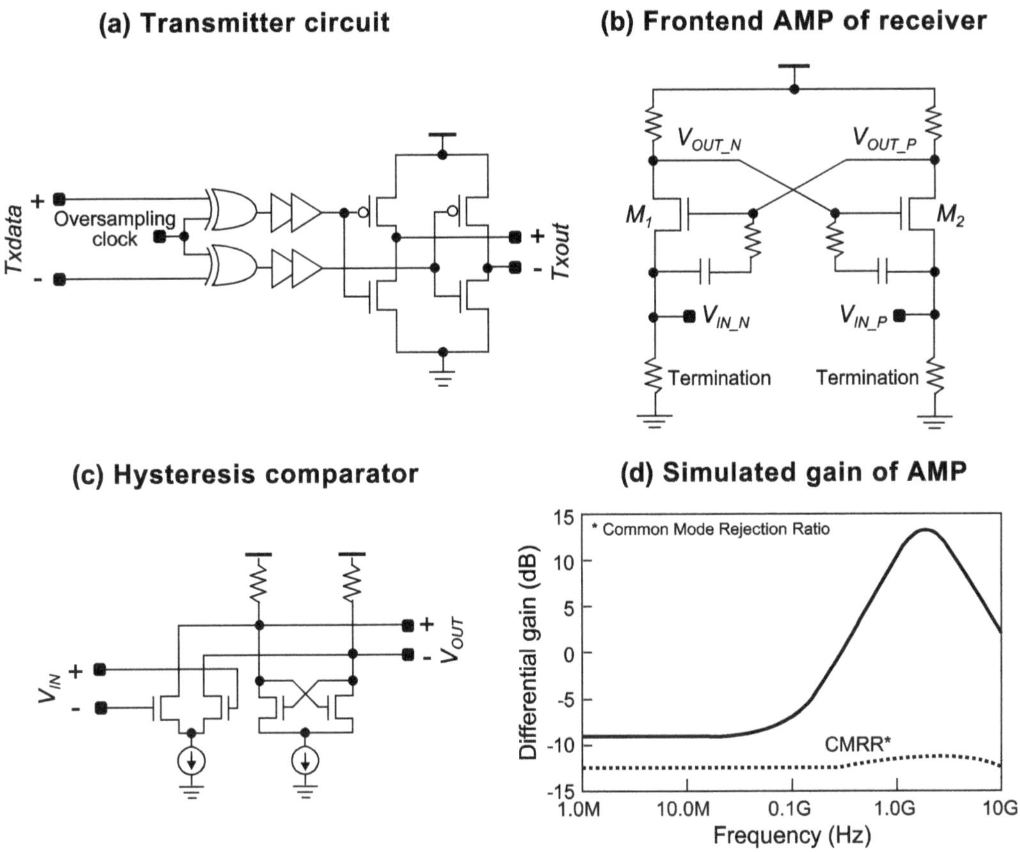

Figure 3.62 Oversampling Manchester coding transceiver design. © 2016 IEEE. Reprinted, with permission, from [36]

increase in energy consumption is much less than proportionally. Indeed, simulation shows that when $N = 5$, the increase in energy per bit is only about 40%.

3.4.2.2 Correlator-Based, High Noise Immunity CDR

While the use of oversampling Manchester coding significantly improves both noise immunity and EMI of data, separate enhancement is required for clock delivery. In automotive LAN, CDR-based data transmission is utilized to remove the clock signal to reduce the number of signal lines and hence wiring weight. Furthermore, removing the constant frequency clock helps reduce EMI. It is thus necessary to adopt a CDR scheme that can recover the clock from a noisy data signal.

One common solution is to insert an averaging filter such as a low-pass filter into the feedback loop of a conventional phase detector-based analog CDR. While this solution has the merit of being easy to implement, it is difficult to design a low-pass filter with low enough cutoff frequency to remove low-frequency noise in the 10–400 MHz range. Low cutoff low-pass filters suffer from poor tracking performance and narrow lock range. Furthermore, large on-chip capacitance is required, resulting in a large area penalty. In addition, large jitter noise can result from injecting a noisy signal directly into an analog phase comparator [46].

Alternatively, an oversampling digital CDR can be used which has high noise immunity as well as fast lock time [46]. Furthermore, a fully digital solution helps limit power consumption and layout area. By inserting an averaging filter into a multilevel digital CDR, it is possible to recover the correct phase from a noisy received signal. Unfortunately, same as its analog counterpart, this digital CDR solution suffers from the drawback of poor tracking performance and narrow lock range.

The solution proposed in [36] is the correlator-based CDR circuit depicted in Figure 3.63a. In this solution, the transmitter sends a preamble signal that is used for clock recovery. The preamble signal consists of a consecutive stream of bits of the same value, such as "11 … 1." The encoded signal is thus identical to the oversampling clock. The received preamble signal is fed into a correlator together with a locally generated template to compute the correlation factor (ΣQ_i). This is done by oversampling and summing the correlator output during the half of the period when

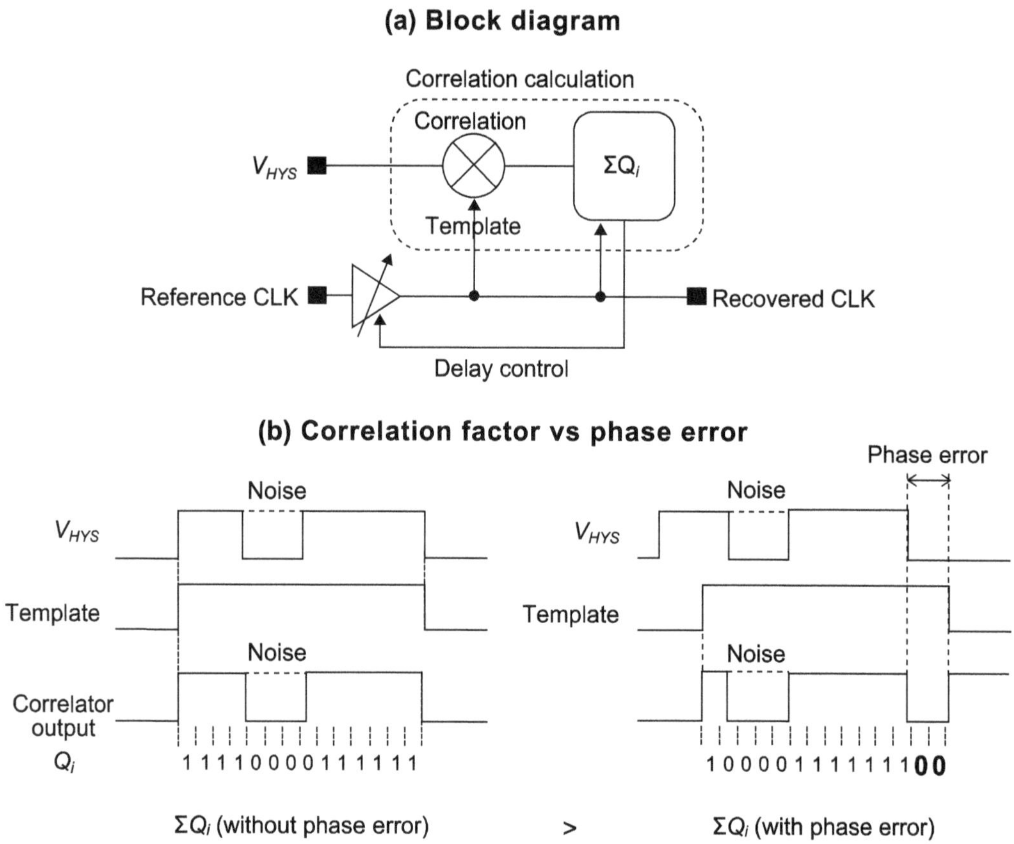

Figure 3.63 Correlator-based CDR block diagram and determination of phase error. © 2016 IEEE. Reprinted, with permission, from [36]

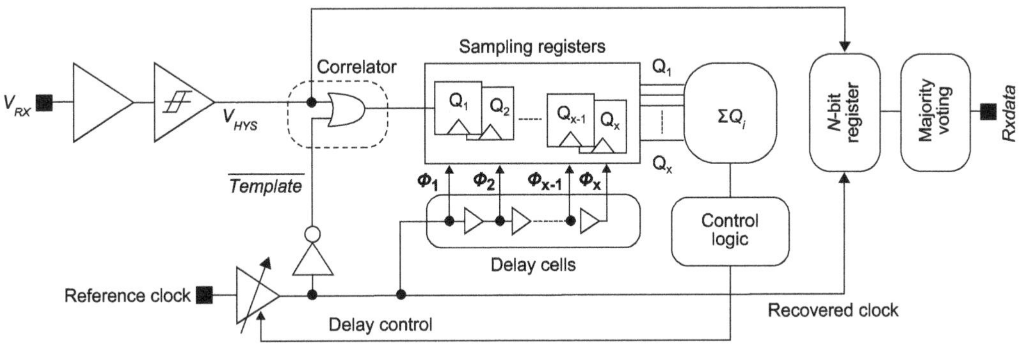

Figure 3.64 Detailed block diagram of correlator-based CDR. © 2016 IEEE. Reprinted, with permission, from [36]

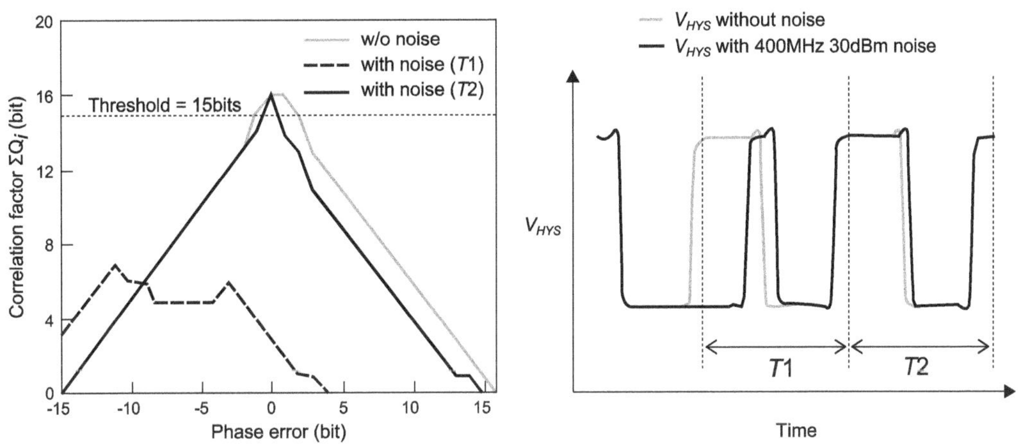

Figure 3.65 Simulated correlation factor as a function of phase error. © 2016 IEEE. Reprinted, with permission, from [36]

the template clock is low. As illustrated in Figure 3.63b, the larger the phase error, the smaller is the correlation factor. By repeatedly computing the correlation factor while shifting the phase of the template, the CDR circuit finds the phase with the largest correlation factor and locks to it. By enforcing a predefined threshold as the lower limit, it is possible to prevent locking to an incorrect phase distorted by large noise that has a small correlation factor lower than the threshold.

Figure 3.64 depicts a detailed block diagram of the correlator-based CDR. Output from the hysteresis comparator and the inverted, locally generated reference clock used as the template is fed into the correlator, which is implemented with an OR-gate. Output of the OR-gate is oversampled, and the sampled values are summed to compute the correlation factor. The multiphase clock used for oversampling is generated by a cascaded array of delay cells.

Figure 3.65 compares the simulated correlation factor between three cases – no noise, large 30 dBm, 400 MHz noise (during time period T1), and small noise (during time period T2). The large noise during T1 causes so much distortion that the correlation factor is below the predefined threshold of 15 bits for all phase values. As a result, the CDR circuit ignores the phase information from this computation. However, when the noise is small during T2, the correlation factor is almost the same as when there is no noise. Therefore, the CDR circuit uses the phase information for its clock recovery. As long as (3.44) is satisfied, time period T2 of low noise should occur more than 50% of the time. Consequently, by performing the correlation factor computation at least twice, the CDR circuit should be able to adequately reject noise and accurately recover the correct clock phase.

3.4.3 Automotive TLC Experimental Results

Figure 3.66 depicts the setup for EMC testing of the proposed automotive LAN TLC, which includes an EM-Clip prototype and a high EMC immunity transceiver test chip. The EM-Clip prototype consists of a FR4 base and 20 mm long electrodes made of copper tape. The network bus is implemented with a 10 m long AWG 24 unshielded twisted pair cable. The test chip is fabricated in a 65 nm process and implements oversampling Manchester coding transceivers with a correlator-based CDR. Using a base data rate of 280 Mb/s and $N = 5$ coding, the encoded data runs at 1.4 GHz. Power consumption is 23.4 mW for TX, 13.6 mW for RX, and 4.8 mW for CDR, for a total of 41.8 mW. Energy efficiency is therefore 0.15 nJ/b. EMI and EMS test setup are based on CISPR and ISO specifications respectively, as shown in Figure 3.66c, d. EMI testing is performed using the PRBS2^7–1 data pattern, while EMS testing is done using a 30 dBm noise source of 1–400 MHz, injected with a bulk current injection (BCI) probe.

Figure 3.67 plots the S parameter measurement results which include coupling gain and EMS. Using a BD-TLC design the EM-Clip is able to achieve close to identical coupling gains from Port 1 onto the bus in both directions toward Port 3 and Port 4 for well beyond 1.4 GHz (Figure 3.67a). Figure 3.67b illustrates that within the ISO regulation band of up to 400 MHz, common mode noise injected into the twisted pair bus creates differential noise at the receiver input with a gain of less than –40 dB, which is low compared to the EM-Clip coupling gain of more than –20 dB.

Signal degradation due to the radiated noise is evaluated and depicted in Figure 3.68. From Figure 3.68b, 30 dBm, 100 MHz noise induces voltage noise on the order of 60 mVpp, which is 5% of the signal amplitude of 1.2 Vpp. The noise is mainly common mode; differential noise is negligible. Meanwhile, because BER is degraded, the number of ECUs that can be attached to the bus decreases. Between noise frequency of 1–400 MHz, the higher the frequency, the larger the BER degradation, and hence the fewer ECUs that can be attached to the LAN at the same BER. But even with a 30 dBm, 400 MHz noise, up to 10 ECUs at a uniform interval of 1 m can be connected to the bus while BER remains below 10^{-11}. At this noise frequency, the noise amplitude at the receiver input is 1.4 Vpp. The minimum received signal amplitude required to achieve BER $< 10^{-11}$ is 130 mVpp.

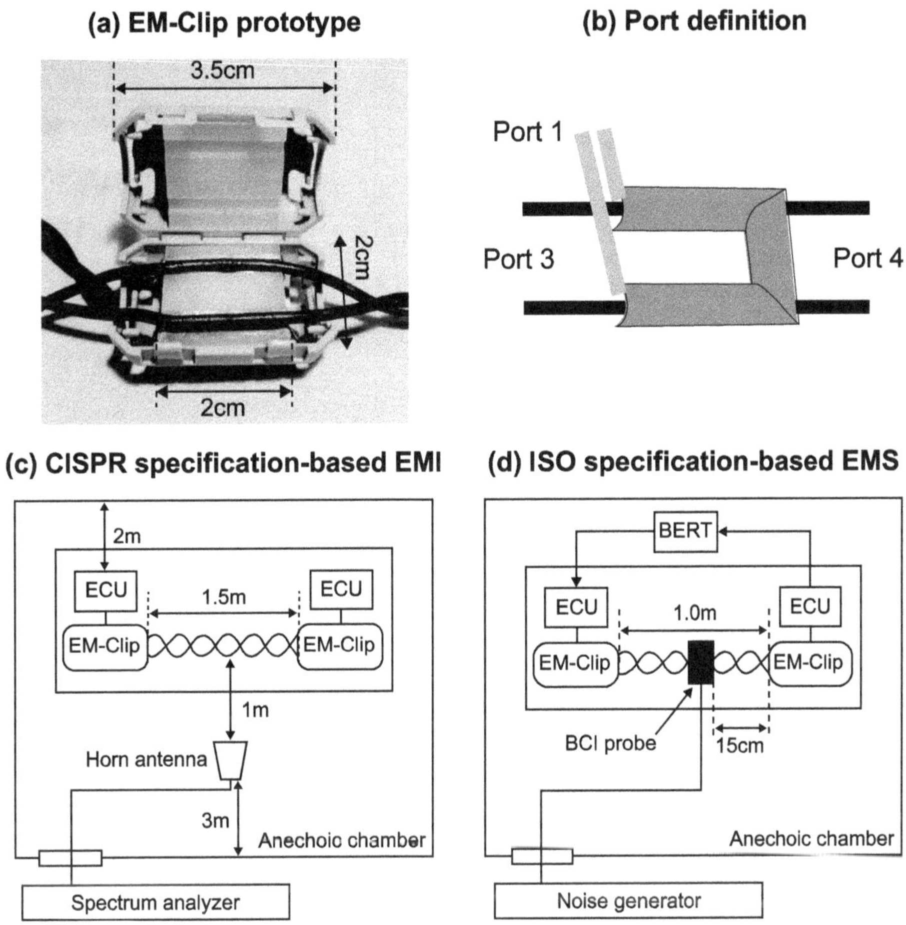

Figure 3.66 Setup for testing of automotive LAN TLC. © 2016 IEEE. Reprinted, with permission, from [36]

EMI measurement results using uncoded NRZ signaling and Manchester coding with $N = 5$ are depicted in Figure 3.69a and b respectively. Without Manchester coding, the signal spectrum peaks around the frequency range where the specification is the strictest. As a result, it is not able to stay entirely within the CISPR-25 class 3 mask. With Manchester coding, however, the signal spectrum is shifted high to 1.4 GHz, where the CISPR specification is more relaxed. Consequently, it is able to meet the specification with margin.

Furthermore, EMI in the GPS bands is measured to be 35.14 dBμV/m at 1.23 GHz and 28.57 dBμV/m at 1.58 GHz respectively. Further reduction is possible by more oversampling such as using $N = 7$. Even with the larger N, the data rate is 200 Mb/s, which is still sufficiently fast for next-generation automotive LAN applications [37].

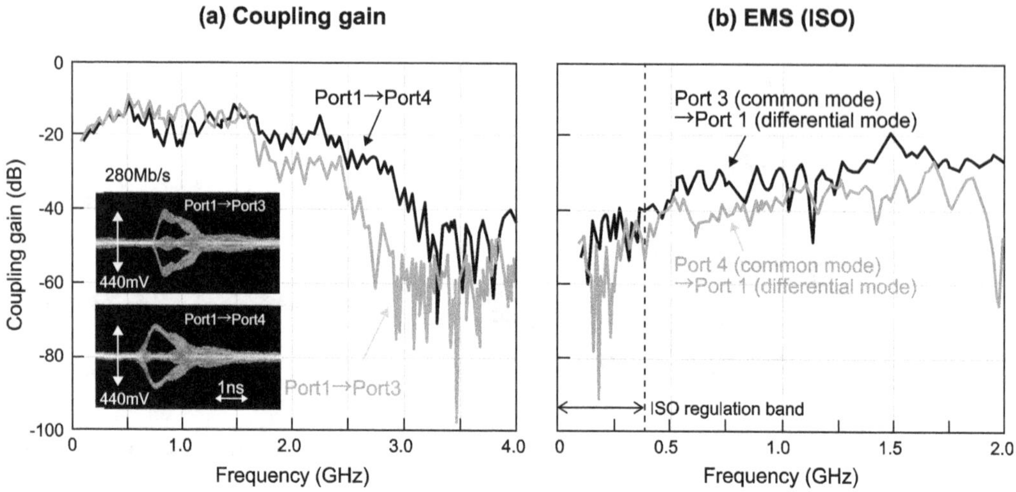

Figure 3.67 S parameter measurement results. © 2016 IEEE. Reprinted, with permission, from [36]

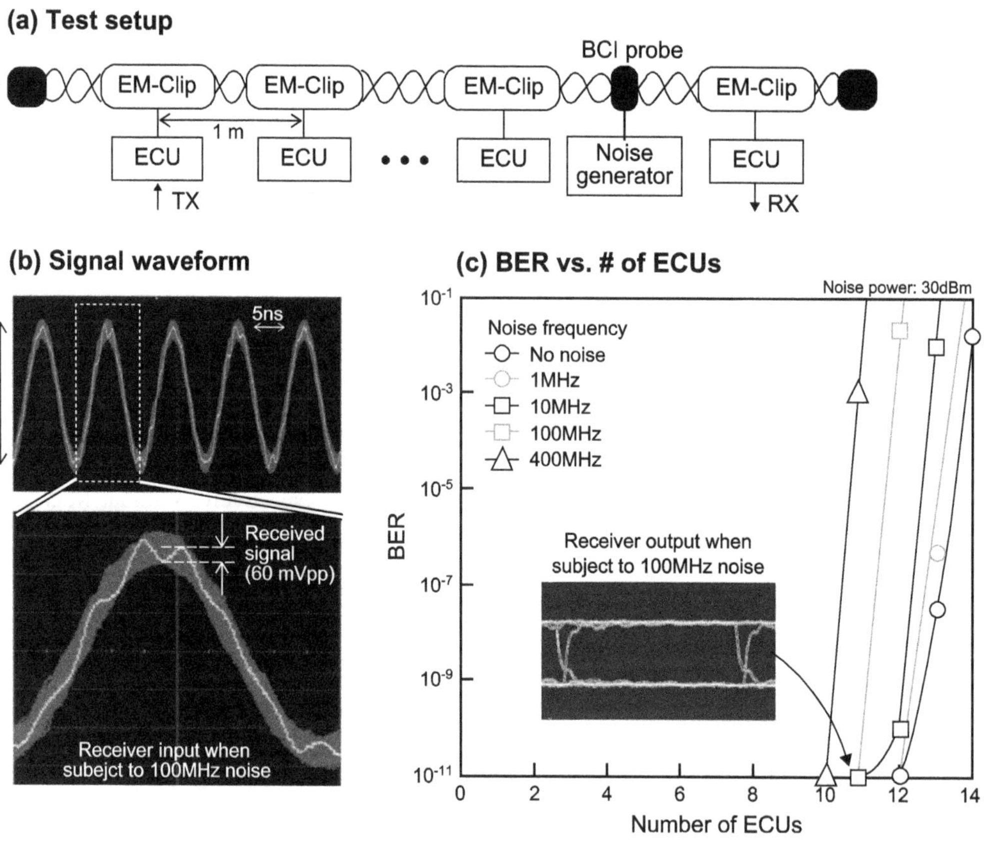

Figure 3.68 Measured signal degradation induced by radiated noise. © 2014 IEEE. Reprinted, with permission, from [47]

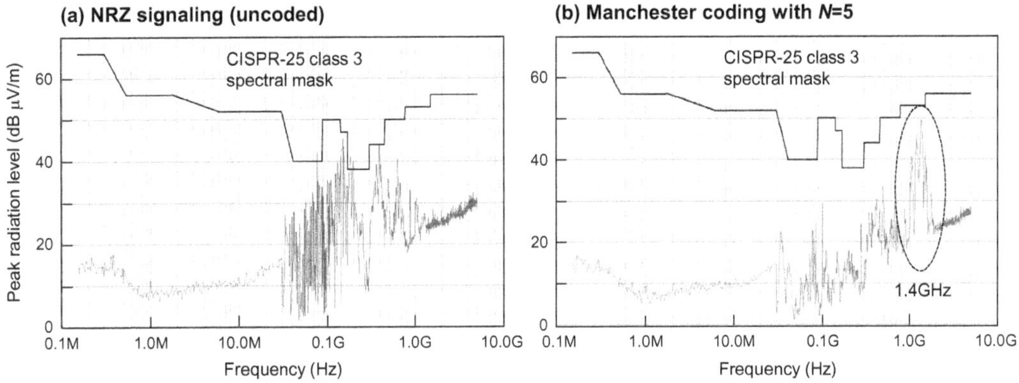

Figure 3.69 EMI measurement results. © 2016 IEEE. Reprinted, with permission, from [36]

3.4.4 Closing Thoughts

Because of the contactless nature of TLC, it offers a superior automotive LAN connector solution in terms of robustness against vibration. As a result, it enables significant simplification of protection required to prevent against vibration-induced disconnection and hence communication failure. This in turn eliminates the need of routing to a junction box, thereby significantly reducing wiring length and weight for improved gas mileage. Furthermore, impedance matching enables implementation of a multidrop bus with minimal reflection at branching points. Together with smaller loading capacitance due to elimination of the need for ESD protection, TLC enables a fast data rate that is a leap from what is required by current applications.

In order to utilize existing infrastructure where automotive LAN uses twisted pair cables, an EM-Clip is developed to enable ECUs to electromagnetically couple with twisted pair cables. In addition, use of BD-TLC termination results in coupling in both directions of the bus to implement a multimaster bus to allow multiple ECUs to communicate with each other.

The weakness of the standard TLC is in EMC, since it operates with a very small signal amplitude. Meanwhile, automotive applications demand strict EMC compliance with ISO and CISPR specifications. Two circuit technologies are developed to achieve compliance – oversampling Manchester coding to improve both EMS and EMI of data transmission, and correlator-based CDR to improve EMS of clock recovery. The resultant transceiver design, together with the EM-Clip, delivers an automotive LAN TLC solution that is lightweight and high speed and has high EMC immunity.

Table 3.6 compares the EM-Clip solution with three advanced automotive LAN transceivers from around the same time. The EM-Clip solution implements a 10-ECU multidrop bus that runs at a higher data rate than all the other three transceivers used in point-to-point communication. In particular, while consuming about the same power, the EM-Clip multidrop bus runs at seven times the data rate of the point-to-point bus in [48].

Table 3.6. Comparison of advanced automotive LAN transceivers. © 2016 IEEE. Reprinted, with permission, from [36].

	[40], 2015	[48], 2010	[49], 2014	EM-Clip, 2016
# of ECUs per bus	1 (Point-to-point)	1 (Point-to-point)	1 (Point-to-point)	10
Data rate (Mb/s)	10	40	100	280
Power consumption (mW)	181.5	43.0	27.0 (TX only)	41.8 (23.4 if TX only)
Power efficiency (nJ/b)	18.15	1.08	0.27 (TX only)	0.15 (0.08 if TX only)
CMOS process (nm)	180	180	28	65

A state-of-the-art 100Base transmitter that is intended for multimedia applications is documented in [49]. Since 100Base transmitters are finding their way into automotive applications [50], it is included here for comparison. Even compared to this state-of-the-art transmitter, the EM-Clip solution has almost three times the data rate and three times better TX power efficiency using a more mature process (65 vs. 28 nm). This is all on top of being lightweight and hence yielding better gas mileage.

3.5 TLC SSD

With the continuous increase in memory density and decrease in cost per bit, NAND flash memory continues to gain acceptance as a low-cost, low-power, lightweight, and fast-access storage solution both in the form of memory cards and solid-state drives (SSDs), in mobile applications on the client side and data center applications on the server side. However, the use of conventional metal contacts for connection means that NAND flash memory solutions are subject to reliability problems that result from electrostatic discharge (ESD), environmental contamination including water corrosion, and wear and tear due to multiple insertions in the case of memory cards. Such reliability problems can also limit the communication data rate and footprint reduction. ESD protection results in higher loading capacitance and hence more signal distortion, while wear and tear degrade high-frequency characteristics of the contacts. On the other hand, since smaller metal contacts are more susceptible to wear and tear, reliability concerns can limit how much the interface footprint can shrink. By adopting TLC as the memory interface, the connection becomes contactless and its communication terminals can be concealed to make them environmentally robust, creating a greatly enhanced memory solution for both mobile applications, which require operation in harsh environments, and data center applications, which require high reliability.

In this section, we will describe an SSD system that utilizes a fully wireless interface between the host controller and the memory system, for both data communication and power delivery. In addition to the interface physical layer (PHY) design, we will explain solutions to address system-level issues, including startup and error correction. Although SD flash memory cards with wireless interface already exist in the market, they use Wi-Fi technology and have limited data rate at 72.2 Mb/s as of 2019 [51]. With TLC, we are able to achieve a high data rate of 1.6 Gb/s/link, representing a more than 20-fold increase.

3.5.1 System Architecture

Figure 3.70 depicts the conceptual view of the wireless SSD system. The memory chips are completely sealed within a plastic housing with no exposed metal contacts. As a result, it offers high resistance to water and robustness against harsh environmental conditions. The entire wireless interface including a $20 \times 20\ mm^2$ power coil and four small TLC couplers placed inside the coil can fit within the $24 \times 32\ mm^2$ footprint of a standard SD card. Two stacks of flash memory chips each measuring 10 mm on the side are mounted on the back of the memory card, providing total capacity of 1 TB. The wireless couplers are implemented in the top layer and connected to an interface chip mounted on the same side of the PCB. The memory card is then flipped and mounted on top of the controller PCB at a communication distance of 1 mm.

Figure 3.71 depicts the block diagram of the wireless SSD system. The connection between the controller and NAND memory card consists of two wireless interfaces, one for power delivery and the other data communication. The 1 W power delivery interface includes a feedback loop to adjust the amount of power being delivered dynamically based on the loading on the memory card supply voltage. The data interface consists of an uplink and a downlink, each with two channels, one for data and the other for clock to

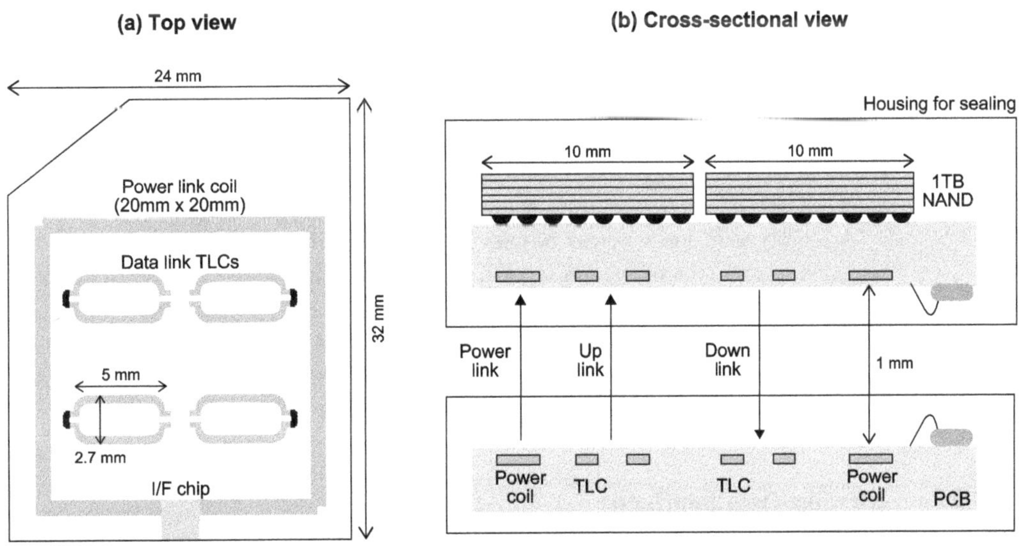

Figure 3.70 Construction of a fully wireless SSD system. © 2015 IEEE. Reprinted, with permission, from [52]

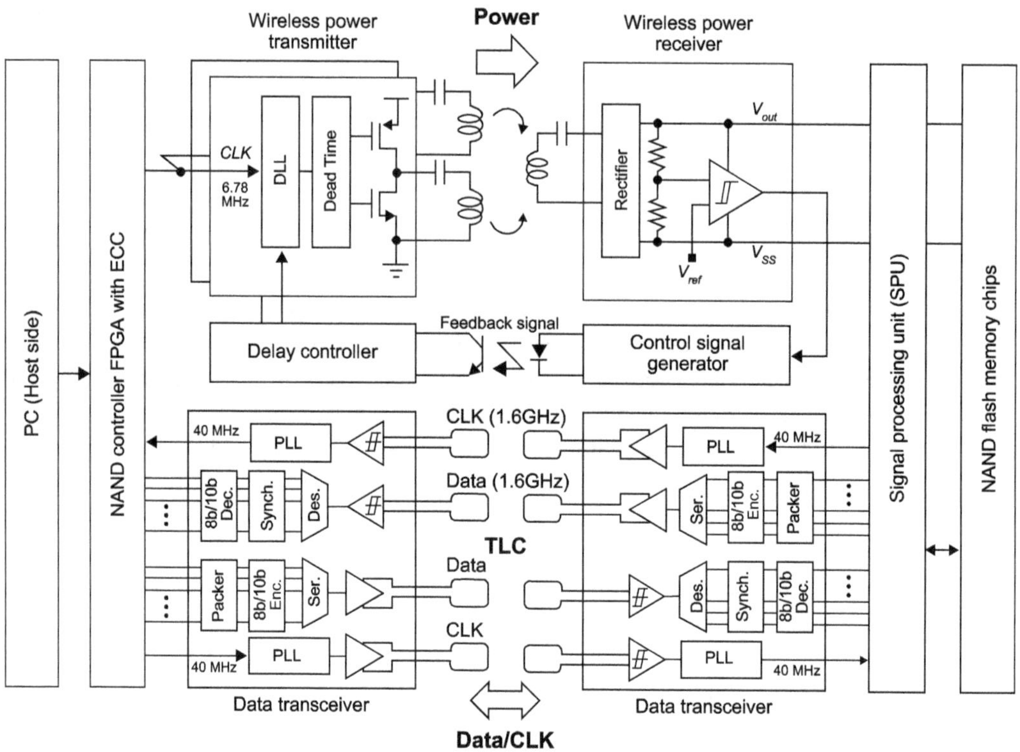

Figure 3.71 Wireless SSD system block diagram. © 2015 IEEE. Reprinted, with permission, from [52]

implement source synchronous communication. SerDes is used to serialize memory address and control signals as well as an 8-bit DQ signal into one 1.6 Gb/s data stream to minimize footprint. The controller includes error correction code (ECC) functionality to correct both memory and interface errors. On the memory card, a signal processing unit (SPU) is included to generate commands for memory access based on the incoming signal. It also controls the startup sequence on the memory side.

A prototype of the wireless SSD system is reported in [53]. While the SPU and transceivers are implemented in separate chips in the prototype, they can be combined into one chip in actual products to save costs and footprint. Similarly, while the controller and ECC function are implemented in an FPGA chip in the prototype, they can be implemented in an ASIC instead to save per-unit costs, footprint, and power consumption.

3.5.2 Wireless Data Interface

Figure 3.72 depicts the TLC transceiver and coupler design for the wireless data interface. The transceiver consists of a standard CML transmitter with 50 Ω source termination and a standard hysteresis comparator receiver, and the coupler also

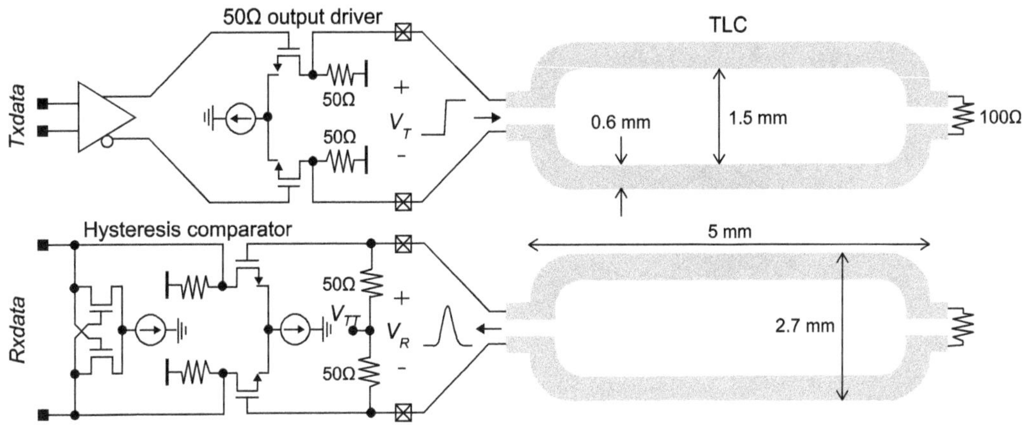

Figure 3.72 TLC wireless data interface. © 2015 IEEE. Reprinted, with permission, from [52]

employs a standard design. Each transmission line pair of the coupler is designed to have a differential characteristic impedance of 100 Ω and terminated at the far end by the same impedance. Since TLC center frequency is inversely proportional to the coupler length based on (3.9), it is generally advantageous to use a short coupler length in order to support a higher-frequency pass band for a higher data rate. Nevertheless, as the pass band is shifted high, the lower cutoff frequency also increases, resulting in loss in signal amplitude. As discussed in Section 3.1, the optimal length l is chosen to achieve saturated coupling, which is determined by the signal rise and fall time t_r using (3.24), where υ_p is the phase velocity of the surrounding dielectric medium. Since t_r of the prototype system is 70 ps, the coupler length l is chosen to be 5 mm.

Figure 3.73a depicts the fabricated TLC coupler and transceiver chip used in the prototype system. The coupler measures $2.7 \times 5\ \text{mm}^2$ and is formed on the surface of a 1.6 mm thick FR4 PCB. The transceiver chip is fabricated in a 90 nm CMOS process and measures $0.8 \times 2.4\ \text{mm}^2$, including two clock and two data channels, serializer, deserializer, and PLLs. Using a 1.6 Gb/s data rate and communicating across a distance of 1 mm, the transceiver consumes 18 mW at a supply voltage of 1.2 V.

Figure 3.73b shows the effect of placing the TLC couplers inside the power interface coil. Turning on the power interface to supply 1 W of power does degrade the wireless signal transmission. Nevertheless, with about 20% increase in signal amplitude, a BER of 10^{-11} can be achieved. The reason for the small interference is because the wireless power interface operates at 6.78 MHz, which is about three orders of magnitude lower than the low cutoff frequency of the TLC channel at 1.2 GHz.

3.5.3 Wireless Power Interface

The challenge in designing the wireless power interface is that it must be able to dynamically adjust itself in response to the fluctuating load of the SSD memory

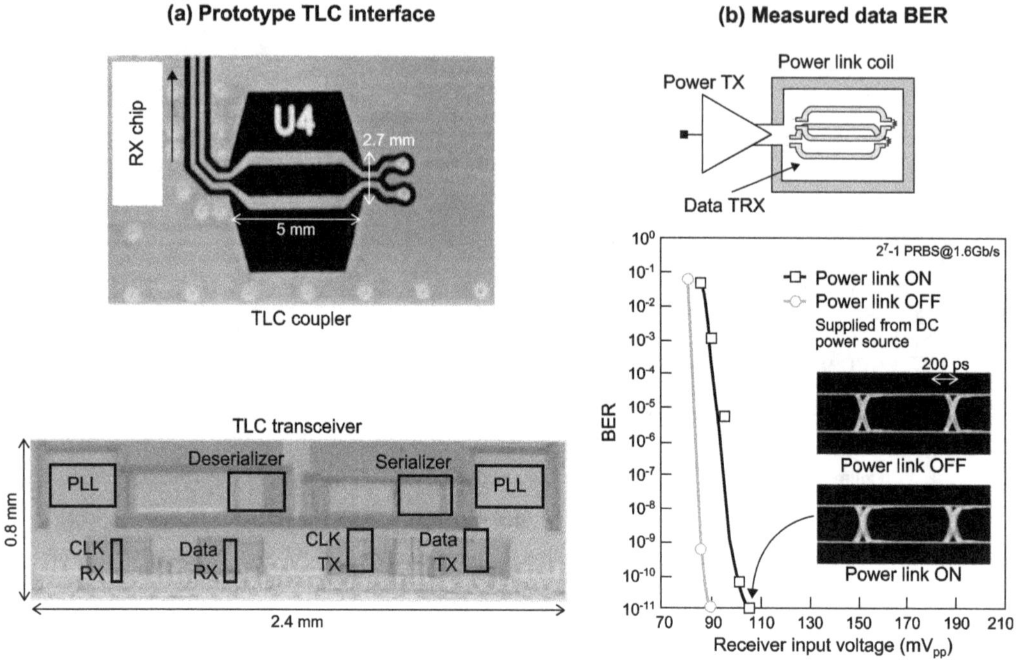

Figure 3.73 Wireless data interface prototype and measured BER. © 2016 IEEE. Reprinted, with permission, from [53]

system as a result of mode transitions between standby and read/write operation. Size and cost prevent the installation of a battery on the SSD card to stabilize the supply voltage. A shunt regulator, or voltage limiter, can be used to suppress voltage fluctuations by consuming surplus power, but that will result in increased power consumption. Instead, the vector summing scheme from [54] is adopted to minimize footprint, cost, and power in our wireless SSD system.

Figure 3.74a depicts the coupled coil design of the power interface. The transmitter coil consists of two concentric coils with a vertical offset. The two coils are driven by two clock signals CLK_1 and CLK_2 of the same frequency and amplitude but with a phase offset of θ in between (Figure 3.74c). The receiver coil, which is aligned concentrically with the transmitter coils, combines the separate power signals P_O and $P_O e^{j\theta}$ induced respectively by the two transmitter coils, resulting in an effective received power given by

$$|P_{RX}| = P_o(1 + \cos\theta). \tag{3.51}$$

Any reactive power is returned to the power supply without loss. The effective received power can therefore be dynamically adjusted by varying the phase difference θ. As shown in Figure 3.71, θ is controlled by a DLL in each of the transmitters. The DLL delays are in turn controlled by feedback from a comparator circuit in the receiver which monitors fluctuations in the memory supply voltage. The reference

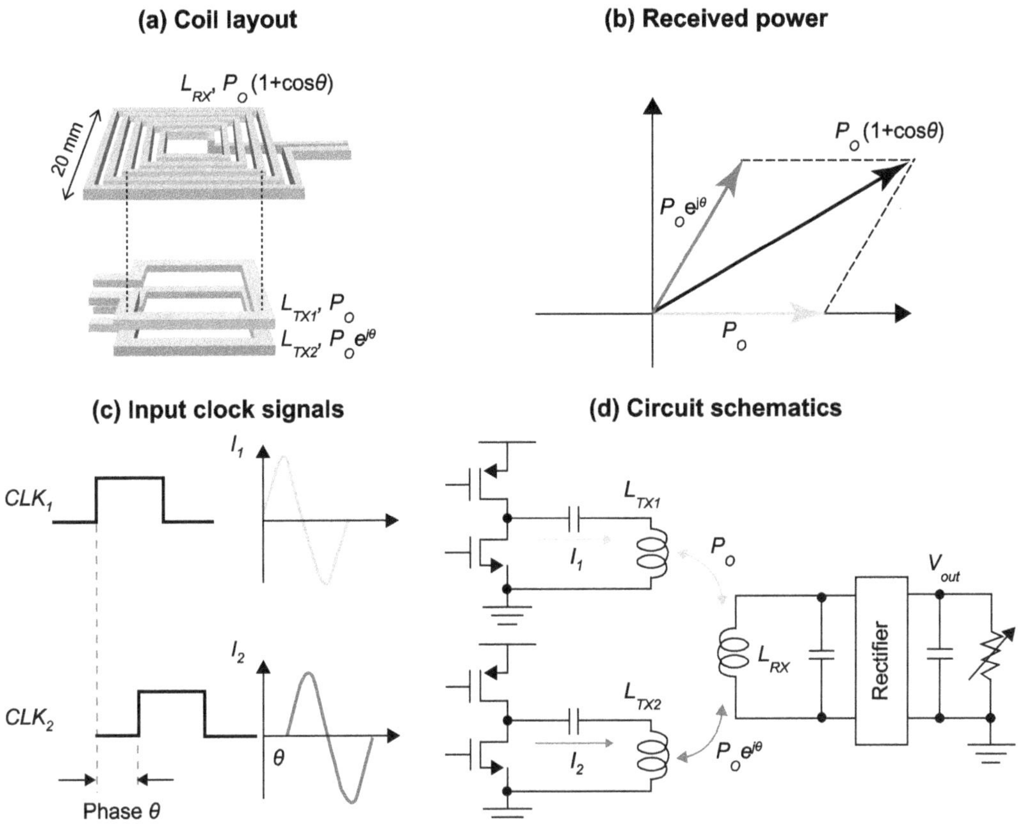

Figure 3.74 Wireless power interface. © 2015 IEEE. Reprinted, with permission, from [52]

voltage for the comparator, V_{ref}, is generated using a band-gap reference circuit. The clock signals are chosen to be 6.78 MHz from the industrial, scientific, and medical (ISM) radio bands designated for radio frequency energy transfer.

Figure 3.75a shows the fabricated power transmitter coil and receiver chip used in the prototype system. The coil measures 20×20 mm^2 and is formed on the surface of a 1.6 mm thick FR4 PCB. It delivers up to 1 W of power across a distance of 1 mm, with an output voltage of 16.3 V and maximum transmission efficiency of 61.8%. The receiver chip is fabricated in a 180 nm high-voltage CMOS process and measures 2.5×2.5 mm^2.

Figure 3.75b plots the measured power transmission efficiency as a function of the environment. The left curve shows power transmission vs. transmitter output voltage and depicts the effect of immersing the interface in water. There is a decrease in transmission efficiency of up to 10% in a water medium, but the interface still functions properly. Furthermore, the degradation is the result of water-induced warpage of the board, which increases the communication distance. Since water has relative permeability that is similar to air, it should not otherwise affect the performance of the power transmission interface that is based on inductive coupling.

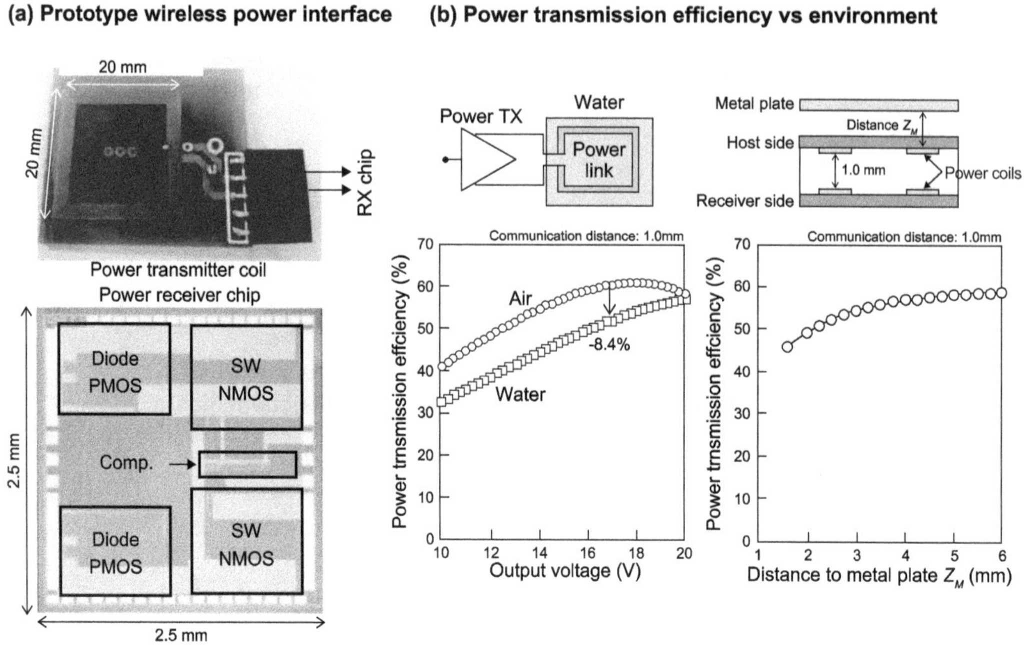

Figure 3.75 Wireless power interface prototype and transmission efficiency. © 2016 IEEE. Reprinted, with permission, from [53]

The right curve of Figure 3.75b plots the effect of a grounded metal plate placed in the vicinity of the transmitter coil. Due to eddy current induced in the metal plate, which weakens inductive coupling, power transmission efficiency drops to around 45% when the plate is in close proximity at a distance of 1.5 mm away. The degradation decreases as the plate moves away and becomes negligibly small when the plate is more than 5 mm away, which is five times the communication distance.

3.5.4 System-Level Challenges and Solutions

3.5.4.1 System Startup Sequence

If wireless power delivery has to be always on to support hot plug functionality, power consumption of the SSD system would become too excessive for a mobile device. Therefore, the system must be able to detect when a memory card is inserted before turning on wireless power delivery. Figure 3.76 illustrates the startup sequence that has been developed to support hot plug. In the absence of a memory card, the power transmitter operates only intermittently. The duty cycle of the intermittent operation is designed to be long enough to power up the memory card for initial communication once insertion is detected. The memory card uses the power to start up the wireless transceiver and SPU to send a predetermined "card inserted" signal to the host. Upon receipt of the signal, the host switches the power transmitter to continuous operation mode. Meanwhile, the host also sends a reset signal to initialize the memory card.

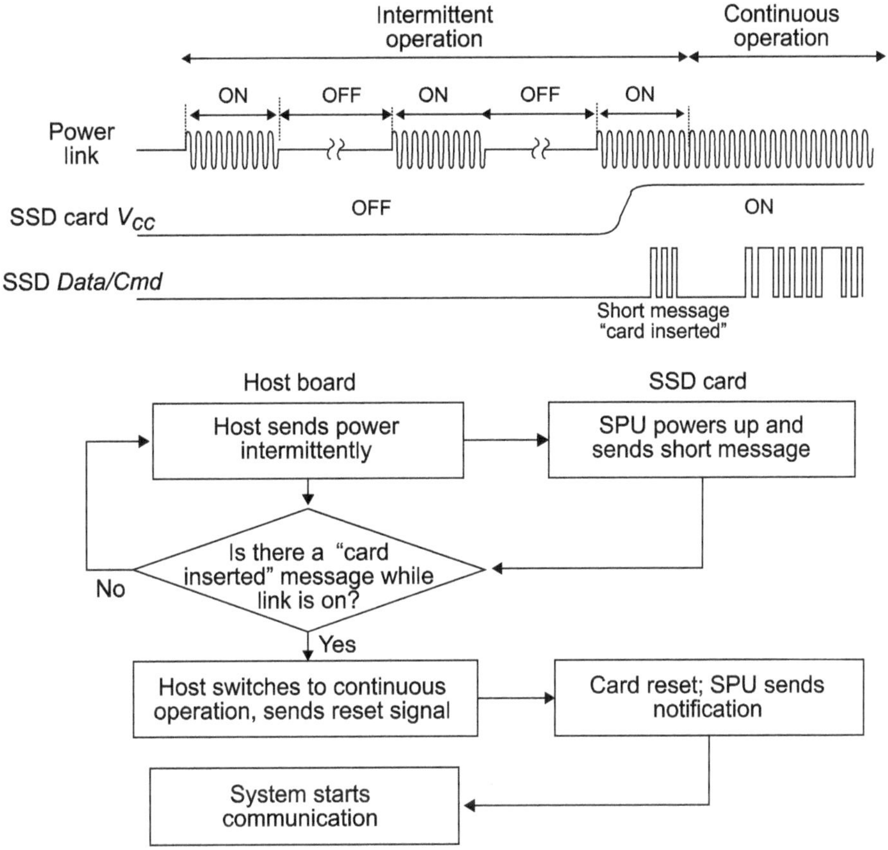

Figure 3.76 SSD system startup sequence. © 2015 IEEE. Reprinted, with permission, from [52]

Upon completion of initialization, the memory card sends a notification to the host, which then activates full system communication.

The amount of power saving from intermittent operation in the absence of a memory card depends on its duty cycle. The on-time has to be long enough to bring up the memory card voltage supply and the wireless transceiver. The former is a function of the capacitive loading on the wireless power receiver that drives the memory card supply voltage. The latter depends on the lock time of the PLLs and startup time of CDR as well as the bias generation circuits. While the latter can be reduced to 20 ns [55], the former is on the order of milliseconds. Hence the former dominates. Figure 3.77a depicts the memory card startup time as a function of its voltage supply decoupling capacitance C_d measured on the prototype system, while Figure 3.77b shows the startup waveforms. Startup time decreases to 3.5 ms when C_d is reduced to 1 μF. NAND flash memory cards on the market use C_d as low as 0.1 μF [56], in which case startup time is extrapolated to be approximately 3 ms. Therefore, an on-time of 5 ms should provide sufficient timing margin for the startup sequence to work. From a user experience standpoint, if a 1 s response time is acceptable when a

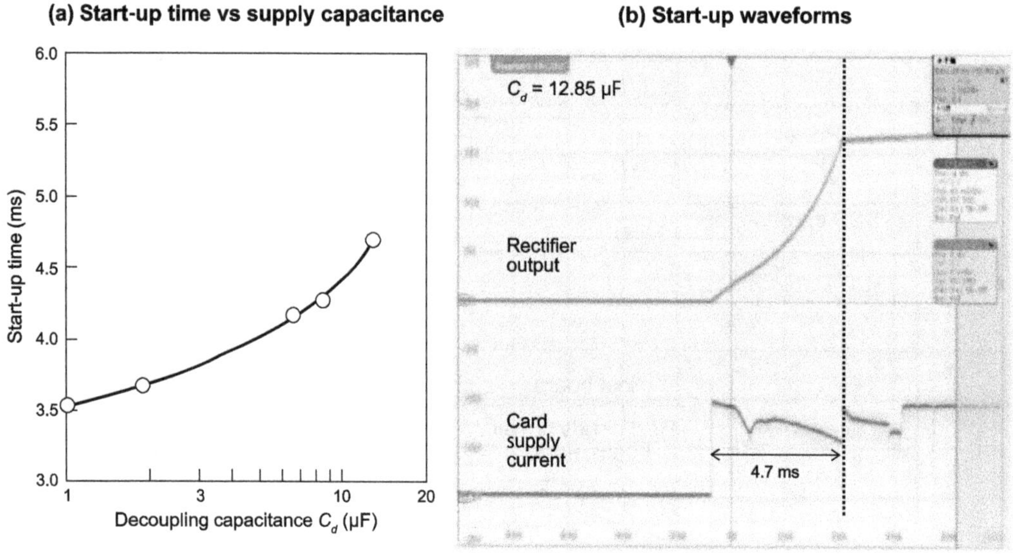

Figure 3.77 Memory card startup time. © 2016 IEEE. Reprinted, with permission, from [53]

memory card is inserted, the intermittent operation cycle can be set at 1 s, resulting in a duty cycle of 0.5% and a 99.5% saving in power compared to an always-on solution.

3.5.4.2 Wireless Interface Error Correction

In an electromagnetically noisy environment, such as inside a smartphone, the small amplitude (~100 mVpp) signal used by the TLC wireless interface is susceptible to interference from large amplitude RF signals. The interference can be suppressed using a circuit solution such as the one described in Section 3.3.2 that consists of a biphase pulse transmitter and a synchronous CDR receiver, but that adds complexity to the circuit design. Alternatively, in the SSD prototype system a postprocessing solution in the form of ECC is adopted to allow a simpler circuit solution and hence lower power. This is because ECC technique is already widely adopted in NAND flash memory for dependable operation due to increasing error rates that result from ever-shrinking memory cell dimensions [57–58]. If separate interface ECC processing is implemented independent of memory ECC processing, it would require processing on both the host and memory card sides, as shown in Figure 3.78a. ECC encoders and decoders consist of tens of thousands of gates and consume tens of mW of power [59]. Therefore, the dedicated interface ECC would require additional chips on the memory card, and that results in power consumption increase on the order of 10%. To avoid cost, form factor, and power consumption increase to the memory card, a merged ECC solution is proposed (see Figure 3.78b). It uses the existing encoder and decoder for NAND flash memory modified with a small number of additional circuits to support an increased number of ECC bits. Hence the added cost is minimized.

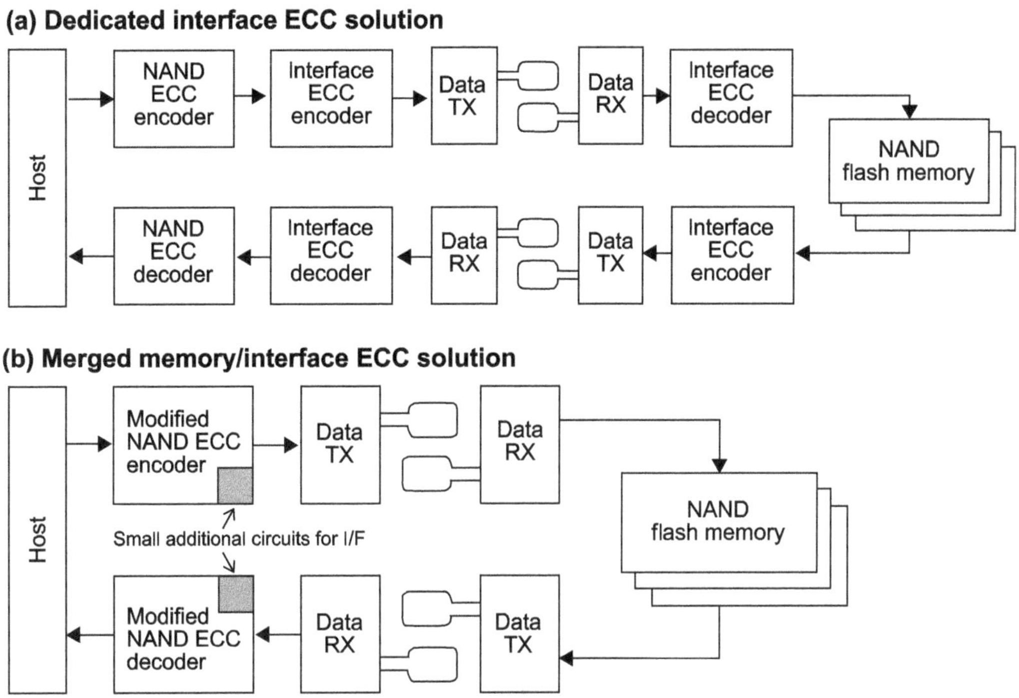

Figure 3.78 Dedicated vs. merged ECC for an SSD wireless data interface. © 2015 IEEE. Reprinted, with permission, from [52]

Figure 3.79 shows the results of an investigation to understand the nature of TLC interface errors that result from wireless interference. Figure 3.79a shows the simulation setup where a 30 dBm, 2.4 GHz RF aggressor signal is generated by a dipole antenna placed at 5 mm away from the TLC coupler. The aggressor signal saturates the TLC receiver amplifier, resulting in distorted output waveform V_{RX}. As a result, the receiver cannot restore the transmitted data correctly, resulting in bit errors. Once the receiver generates a wrong bit, it continues to hold the erroneous value until the transmitted data transitions again. Consequently, interface errors occur in bursts, as opposed to memory errors, which occur at random. The longer the transmitted data hold their value, the more continuous error bits there are, requiring more ECC bits to correct. Therefore, to reduce the number of continuous error bits, 8b/10b encoding and data scrambling are adopted to increase the number of data transitions.

The commonly used ECC schemes for NAND flash memory are the Bose–Chaudhuri–Hocquenghem (BCH) and Reed–Solomon (RS) coding schemes [60]. They offer similar performances in correcting the random errors of flash memory. But since BCH is easier to implement than RS, the former is the usual choice. However, the burst errors of the TLC interface require a different solution. Figure 3.80 shows the comparison results between the two schemes in correcting TLC interface errors as the number of ECC bits is increased, based on MATLAB simulations. BER is set to 10^{-3} for memory and 10^{-4} for interface respectively, while

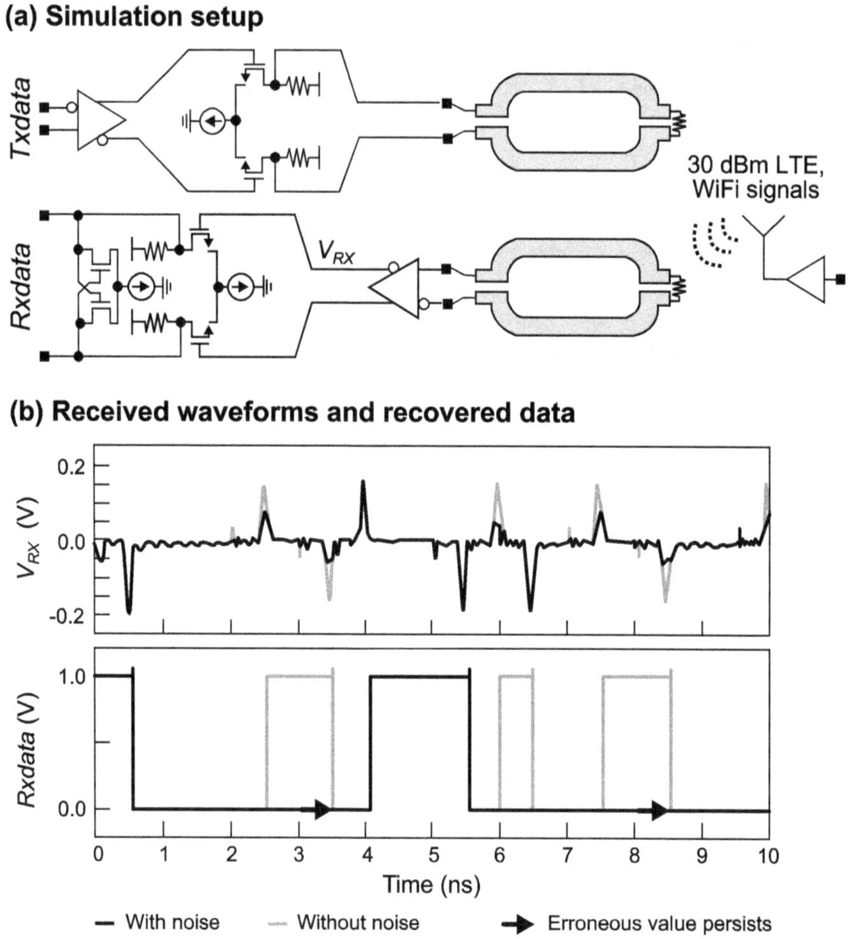

Figure 3.79 Interface errors due to wireless interference. © 2016 IEEE. Reprinted, with permission, from [53]

maximum burst error length is set to 5 bits. The data reveal that RS is much more effective in correcting burst errors. As a result, it is adopted in the prototype system. Efficient coding and decoding techniques such as those in [57] can be employed to alleviate the complexity problem of RS implementation.

Figure 3.81 plots and compares page error rates measured on the prototype system using three different interface solutions: (1) wireless interface with normal V_{DD},; (2) wireless interface with 20% lower V_{DD}; and (3) wireline interface. Error rates are compared for different numbers of bits added for interface ECC. Since the wireline interface error rate is much lower than that of NAND flash memory, the measured error rates in case (3) correspond to memory error rates. Meanwhile, the lower V_{DD} of case (2) increases BER. As a result, when the ECC length is short, its error rate is significantly

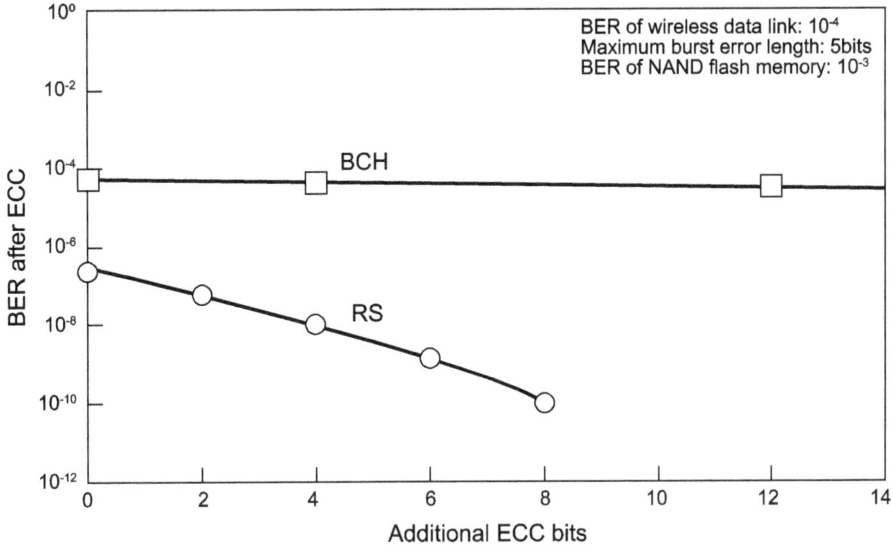

Figure 3.80 ECC schemes comparison in burst error correction. © 2016 IEEE. Reprinted, with permission, from [53]

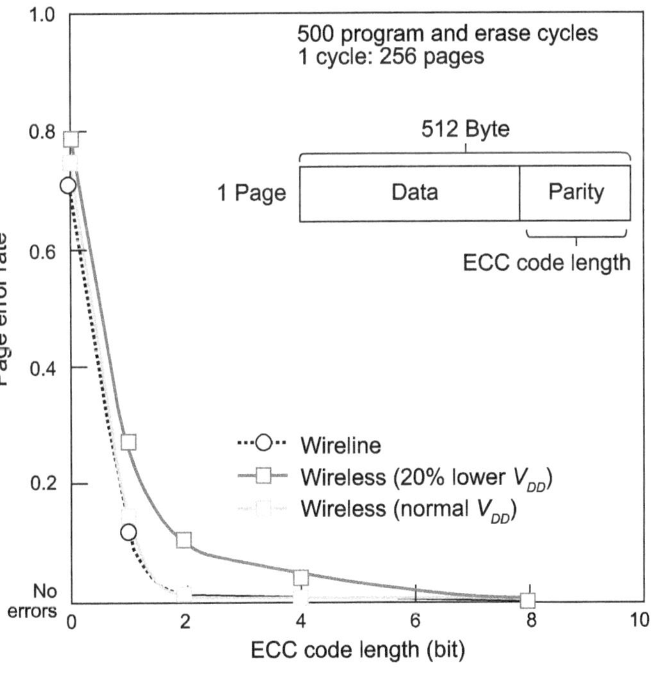

Figure 3.81 Page error rate reduction using merged ECC. © 2016 IEEE. Reprinted, with permission, from [53]

higher than cases (1) and (3). However, when ECC length is increased to 8 bits, the case (2) error rate converges with cases (1) and (3). In other words, by adding 8 bits to the NAND flash ECC, the wireless interface supply voltage can be lowered by 20% without increasing the error rate, resulting in significant power saving. Since typical NAND flash memory uses 40 bits [57], the code length increase is 20%.

3.5.5 Closing Thoughts

A TLC data interface combined with a power transmission interface based on inductively coupled coils can be used to provide a fully wireless interface between a host controller and a NAND flash SSD system. Using four TLC couplers measuring 2.7×5 mm^2 each placed inside a power coil measuring 20×20 mm^2, the combined interface can fit into a standard SD flash memory card footprint of 24×32 mm^2. Using PCB thickness of 1.6 mm and communication distance of 1 mm, the total height profile measures 4.2 mm. Even with a more mature 90 nm process for the data and a 180 nm high-voltage process for the power interface, a TLC data rate of 1.6 Gb/s/link, which is more than 20 times the 72.2 Mb/s currently available on the market, and a 1 W power transmission of up to 62% efficiency are achievable.

While the small TLC signal is susceptible to RF interference such as those that exist in a smartphone, robust data transmission can still be achieved by extending ECC correction widely adopted in flash memory to correct interface errors. The cost can be minimized by adding a small amount of circuitry into existing ECC encoder and decoder to support extra ECC bits for interface errors.

Wireless power transmission is possible even when the SSD system is immersed in water or in the proximity of metal objects. However, transmission efficiency may degrade down to 30–45%. Hence such effects must be accounted for to ensure enough power is delivered.

To support hot plug, intermittent power transmission can be used to significantly reduce power consumption while the host is waiting for the memory card to be inserted. The amount of power reduction depends on how fast the memory card can be powered up once it is attached.

We have thus demonstrated the feasibility of making the SSD memory interface completely wireless while increasing the communication data rate. This should increase memory operation reliability and robustness against harsh environmental conditions while enhancing data access performance, hence improving the versatility of flash memory as a storage solution.

References

[1] A. Kosuge. "伝送線路結合器を用いた高信頼非接触インタフェース [High reliability contactless interface using transmission line coupler]," in Japanese, Ph.D. dissertation, Department of Electronics and Electrical Engineering, Keio University, Yokohama, Kanagawa Prefecture, Japan, 2016.

[2] N. Kinayman and M. I. Aksun. *Modern Microwave Circuits*. London: Artech House, 2005.

[3] M. K. Krage and G. I. Haddad. (1970, Apr.). "Characteristics of coupled microstrip transmission lines-I: coupler-mode formulation of inhomogeneous lines." *IEEE Transactions on Microwave Theory and Techniques. 18*(4), pp. 217–222.

[4] D. M. Pozar. *Microwave Engineering,* 4th ed. Hoboken: Wiley, 2012.

[5] A. Kosuge, S. Ishizuka, M. Taguchi, H. Ishikuro, and T. Kuroda. (2015, Aug.). "Analysis and design of an 8.5-Gb/s/link multi-drop bus using energy-equipartitioned transmission line couplers." *IEEE Transactions on Circuits and Systems I: Regular Papers. 62*(8), pp. 2122–2131.

[6] N. Miura, D. Mizoguchi, M. Inoue, K. Niitsu, Y. Nakagawa, M. Tago, M. Fukaishi, T. Sakurai, and T. Kuroda. (2007, Jan.). "A 1 Tb/s 3 W inductive-coupling transceiver for 3D-stacked inter-chip clock and data link." *IEEE Journal of Solid-State Circuits. 42*(1), pp. 111–122.

[7] T. Takeya. "伝送線路結合器を用いた非接触1対1通信インタフェース [Contactless point-to-point communication interface using transmission line coupler]," in Japanese, Ph.D. dissertation, Department of Electronics and Electrical Engineering, Keio University, Yokohama, Kanagawa Prefecture, Japan, 2013.

[8] J. Poulton, R. Palmer, A. M. Fuller, T. Greer, J. Eyles, W. J. Dally, and M. Horowitz. (2007, Dec.). "A 14-mW 6.25-Gb/s transceiver in 90-nm CMOS." *IEEE Journal of Solid-State Circuits. 42*(12), pp. 2745–2757.

[9] K. Fukuda, H. Yamashita, G. Ono, R. Nemoto, E. Suzuki, N. Masuda, T. Takemoto, F. Yuki, and T. Saito. (2010, Dec.). "A 12.3-mW 12.5-Gb/s complete transceiver in 65-nm CMOS process." *IEEE Journal of Solid-State Circuits. 45*(12), pp. 2838–2849.

[10] K. Suzuki, Y. Tomita, H. Yamaguchi, T. Cheung, T. Yamamoto, and H. Tamura. "A 24-Gb/s source-series terminated driver with inductor peaking in 28-nm CMOS," *2012 IEEE Asian Solid-State Circuits Conference*, pp. 137–140, Nov. 2012.

[11] M. Hossain and A. C. Carusone. (2010, Jan.). "Multi-Gb/s bit by bit receiver architectures for 1-D partial-response channels." *IEEE Transactions on Circuits and Systems I: Regular Papers. 57*(1), pp. 270–279.

[12] H. J. Chi, J. S. Lee, S. H. Jeon, S. J. Bae, Y. S. Sohn, J. Y. Sim, and H. J. Park. (2011). "A single-loop SS-LMS algorithm with single-ended integrating DFE receiver for multi-drop DRAM interface." *IEEE Journal of Solid-State Circuits. 46*(9), pp. 2053–2063.

[13] H. Fredriksson and C. Svensson. (2009, Aug.). "Improvement potential and equalization example for multidrop DRAM memory buses." *IEEE Transactions on Advanced Packaging. 32*(3), pp. 675–682.

[14] S. J. Bae, H. J. Chi, Y. S. Sohn, and H. J. Park. "A 2 Gb/s 2-tap DFE receiver for multi-drop single-ended signaling systems with reduced noise." *2004 IEEE International Solid-State Circuits Conference*, pp. 244–525, Feb. 2004.

[15] W. Y. Shin, G. M. Hong, H. Lee, J. D. Han, S. Kim, K. S. Park, D. H. Lim, J. H. Chun, D. K. Jeong, and S. Kim. "A 4.8 Gb/s impedance-matched bidirectional multi-drop transceiver for high-capacity memory interface." *2011 IEEE International Solid-State Circuits Conference*, pp. 494–495, Feb. 2011.

[16] A. Kosuge, S. Ishizuka, M. Abe, S. Ichikawa, and T. Kuroda. "A 6.5Gb/s shared bus using electromagnetic connectors for downsizing and lightening satellite processor system by 60%." *2015 IEEE International Solid-State Circuits Conference*, pp. 434–436, Feb. 2015.

[17] W. Mizuhara, T. Shidei, A. Kosuge, T. Takeya, N. Miura, M. Taguchi, H. Ishikuro, and T. Kuroda. "A 0.15mm-thick non-contact connector for MIPI using vertical directional

coupler." *2013 IEEE International Solid-State Circuits Conference*, pp. 200–201, Feb. 2013.

[18] A. Kosuge, W. Mizuhara, T. Shidei, T. Takeya, N. Miura, M. Taguchi, H. Ishikuro, and T. Kuroda. (2014, Jan.). "A 0.15-mm-thick noncontact connector for MIPI using a vertical directional coupler." *IEEE Journal of Solid-State Circuits. 49*(1), pp. 223–231.

[19] T. Takeya and T. Kuroda. (2013, May). "Transmission line coupler design and mixer-based receiver for dicode partial response communications." *IEICE Transactions on Fundamentals of Electronics, Communications and Computer Science. 96*(5), pp. 940–946.

[20] A. Kosuge, S. Ishizuka, J. Kadomoto, and T. Kuroda. "A 6Gb/s 6pJ/b 5mm-distance non-contact interface for modular smartphones using two-fold transmission-line coupler and EMC-qualified pulse transceiver." *2015 IEEE International Solid-State Circuits Conference*, pp. 176–177, Feb. 2015.

[21] A. Kosuge, J. Kadomoto, and T. Kuroda. (2016, Jun.). "A 6 Gb/s 6 pJ/b 5 mm-distance non-contact interface for modular smartphones using two-fold transmission line coupler and high EMC tolerant pulse transceiver." *IEEE Journal of Solid-State Circuits. 51*(6), pp. 1446–1456.

[22] S. H. Lee, M. S. Hwang, Y. Choi, S. Kim, Y. Moon, B. J. Lee, D. K. Jeong, W. Kim, Y. J. Park, and G. Ahn. (2002, Dec.). "A 5-Gb/s 0.25-μm CMOS jitter-tolerant variable-interval oversampling clock/data recovery circuit." *IEEE Journal of Solid-State Circuits. 37*(12), pp. 1822–1830.

[23] J. G. Maneatis and M. A. Horowitz. (1993, Dec.). "Precise delay generation using coupled oscillators." *IEEE Journal of Solid-State Circuits. 28*(12), pp. 1273–1282.

[24] M. Mansuri, J. E. Jaussi, J. T. Kennedy, T. C. Hsueh, S. Shekhar, G. Balamurugan, F. O'Mahony, C. Roberts, R. Mooney, and B. Casper. (2013, Dec.). "A scalable 0.128–1 Tb/s, 0.8–2.6 pJ/bit, 64-lane parallel I/O in 32 nm CMOS." *IEEE Journal of Solid-State Circuits. 48*(12), pp. 3229–3242.

[25] J. Lee and M. Liu. (2008, Mar.). "A 20-Gb/s burst-mode clock and data recovery circuit using injection-locking technique." *IEEE Journal of Solid-State Circuits. 43*(3), pp. 619–630.

[26] J. Ko, J. Kim, S. Cho, and K. Lee. (2005, Jul.). "A 19-mW 2.6-mm^2 L1/L2 dual-band CMOS GPS receiver." *IEEE Journal of Solid-State Circuits. 40*(7), pp. 1414–1424.

[27] C. R. Paul. *Introduction to Electromagnetic Compatibility*, 2nd ed. Hoboken: Wiley, 2006.

[28] T. Takeya, L. Nan, S. Nakano, N. Miura, H. Ishikuro, and T. Kuroda. "A 12Gb/s non-contact interface with coupled transmission lines." *2011 IEEE International Solid-State Circuits Conference*, pp. 492–493, Feb. 2011.

[29] S. Kawai, H. Ishikuro, and T. Kuroda. "A 2.5Gb/s/ch inductive-coupling transceiver for non-contact memory card." *2010 IEEE International Solid-State Circuits Conference*, pp. 264–265, Feb. 2010.

[30] H. Cho, U. Ha, T. Roh, D. Kim, J. Lee, Y. Oh, and H. J. Yoo. "A 1.2 Gb/s 3.9 pJ/b mono-phase pulse-modulation inductive-coupling transceiver for mm-range board-to-board communication." *2013 IEEE International Solid-State Circuits Conference*, pp. 202–203, Feb. 2013.

[31] K. Hijioka, M. Matsudaira, K. Yamaguchi, and M. Mizuno. "A 5.5 Gb/s 5mm contactless interface containing a 50Mb/s bidirectional sub-channel employing common-mode OOK signaling." *2013 IEEE International Solid-State Circuits Conference*, pp. 406–407, Feb. 2013.

[32] W. J. Yun, S. Nakano, W. Mizuhara, A. Kosuge, N. Miura, H. Ishikuro, and T. Kuroda. "A 7Gb/s/link non-contact memory module for multi-drop bus system using energy-equipartitioned coupled transmission line." *2012 IEEE International Solid-State Circuits Conference*, pp. 52–53, Feb. 2012.

[33] W. Mizuhara, T. Shidei, A. Kosuge, T. Takeya, N. Miura, M. Taguchi, H. Ishikuro, and T. Kuroda. "A 0.15mm-thick non-contact connector for MIPI using vertical directional coupler." *2013 IEEE International Solid-State Circuits Conference*, pp. 200–201, Feb. 2013.

[34] R. Inti, S. Shekhar, G. Balamurugan, J. Jaussi, C. Roberts, T. C. Hsueh, and B. Casper. "A 0.5-to-0.75V, 3-to-8 Gbps/lane, 385-to-790 fJ/b, bi-directional, quad-lane forwarded-clock transceiver in 22nm CMOS." *2015 IEEE Symposium on VLSI Circuits*, pp. C346–C347, Jun. 2015.

[35] L. Reger. "The road ahead for securely-connected cars." *2016 IEEE International Solid-State Circuits Conference*, pp. 29–33, Feb. 2016.

[36] A. Kosuge, A. Okada, M. Taguchi, H. Ishikuro, and T. Kuroda. (2016, Feb.). "A 280Mb/s in-vehicle LAN system using electromagnetic clip connector and high-EMC transceiver." *IEEE Transactions on Circuits and Systems I: Regular Papers. 63*(2), pp. 265–275.

[37] Y. Kim and M. Nakamura. (2011, Mar.). Automotive ethernet network requirements. IEEE 802.1 AVB Task Force Meeting. Available: www.ieee802.org/1/files/public/docs2011/new-avb-KimNakamura-automotive-network-requirements-0311.pdf.

[38] Robert Bosch GmbH. (1991, Sep.). CAN Specification version 2.0.

[39] FlexRay Consortium. (2010, Oct.). FlexRay Communications System Protocol Specification version 3.0.1.

[40] C. C. Wang, C. L. Chen, Z. Y. Hou, Y. Hu, J. W. Lee, W. Y. Lin, Y. F. Chang, C. W. Hsu, and M. H. Song. (2015, Mar.). "A 60 V tolerance transceiver with ESD protection for FlexRay-based communication systems." *IEEE Transactions on Circuits and Systems I: Regular Papers. 62*(3), pp. 752–760.

[41] A. Gendron, C. Gill, C. Zhan, M. Kaneshiro, B. Cowden, C. Hong, R. Ida, and D. Nguyen. "New high voltage ESD protection devices based on bipolar transistors for automotive applications." *2011 IEEE Electrical Overstress/Electrostatic Discharge Symposium*, pp. 1–10, Sep. 2011.

[42] デンソーカーエレクトロニクス研究会 [Denso Car Electronics Study Group]. *図解カーエレクトロニクス 下 要素技術編 [An illustrative guide to car electronics, part 2 – underlying technologies]*. Tokyo: Nikkei Business Publications, 2010.

[43] A. Casadei and R. Broda. "Impact of vehicle weight reduction on fuel economy for various vehicle architectures." Ricardo, Inc., Van Buren Twp, MI, Research Report RD.07/71602.2, Apr. 2008. [Online]. Available: www.h3xed.com/blogmedia/Ricardo_FE_MPG_Study.pdf

[44] ISO. "Road vehicles – component test methods for electrical disturbances from narrow-band radiated electromagnetic energy – Part 4: Harness excitation methods." ISO 11452-4, Nov. 2011.

[45] CISPR. "Vehicles, boats and internal combustion engines – Radio disturbance characteristics – Limits and methods of measurement for the protection of on-board receivers." CISPR 25 Ed. 3.0, Jan. 2009.

[46] J. Kim and D. K. Jeong. (2003, Dec.). "Multi-gigabit-rate clock and data recovery based on blind oversampling." *IEEE Communications Magazine. 41*(12), pp. 68–74.

[47] A. Kosuge, S. Ishizuka, L. Liu, A. Okada, M. Taguchi, H. Ishikuro, and T. Kuroda. "An electromagnetic clip connector for in-vehicle LAN to reduce wire harness weight by 30%." *2014 IEEE International Solid-State Circuits Conference*, pp. 496–498, Feb. 2014.

[48] C. C. Wang, G. N. Sung, P. C. Chen, and C. L. Wey. (2010, Feb.). "A transceiver front end for electronic control units in FlexRay-based automotive communication systems." *IEEE Transactions on Circuits and Systems I: Regular Papers*. *57*(2), pp. 460–470, Feb. 2010.

[49] H. Pan, Y. Yao, M. Hammad, J. Tan, K. Abdelhalim, E. Wang, R. Hsu, J. Yu, J. Aziz, D. Tam, and I. Fujimori. "A full-duplex line driver for Gigabit Ethernet with rail-to-rail class-AB output stage in 28 nm CMOS." *2014 IEEE International Solid-State Circuits Conference*, pp. 148–149, Feb. 2014.

[50] Broadcom Corporation. (2014, May). BroadR-Reach® physical layer transceiver specification for automotive applications V3.0. [Online]. Available: www.ieee802.org/3/1TPCESG/public/BroadR_Reach_Automotive_Spec_V3.0.pdf.

[51] KIOXIA (formerly Toshiba). FlashAir™ product information. In Japanese. [Online]. Retrieved on Jun. 28, 2019. Available: https://jp.toshiba-memory.com/endproduct/flashair/sdwe/index_j.htm.

[52] A. Kosuge, J. Hashiba, T. Kawajiri, S. Hasegawa, T. Shidei, H. Ishikuro, T. Kuroda, and K. Takeuchi. "Inductively-powered wireless solid-state drive (SSD) system with merged error correction of high-speed non-contact data links and NAND flash memory." *2015 Symposium on VLSI Circuits*, pp. c218–c219, Jun. 2015.

[53] A. Kosuge, J. Hashiba, T. Kawajiri, S. Hasegawa, T. Shidei, H. Ishikuro, T. Kuroda, and K. Takeuchi. (2016, Apr.). "An inductively powered wireless solid-state drive system with merged error correction of high-speed wireless data links and NAND flash memories." *IEEE Journal of Solid-State Circuits*. *51*(4), pp. 1041–1050.

[54] K. Tomita, R. Shinoda, T. Kuroda, and H. Ishikuro. (2012, Nov.). "1-W 3.3-16.3-V boosting wireless power transfer circuits with vector summing power controller." *IEEE Journal of Solid-State Circuits*. *47*(11), pp. 2576–2585.

[55] T. Anand, M. Talegaonkar, A. Elkholy, S. Saxena, A. Elshazly, and P. K. Hanumolu. "A 7Gb/s rapid on/off embedded-clock serial-link transceiver with 20ns power-on time, 740 μW off-state power for energy-proportional links in 65nm CMOS." *2015 IEEE International Solid-State Circuits Conference*, pp. 64–65, Feb. 2015.

[56] Spansion. "Spansion® SLC NAND flash memory for embedded - S34MS01G2, S34MS02G2, S34MS04G2 data sheet." Sep. 2014. [Online]. Available: http://natisbad.org/NAS8/refs/S34MS01G2_04G2.pdf

[57] S. Tanakamaru, C. Hung, A. Esumi, M. Ito, K. Li, and K. Takeuchi. "95%-lower-BER 43%-lower-power intelligent solid-state drive (SSD) with asymmetric coding and stripe pattern elimination algorithm." *2011 IEEE International Solid-State Circuits Conference*, pp. 204–205, Feb. 2011.

[58] S. Tanakamaru, Y. Yanagihara, and K. Takeuchi. "Over-10x-extended-lifetime 76%-reduced-error solid-state drives (SSDs) with error prediction LDPC architecture and error-recovery scheme." *2012 IEEE International Solid-State Circuits Conference*, pp. 424–425, Feb. 2012.

[59] Y. M. Lin, C. H. Hsu, H. C. Chang, and C. Y. Lee. (2014, Jul.). "A 2.56 Gb/s Soft RS (255, 239) decoder chip for optical communication systems." *IEEE Transactions on Circuits and Systems I: Regular Papers*. *61*(7), pp. 2110–2118.

[60] Spansion. "Application Note: What Types of ECC Should Be Used on Flash Memory?" Publication Number: Types_of_ECC_Used_on_Flash_AN, Mar. 2011.

4 The Future of Computing
3D SRAM for Neural Network, eBrain

The test chip results and application examples described in Chapters 2 and 3 demonstrate the performance leaps enabled by the application of TCI and TLC near-field coupling-based wireless interface technologies to IC and module integration respectively. This is More than Moore performance scaling in the fourth dimension, beyond two-dimensional process scaling as well as process and device innovations in the third dimension, such as FinFET and 3D NAND flash technologies. Such performance leaps not only extend performance scaling in existing applications but also enable paradigm shifts in computing to address emerging applications that were not possible before.

One such paradigm shift is driven by TCI stacking of SRAM chips. Because of lower cost and higher density, DRAM has been the conventional solution for applications that require large memory capacity, even though SRAM has faster access speed and consumes less power. While the conventional DRAM bit cell is constructed using one transistor and one capacitor, the common SRAM bit cell consists of six transistors. As a result, generally SRAM memory density is only a fraction of that of DRAM. Consequently, SRAM memory is commonly used in small quantites as cache memory for fast data and instruction access and integrated with the processor on the same chip. However, using TCI to wirelessly connect multiple SRAM dice and the processor all stacked in 3D, it is possible to offer SRAM capacity that is more than an order-of-magnitude higher to dramatically expand the availability of fast memory and potentially alter how computation is done.

In the next two sections, we will explore how this TCI SRAM stack can deliver new levels of performance in the field of AI deep learning using dedicated neural network accelerators and reconfigurable FPGA respectively for processing.

4.1 Neural Net Accelerator on TCI 3D-SRAM

Figure 4.1 depicts the basic structure of a convolutional neural network (CNN), which is a class of neural network widely used in image recognition. Its key computational layers are convolutional (CONV) layers for low-level feature extraction and fully connected (FC) layers for high-level image classification. The figure also defines the different parameters of each layer type that determine the computational complexity and data storage requirement. In a CONV layer, N_i input activation maps are

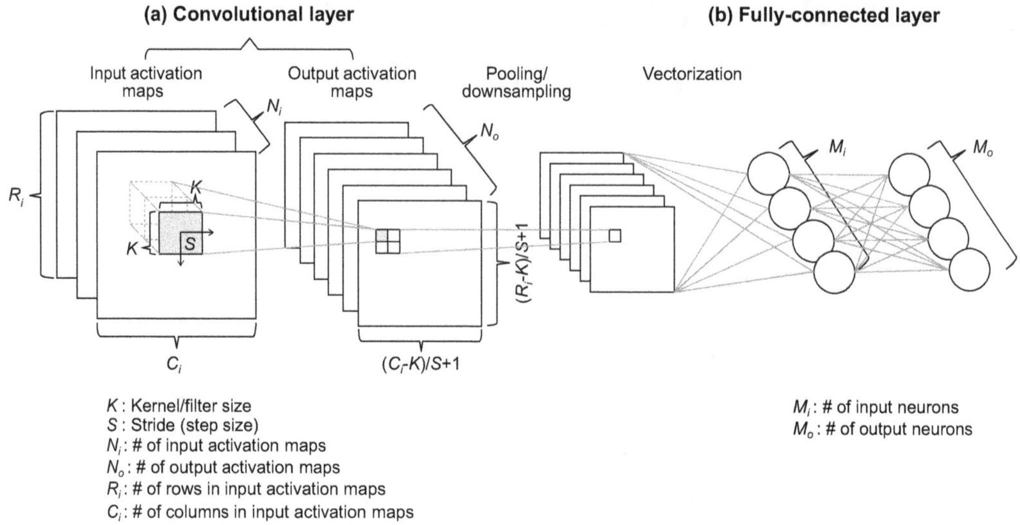

Figure 4.1 Basic structure of a convolutional neural network. © 2019 IEEE. Reprinted, with permission, from [1]

processed in parallel by applying a filter to a window of $K \times K$ in size at a time. The filter weights are multiplied with values on all N_i input activation maps and summed to create one pixel in the output activation map. The $K \times K$ window is stepped through the input activation maps with a step size of S (known as stride) to generate one output activation map. The process is repeated for N_o different filters, for example, each for a different image feature, resulting in N_o output activation maps. On the other hand, in an FC layer weights are multiplied with values on the activation maps to determine if features correlated to a particular class are present. It is called a fully connected layer because every input neuron is connected to every output neuron. A typical CNN has multiple CONV and FC layers as well as pooling (or downsampling) layers to reduce the activation map size. The computation of a CONV layer can be represented by

$$O = f\left\{\alpha \sum (W \times A) + B\right\}, \tag{4.1}$$

where O is the output activation, W the weight, A the input activation, B the bias, and $f\{\}$ a nonlinear activation function. The computation of an FC layer can be represented by a similar equation. For an inference engine, W and B are static data while A and O are dynamic intermediate data.

During image inference processing, all data are stored in memory where prestored static data W and B are first retrieved using long burst access, followed by dynamic read/write of intermediate data A and O through short burst access. A deep neural network (DNN) with many layers therefore requires large memory capacity that necessitates the use of DRAM memory. However, if memory access latency is high, retrieval of intermediate activations can take longer than the processing time of the processing element (PE) array, and computation will stall as a result (Figure 4.2). The

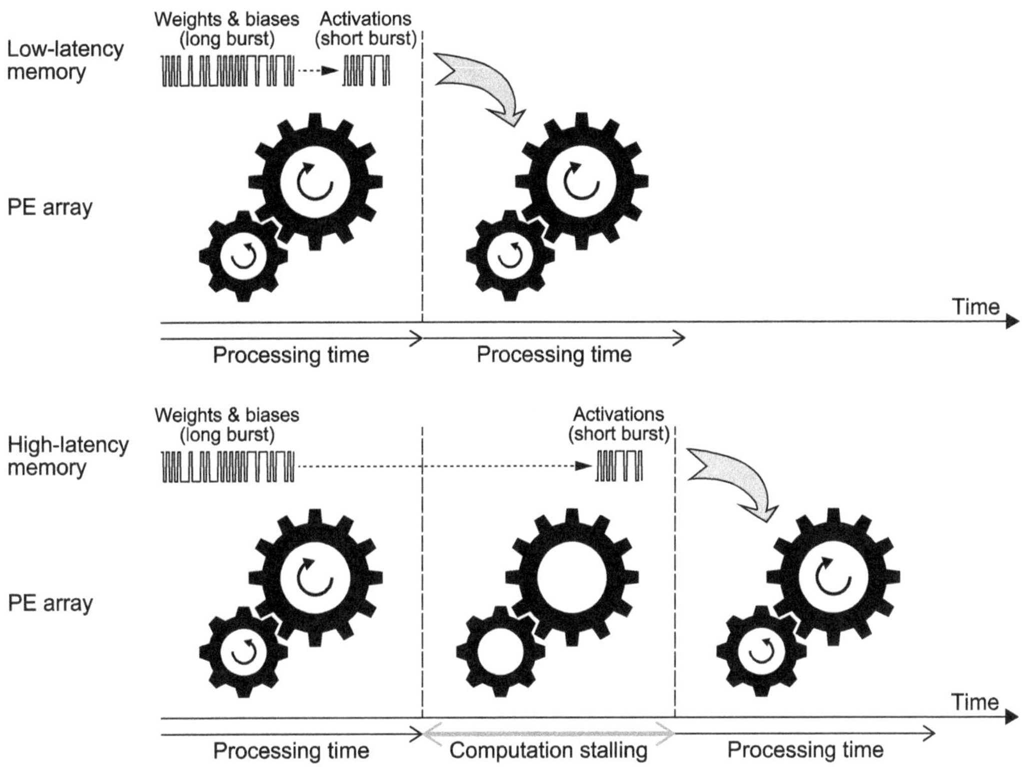

Figure 4.2 Neural network computation vs. memory access latency

rapid increase in the number of layers and model complexity of DNNs driven by their application to increasingly large and complex problems with demand for increasing recognition accuracy is creating demand for access to not only large capacity but also low latency memory. While there are proposals for 3D DRAM memory solutions – stacking multiple DRAM dice on top of a DNN accelerator – to provide large memory capacity with high bandwidth and low power [2–3], the long DRAM latency remains a challenge. By using TCI to stack SRAM chips instead on top of a DNN accelerator, the low latency of SRAM and high bandwidth of TCI can address the needs of both short and long burst access, while delivering large memory capacity at high energy efficiency to handle large neural networks with complex models. The idea has been demonstrated in a prototype system that consists of a stack of eight SRAM dice and a general-purpose DNN inference engine known as QUEST [1].

4.1.1 The DNN Inference Engine

Figure 4.3 depicts the block diagram of QUEST, which is a programmable, multiple instruction, multiple data (MIMD) parallel accelerator for general-purpose DNN inference applications. It consists of 24 processing cores connected by two on-chip

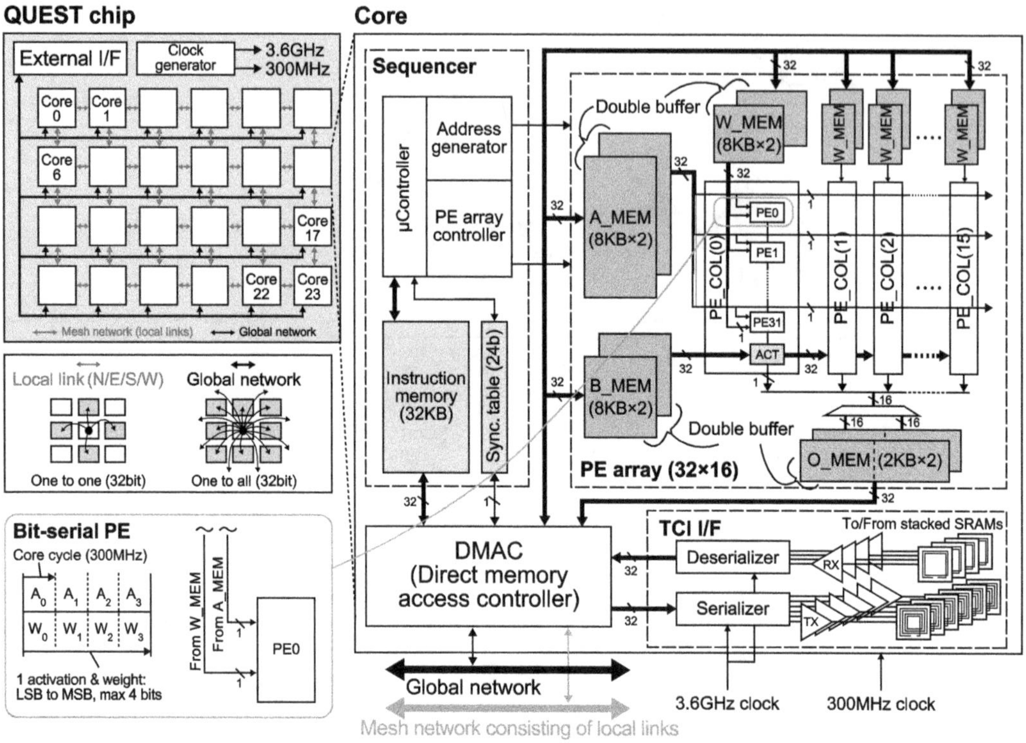

Figure 4.3 Block diagram of QUEST, a DNN inference engine. © 2018 IEEE. Reprinted, with permission, from [4]

networks – a mesh network where each core has local links to communicate directly with its immediate neighbors, and a global network that connects each core directly to all the other cores. The global network is also connected to an external interface to communicate with the outside world. Each core consists of a 32 row × 16 column PE array where computation takes place, a sequencer to control processing and data flow that includes a synchronization table to synchronize processing across cores, a TCI interface connected to its local SRAM memory vault, and a direct memory access controller (DMAC). The PE array includes internal memory blocks (shown darkly shaded in Figure 4.3) dedicated separately to *W*, *B*, *A*, and *O*. Input activations and weights are retrieved from A_MEM and W_MEM respectively for processing by the PEs. Biases are then retrieved from B_MEM and applied by the activation (ACT) block at the bottom of the PE column. Finally, the ACT block applies scaling and the activation function before writing the output activations to O_MEM. Meanwhile, the DMAC controls access to both on-chip memory and off-chip SRAM memory local to the core, as well as access to memory external to the core through the two on-chip networks.

To reduce the required memory footprint and computing hardware complexity, quantization is used to convert the numerical values of weights and activations into a

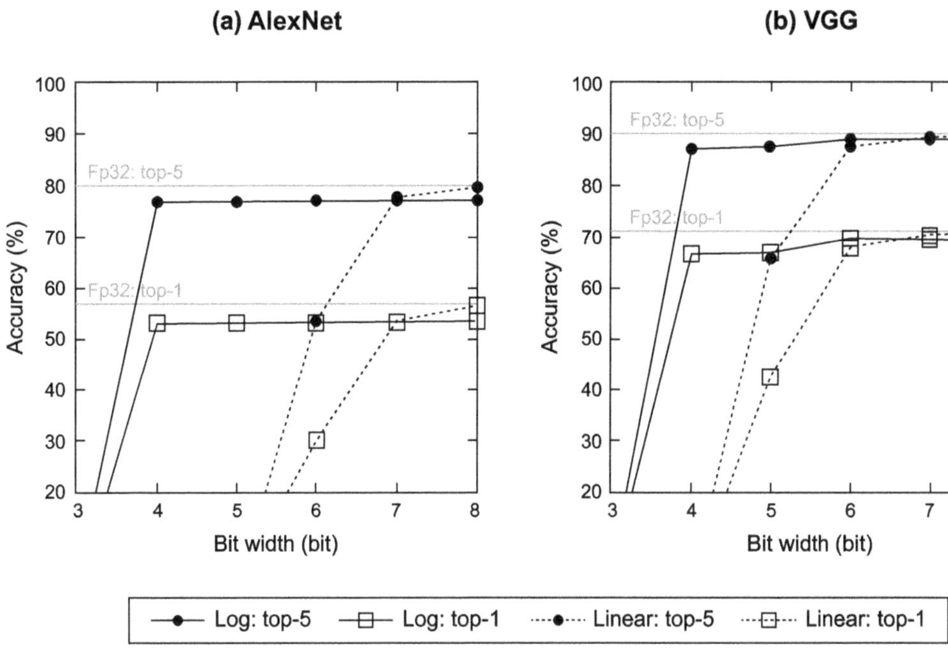

Figure 4.4 ImageNet classification accuracy of log quantization on AlexNet and VGG. © 2019 IEEE. Reprinted, with permission, from [1]

simpler format with fewer bits. In QUEST, the PE array implements a log-quantized neural datapath, which, compared to conventional linear quantization, is better in representing weight and activation distributions and converts resource-consuming multiplication into addition to boost computational performance. A 4-bit log-quantized neural computation can be completed in four cycles. ImageNet classification evaluation on AlexNet and VGG shows that accuracy degradation is small compared to the benchmark FP 32-bit (Log vs Fp32 in Figure 4.4), which achieved the best accuracy in ImageNet large-scale visual recognition challenge in 2015 [5]. Furthermore, log-quantization achieves much better accuracy than linear quantization at low bit width. In addition, since log-quantization converts all multiplications to additions, it results in a low-cost hardware implementation.

4.1.2 The TCI SRAM Stack

Figure 4.5a illustrates the TCI SRAM stack used in the prototype DNN system. The QUEST accelerator chip is placed on top of a stack of eight SRAM dice, each with 12 MB for a total of 96 MB of memory. This memory space is divided vertically into 24 independent vaults each 32-bit wide, 4 MB in capacity, and is connected directly to a matching processing core of the same physical size through a dedicated TCI channel. Figure 4.5b depicts the TCI coil layout and placement on each core and SRAM vault.

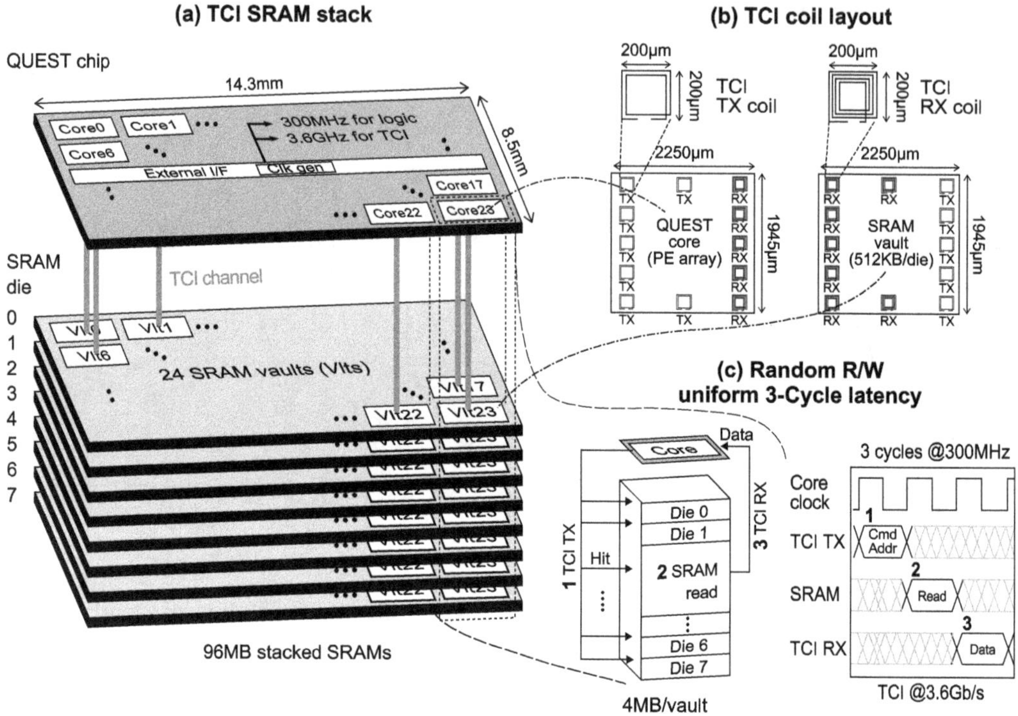

Figure 4.5 TCI SRAM stack in a prototype DNN system. © 2018 IEEE. Reprinted, with permission, from [4]

The coil diameter D is 200 μm. Given that each SRAM is thinned to 10 μm, and the use of a bumpless interchip connection makes the air gap negligible, the maximum communication distance Z across the SRAM stack is 80 μm, resulting in $D = 2.5\ Z$. Each channel consists of seven links for write and five links for read, operating at 3.6 GHz. With a 12:1 serialization ratio, the internal clock runs at 300 MHz. Since internal data width is 32-bit, the channel delivers a data bandwidth of 32*0.3 = 9.6 Gb/s/vault in a source-synchronous manner. As a result, total SRAM memory bandwidth for the entire module is 24*9.6/8 = 28.8 GB/s. With the cores running at 300 MHz, read/write (R/W) latency is three cycles and uniform across the SRAM stack as shown in Figure 4.5c.

Figure 4.6 depicts the TCI interface block diagram for both the controller on a QUEST core and SRAM vault. Command and data are serialized for transmission. The write channel includes a source-synchronous clock link, a chip select (CS), a command (Cmd), and an address (Addr) link, and three data links for a total of seven links. Meanwhile, the read channel includes a DQS and four data links for a total of five links. The variable delay on the DQS receiver can be used to optimize setup and hold time. The interface achieves an energy efficiency of 1.6 pJ/b for TX and 0.4 pJ/b for RX, for a total of 2 pJ/b in a 40 nm CMOS process.

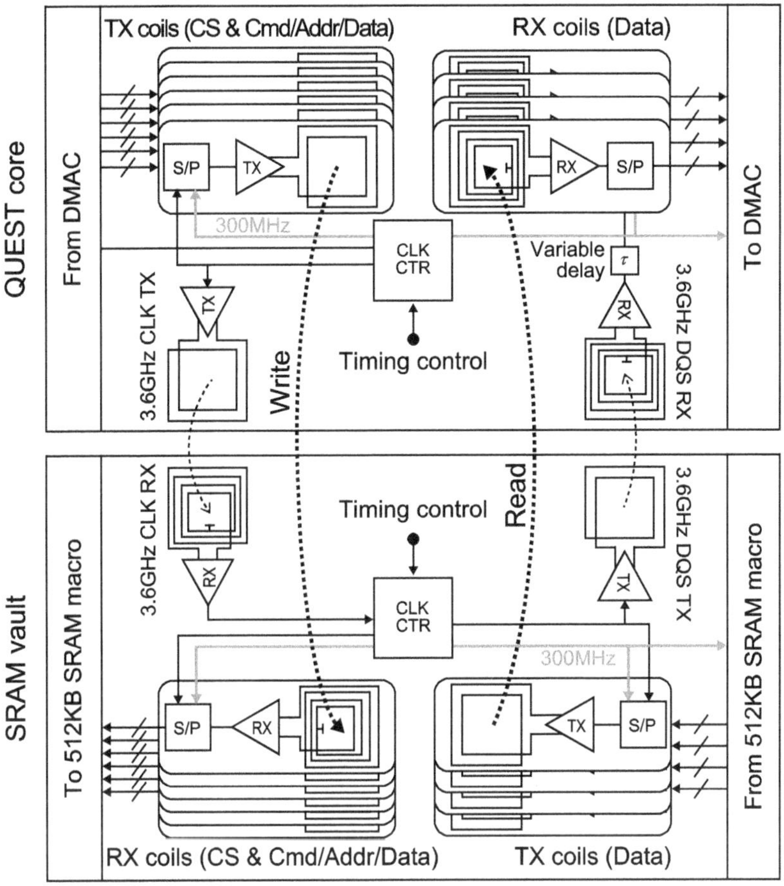

Figure 4.6 TCI interface block diagram. © 2019 IEEE. Reprinted, with permission, from [1]

4.1.3 Performance Analysis

Figure 4.7 summarizes simulation results comparing performance of the TCI SRAM stack with a DRAM solution. While the SRAM stack provides a uniform, three-cycle latency, the DRAM solution is assumed to have an initial latency of 30 cycles in burst memory access, which is an order of magnitude higher. Figure 4.7a shows that while the improvement from using SRAM is small for AlexNet, it is significant for VGG and drastic for LeNet. This is due to the smaller activation and weight bit width used by LeNet, which is one vs. four for AlexNet and VGG. The smaller bit width results in more frequent short burst access, which takes better advantage of the low latency of SRAM. There are, however, large layer-to-layer variations. With 4-bit log-quantization, the improvement can vary from 0 to >50% when using VGG for CIFAR-10 (Figure 4.7b). With sufficient on-chip memory capacity, the SRAM stack can reduce prefetch time to less than the PE execution time to conceal memory access and avoid computation stalls (Figure 4.7c).

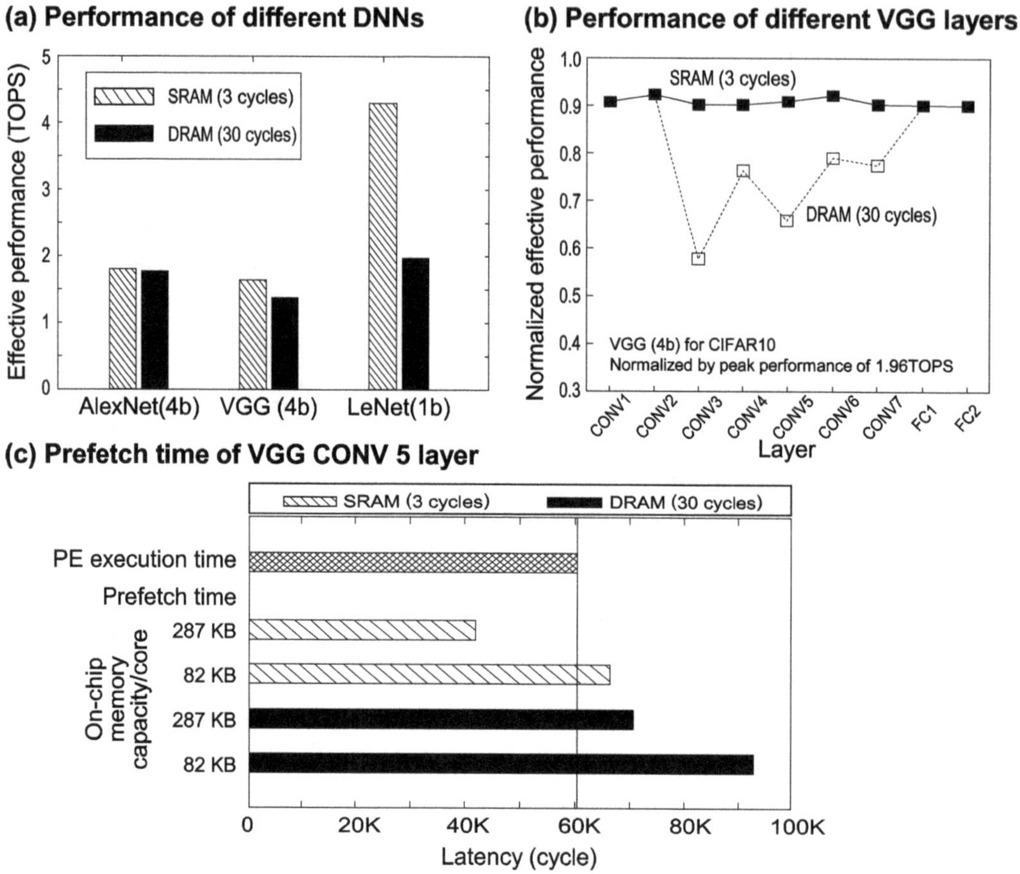

Figure 4.7 Inference performance as a function of memory latency. © 2019 IEEE. Reprinted, with permission, from [1]

Table 4.1 compares the prototpe QUEST system with three recently published DNN accelerators that use look-up table (LUT)-based and linear quantization. One is targeted at CONV, while the other two are general purpose and applicable to CONV, FC, and the recurrent neural network (RNN). The distinguishing features of the prototype QUEST system are its use of log-quantization and the TCI SRAM stack. The prototype system is shown to deliver much higher performance both in terms of peak performance and AlexNet performance. In particular, using the AlexNet benchmark, its performance is between 1.7 and 24× higher for the same 4-bit precision.

By using TCI for 3D integration of SRAM chips, an alternate high-capacity memory solution to DRAM can be created for deep neural network applications. In addition to lower power and higher bandwidth, the order-of-magnitude lower latency offered by SRAM is expected to reduce computation stalls to improve processing performance. Unfortunately, it is hard to do a meaningful direct comparison of power and energy efficiencies between this memory solution and state-of-the-art alternatives since power

Table 4.1. Comparison of recent DNN accelerators. © 2019 IEEE. Reprinted, with permission, from [1].

	[6]	[7]	[8]	QUEST
Year published	2017	2017	2017	2019
Process (nm)	65	28	65	40
Core area (mm^2)	4.0 × 4.0	1.29 × 1.45	3.8 × 3.8	14.3 × 8.5
Bit precision (bit)	4/8/16	4/8/16	8/16	1–4
Target application	CONV/FC/RNN	CONV	CONV/FC/RNN	CONV/FC/RNN
Quantization	LUT-based	Linear	Linear	**Logarithmic**
Clock freq. (MHz)	50–200	25–200	10–200	75–330
External memory	Not discussed			**TCI 3D-SRAM**
Peak performance (TOPS)	1.2@4b CONV 0.025@4b FC	0.102@16b 0.408@4b	0.410@8b	1.96@4b 7.49@1b
AlexNet benchmark				
Voltage (V)	1.1	1.0	1.2	1.1
Clock freq. (MHz)	200	200	200	300
Bit precision (bit)	4	4	4	4
Performance (TOPS)	1.088	0.076	0.368	1.825

Note: Bold indicates the distinguishing features of QUEST.

Table 4.2. Comparison in access energy between on-chip SRAM and DRAM. © 2019 IEEE. Reprinted, with permission, from [1].

	Energy (pJ/access)
100 MB SRAM	86.9
100 MB DRAM	204.9

consumption reported for the other solutions does not include external memory. Instead, access energy consumptions are estimated using CACTI 6.5 [9] and compared in Table 4.2. For the same 100 MB capacity and 32 nm process, on-chip SRAM access is shown to consume less than half the energy as DRAM. The shift of capacity from external memory to on-chip memory can thus contribute to significant saving in access energy. However, increase in on-chip memory comes at the expense of the PE array area, resulting in a trade-off between memory and processing performance. By adopting TCI 3D-SRAM, a memory solution with comparable performance that does not compromise on PE area can be achieved, which can potentially drive a paradigm shift in memory hierarchy.

4.2 FPGA on TCI 3D-SRAM: An Electronic Right Brain

4.2.1 Structural Comparison between an Electronic Left and Right Brain

As noted in Section 1.1.1, one of the two key groundbreaking inventions in the early history of the computer that drove the explosive growth of the industry was the

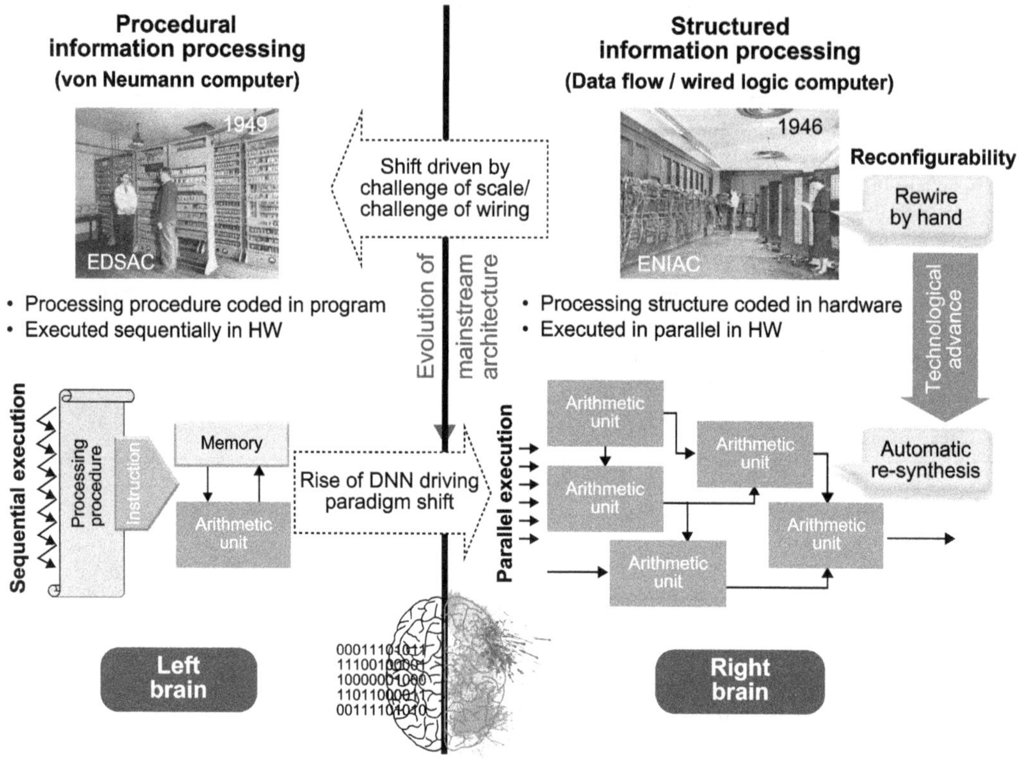

Figure 4.8 Major shifts in computer architecture. © 2018 Masato Motomura [10]. EDSAC photo: Copyright Department of Computer Science and Technology, University of Cambridge. Reproduced under CC BY 2.0. ENIAC photo: By U.S. Army, public domain

invention of the von Neumann computer. It is a general-purpose computer that addresses different applications by executing stored programs written specifically for individual applications while using the exact same hardware. Specifically, as explained in Figure 4.8, a von Neumann computer processes information following a prescribed procedure that is coded in a program. Furthermore, the procedural steps are executed one after another in sequence. Nevertheless, the von Neumann computer, with the Electronic Delay Storage Automatic Calculator (EDSAC) from 1949 as one of its earliest implementations, was actually predated by the Electronic Numerical Integrator and Computer (ENIAC), which was first introduced in 1946. Although the ENIAC was also a general-purpose computer, its programming was done through physically altering the wiring of its components – the computation executed is controlled by how its components are connected together. In other words, the hardware was based on wired logic. How information is processed is determined by the physical structure of the computer as data flows through it. It is therefore also known as a data flow or wired logic computer.

Table 4.3. Energy efficiency comparison between an HPC and the human brain. © 2018 Masato Motomura [10]. HPC (K computer) photo: by Toshihiro Matsui from Tsukuba and Yokohama, Japan - 京コンピュータ, CC BY 2.0, https://commons.wikimedia.org/w/index.php?curid=61879389, photo trimmed.

		HPC	**Human Brain**
			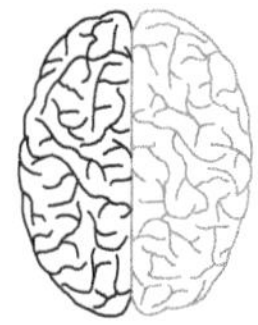
Power consumption		100 kW	10 W
Relative energy efficiency	Shogi (Japanese chess)/Go	1	10^4
	Arithmetic operations	10^8	1
Architecture		Von Neumann arithmetic operator	Data flow/wired logic neural network
		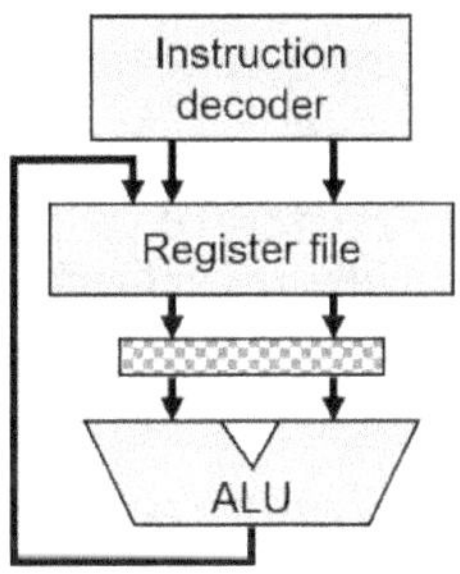	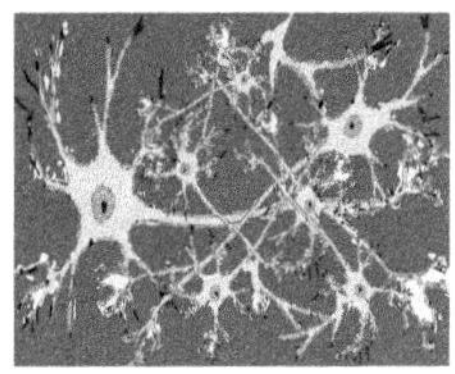

In a data flow computer, programs are mapped onto the structure of the hardware. Consequently, the size of problems it can solve is limited by the scale of the hardware, resulting in the Challenge of Scale. Furthermore, increasing scale of the hardware results in the Challenge of Wiring, where increasing wiring complexity makes it impractical to program the computer. Because of these challenges, the later-born stored program von Neumann computer became the mainstream architecture instead, and has remained so until today.

In the Chinese language, a computer is known as a Diànnǎo, which literally means "electrical brain." Indeed, a modern-day computer can perform a lot of the same tasks as and much better than the human brain. This is illustrated in Table 4.3, which shows a high-performance computer (HPC) as having eight orders-of-magnitude better energy efficiency in performing arithmetic operations than the human brain. Nevertheless, in playing Japanese chess and Go, the same HPC is four orders-of-magnitude lower in energy efficiency than the human brain. Arithmetic operations are logical processing of information, which is the domain of the left brain. Meanwhile, playing chess and Go requires intuition and visualization, which is more the domain of the right brain. Because the von Neumann computer excels only in left-brain tasks, regardless of how powerful it is, it can only mimic the left brain. But with the recent advances in artificial intelligence, we are starting to see AI computers approaching the human brain in certain right-brain tasks such as image classification. Indeed, DNNs, which have been responsible for the advances in AI, are originally modeled after the neural network in the human brain, as opposed to the arithmetic operator-based architecture of a von Neumann computer.

In a DNN, how input data are processed is determined by how the neurons are physically connected together. Furthermore, processing occurs in parallel. In other words, it operates like a data flow computer, like the ENIAC that predated it by almost seven decades. The recent rise of DNN is therefore driving a paradigm shift in computer architecture back from the von Neumann computer that performs like the left brain, to the data flow/wired logic computer that performs like the right brain. History seems to be coming full circle. One important difference, though, between data flow computer then and now is in its reconfigurability. With the ENIAC, reconfiguration to solve a different class of problems would require more than weeks of work to map the new program onto the machine, rewire the machine, and verify and debug. It was impractical and certainly not applicable to problems that require real-time reconfigurability. With technological advance, real-time, automatic resynthesis to create a new hardware configuration has become a possibility. It may finally enable us to build an electronic right brain that rivals the human brain.

4.2.2 Construction of an Electronic Right Brain

As discussed in Section 1.1.2.6, the semiconductor industry is entering its second spring, where innovations come primarily not from increasing quantity but from enhancing quality – from how components are assembled to best solve specific classes of problems. A representative example of the idea can be found in domain-specific processors that are designed to achieve optimal processing performance in a specific application domain. In particular, in the field of DNN, which is increasingly being applied to solve right-brain problems, the shift to domain-specific processors has given rise to accelerators, which offload specific types of neural network processing from the host processor in order to complete the processing in a more cost-effective way, especially in terms of energy efficiency.

Chapter 7 of [11] provides a comprehensive discussion of domain-specific architecture (DSA) in general, and DNN accelerators in particular. DNN has become a prime example domain of DSA because of its application to a widening range of problems, including image processing, speech recognition, automatic translation, and search, which are all killer applications of the day. Furthermore, the wide range of applications helps to enhance the reusability of DNN hardware. DSAs are by nature expensive because they require hardware that is built to solve only a small subset of all the problems a general-purpose computer can solve. With rapidly rising nonrecurring

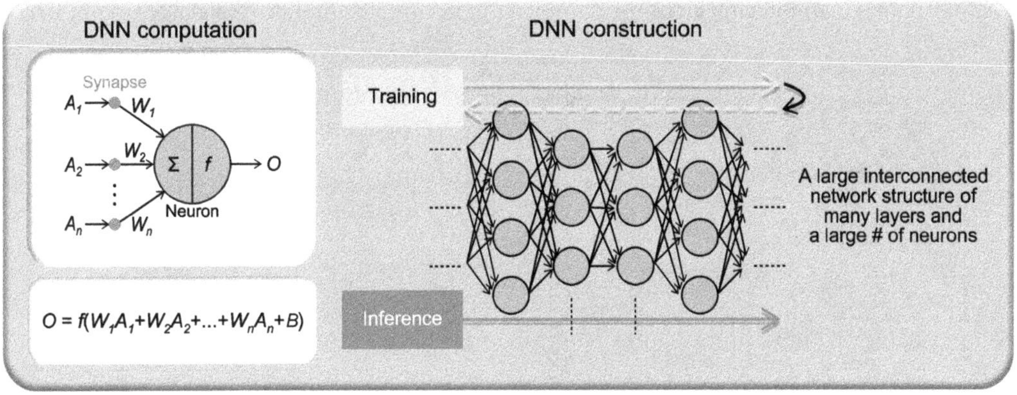

Figure 4.9 DNN computation and basic construction. © 2018 Masato Motomura [10].

engineering (NRE) costs, which require an increasingly larger volume of chips to amortize, the widening range of DNN applications is a welcome trend.

Figure 4.9 presents an abstract view of the operation of DNN. The key feature is that each neuron performs a simple computation that is fixed, although the coefficients W and B can vary. The relationship between the outputs and inputs – in other words, the type of overall processing executed by the DNN – is therefore determined entirely by how the neurons are connected together. As such, it has the construction of a data flow computer and performs structured information processing.

The direction of data flow depends on which stage the DNN is in. In the training stage, different sets of sample data are fed into the network and flow from the input to the output side. The resultant outputs are then used to compute errors that are propagated back through the network in the opposite direction. These errors are used to tune the network parameters to improve accuracy. In the inference stage, input data are processed as they flow through the network, generating the requested output.

To provide specific examples, [11] presents details of four state-of-the-art DNN accelerators, two implemented in an application-specific integrated circuit (ASIC), one as an ASIC/SoC IP core, and one in an FPGA. One important motivation for using an FPGA is to provide cost-performance improvement for a larger range of applications than an ASIC, through its configurability. In particular, the example FPGA-based accelerator has been configured for both CNN and search applications.

The difference between the FPGA-based accelerator and the ASIC-based accelerators are similar to the difference between the ENIAC and the computer that Turing built to break the Enigma machine. While both are data flow machines performing structured information processing in a way defined by the hardware wiring, the former is designed to be programmable while the latter optimized for one specific set of problems. The trade-off is between more versatility for the FPGA-based accelerator, and better performance optimization for the ASIC-based accelerators.

The configurability of the FPGA can be exploited both statically and dynamically. Different accelerators deployed in different servers can be configured differently to

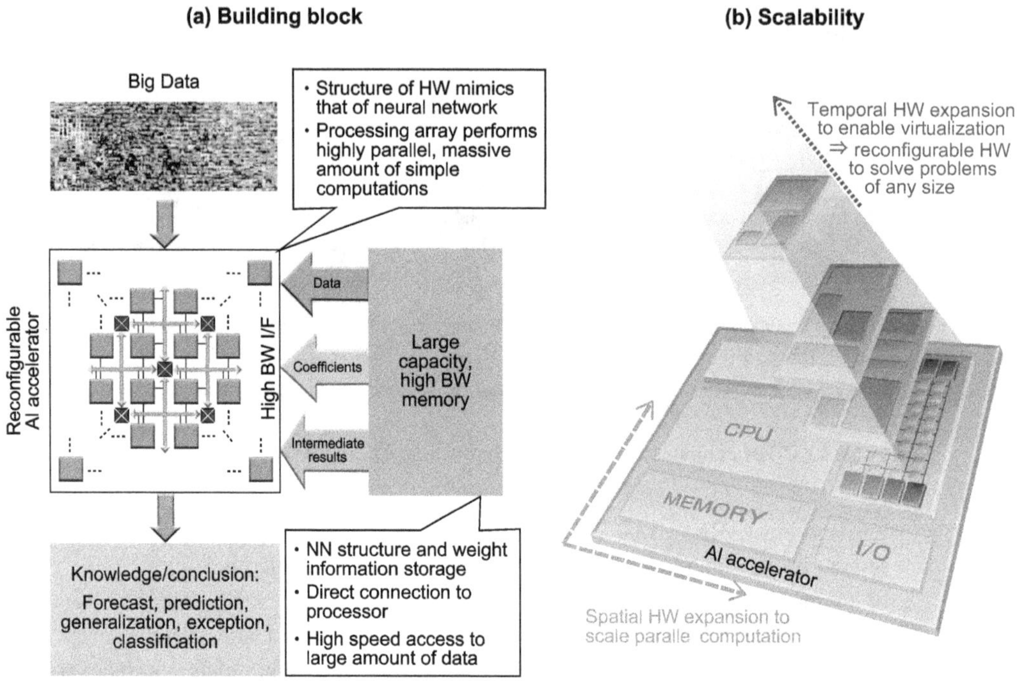

Figure 4.10 Construction of an electronic right brain. © 2018 Masato Motomura [10]

accelerate processing of different applications. At the same time, accelerators in a particular server can be reconfigured differently over time to accelerate new applications. Taken to the extreme, the FPGA can be reconfigured in real time dynamically to step through different processing structures to perform different classes of subcomputation that are required for solving a complex problem. Such is the concept of a time-multiplexed FPGA as proposed in [12].

In [12], the authors propose an architecture where a small number of FPGA configurations are stored in SRAM that can be retrieved to reconfigure the chip. It is possible to switch between configurations dynamically in as little as 30 ns using technology from two decades ago when the paper was published. The architecture supports three operating modes. In particular, in time-share mode, a different, independent FPGA on the same virtual network is realized in each time slot. As a result, one physical FPGA can be used to realize a network of independent FPGAs each using the same computer resource but configured for solving a different part of a complex problem and activated in a different time slot.

A high-performance electronic right brain can thus be implemented by combining a time-multiplexed FPGA with TCI 3D-SRAM that has large capacity and access bandwidth (Figure 4.10). The FPGA is configured to mimic the structure of a neural network, where the processing array performs a massive amount of highly parallel, simple computations at high speed, supported by fast, low-latency access to a large amount of data, coefficients, and intermediate values from memory. Parallel computation can be scaled up through physical expansion by increasing the processing array size and memory capacity. At the same time, virtualization through temporal expansion using time-multiplexing scales the size of problems that can be solved. Real-time dynamic reconfiguration is supported by fast access to configuration parameters stored in memory. By expanding virtualization, hardware of a fixed size can be used to solve problems that are many times bigger in size.

4.2.3 FPGA Right-Brain Performance Leaps Enabled by TCI 3D-SRAM

Two of the main challenges in adopting FPGA are system development costs and power. TCI 3D-SRAM can be used to address both. Figure 4.11 compares the memory capacity available to a FPGA on TCI 3D-SRAM with that from Xilinx's UltraScale+ series. A 40 nm TCI 3D-SRAM stack offers 12.7 kb/LUT of memory capacity, which is more than 40 times that provided by the UltraScale+ series fabricated in a 16 nm process. Such orders-of-magnitude larger SRAM capacity can be used to achieve two important benefits for FPGA – shorter development cycle and higher energy efficiency.

Given that random access to internal SRAM memory is an order of magnitude faster than to external DRAM memory, if there is enough SRAM capacity to replace DRAM, developers can leave the details of memory access to the compiler, which can significantly reduce the complexity of programming and hence shorten development time.

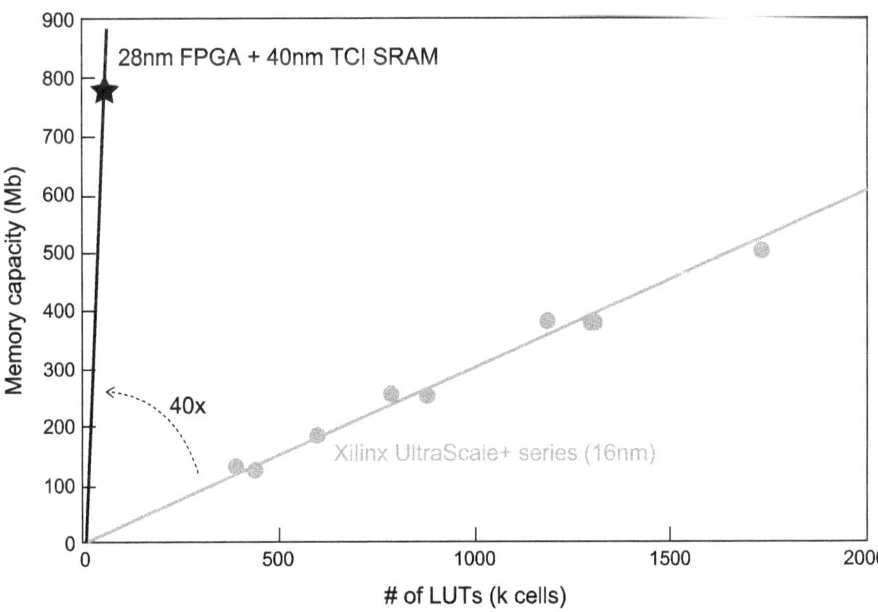

Figure 4.11 Memory capacity of FPGA on TCI SRAM.

Table 4.4. Complexity of six popular DNN applications. This table is based on a table published in *Computer Architecture*: A *Quantitative Approach,* J. L. Hennessy and D. A. Patterson, figure 7.5, p. 547. Copyright Elsevier 2017.

Application	# of weights	Operations/weight	Description
MLP0	20 M	200	MLP = multilayer perception
MLP1	5 M	168	
LSTM0	52 M	64	LSTM = long short-term memory
LSTM1	34 M	96	
CNN0	8 M	2,888	CNN = convolutional neural network
CNN1	100 M	1,750	

Table 4.5. Relative energy costs per operation. This table is based on a table published in *Computer Architecture*: A *Quantitative Approach*, J. L. Hennessy and D. A. Patterson, figure 7.32, p. 584, Copyright Elsevier 2017.

Operation	Energy consumption	Remark
8b LPDDR3 DRAM access	125 pJ	
8b SRAM access	17.1 pJ	High-end estimate to account for overhead due to large capacity
32b floating point arithmetic	2.7 pJ	Assuming use of DSP

On the other hand, the higher energy efficiency of TCI 3D-SRAM can alleviate the FPGA power problem. To quantify the potential improvement, energy efficiencies are estimated for six popular DNN applications, and compared between a conventional FPGA with 300 Mb of SRAM, and a theoretical FPGA with 8 GB of SRAM capacity. The assumptions used for the estimation are listed in Tables 4.4 and 4.5 respectively. Energy costs per operation are based on a TSMC 28 nm HPM process. The DNN architecture assumed is Google TPU and all weights are 32b.

The estimated energy consumption per weight for the different applications and memory capacities are summarized in Table 4.6. Since the comparison is between two different SRAM capacities, the most improvement is derived when memory access dominates total energy consumption, and when 8 GB provides sufficient capacity to avoid external memory access while 300 Mb does not. Under such a scenario, energy efficiency is estimated to be more than doubled (LSTM0 and LSTM1).

Figure 4.12 illustrates an example FPGA on a TCI 3D-SRAM system that is constructed in a similar way as the prototype system described in Section 4.1. The FPGA chip is segmented into 24 independent cores, each with a TCI channel for access to a local SRAM vault.

Figure 4.13 illustrates the block diagram for each FPGA core. In addition to being connected to a TCI interface for access to the stacked SRAM, each FPGA core is also connected to a switch for communication with its immediate neighbors, as well as with external chips through another TCI interface.

Table 4.6. Improvement in energy efficiency due to increased SRAM capacity.

	Energy consumption/weight			
Application	**300 Mb SRAM**	**8 GB SRAM**	**Improvement**	**Remark**
MLP0	837.7	608.4	1.4×	Power dominated by arithmetic operations
MLP1	522.0	522.0	1.0×	300 Mb is sufficient for the small # of weights
LSTM0	595.0	241.2	2.5×	
LSTM1	640.2	327.6	2.0×	
CNN0	7866.0	7866.0	1.0×	300 Mb is sufficient for the small # of weights
CNN1	5184.5	4793.4	1.1×	Even 8 GB is insufficient for the large # of weights

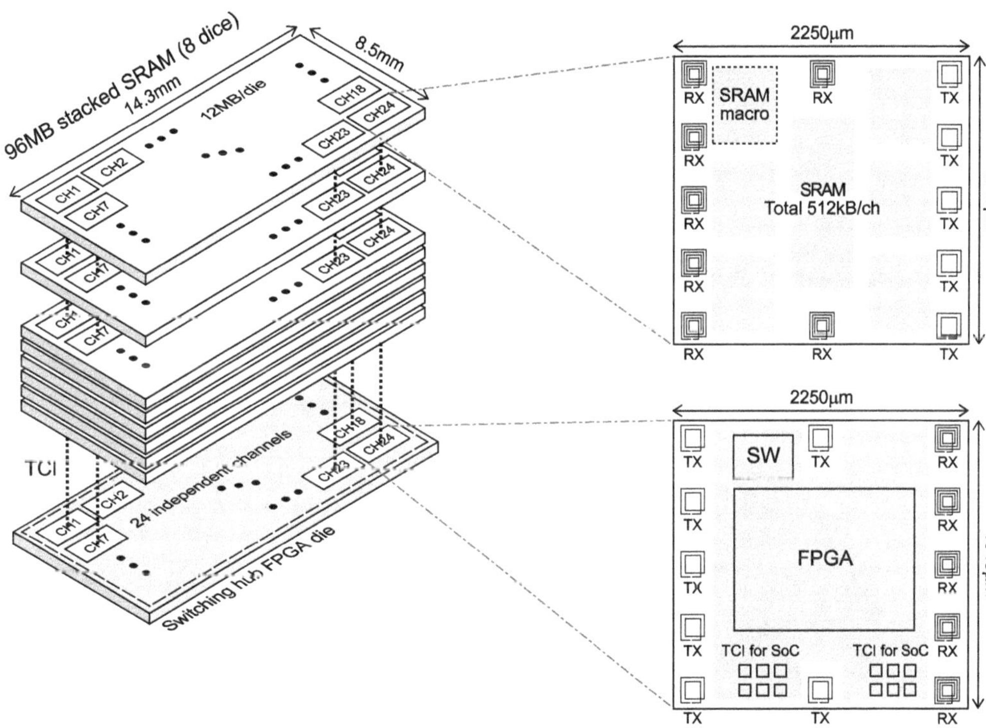

Figure 4.12 Example FPGA on a TCI 3D-SRAM system.

4.3 eBrain: A True Electronic Brain

Figure 4.14 illustrates the requirements for a true electronic brain, a true Diànnǎo. It has two halves, with the left half delivering comparable or better performance (in

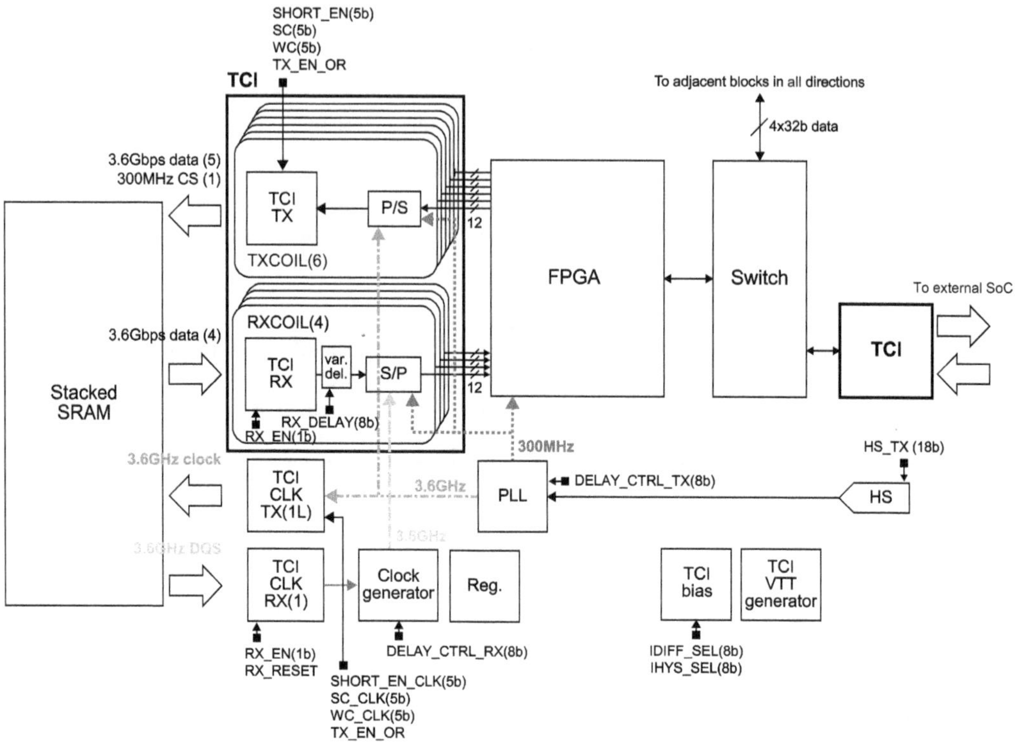

Figure 4.13 Block diagram for each FPGA core.

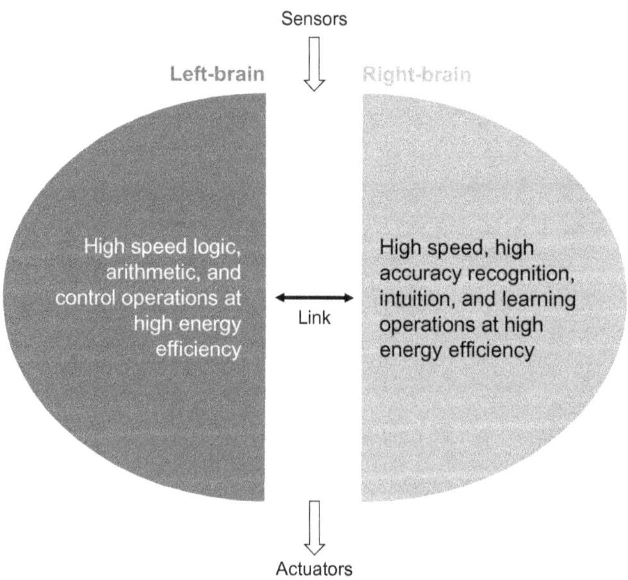

Figure 4.14 Requirements for an electronic brain.

speed and energy efficiency) in logic, arithmetic, and control operations, and the right half delivering comparable or better performance (in speed, energy efficiency, and accuracy) in recognition, intuition, and learning tasks than the human brain. Furthermore, there needs to be a link between the two halves so they can work together to solve problems that require skills from both, and the ability to interact with the outside world – accepting data from sensors and driving actuators.

4.3.1 Construction of an eBrain

As depicted in Table 4.3, the explosive growth in performance of the IC in the last 60 years has created high-performance von Neumann machines that perform arithmetic operations with eight orders-of-magnitude better energy efficiency than the human brain. Nevertheless, they still consume too much power, have too large a form factor, and are constrained by the von Neumann bottleneck in memory access. By adopting TCI, memory and processing chips can be integrated in 3D to alleviate the von Neumann bottleneck while improving energy efficiency and form factor, resulting in a better electronic left brain (Figure 4.15a). The concept has been proven in the wideband DRAM interface introduced in Section 2.6. Using a conservative data rate of 2.22 Gb/s/link, the prototype achieves more than double the 256 GB/s bandwidth of standard HBM2 and reduces footprint and module height through 2.9D integration.

Meanwhile, using the neural network accelerator on TCI 3D-SRAM introduced in Section 4.1, a right brain can be constructed with performance that is an order of magnitude better, and energy efficiency that is 100% better than alternate DNN

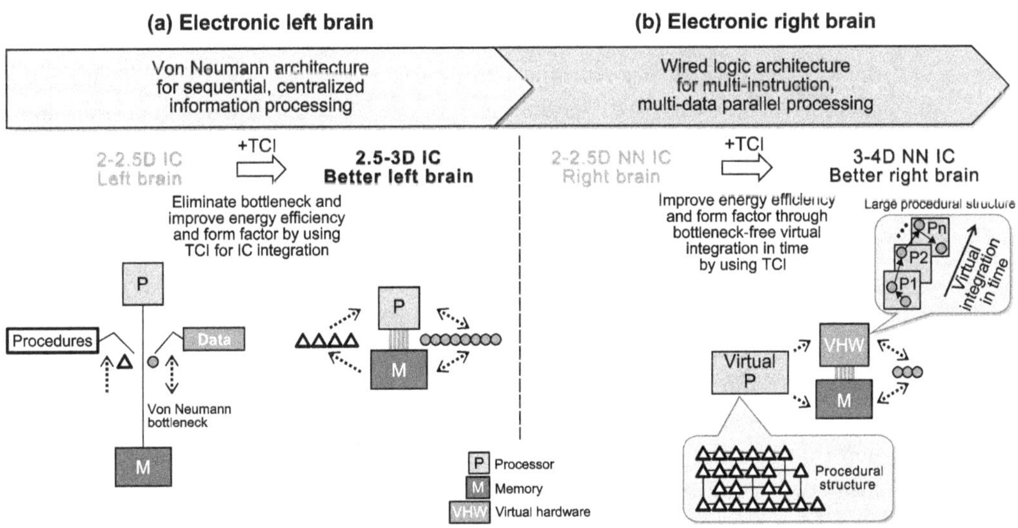

Figure 4.15 TCI-enhanced electronic left and right brains.

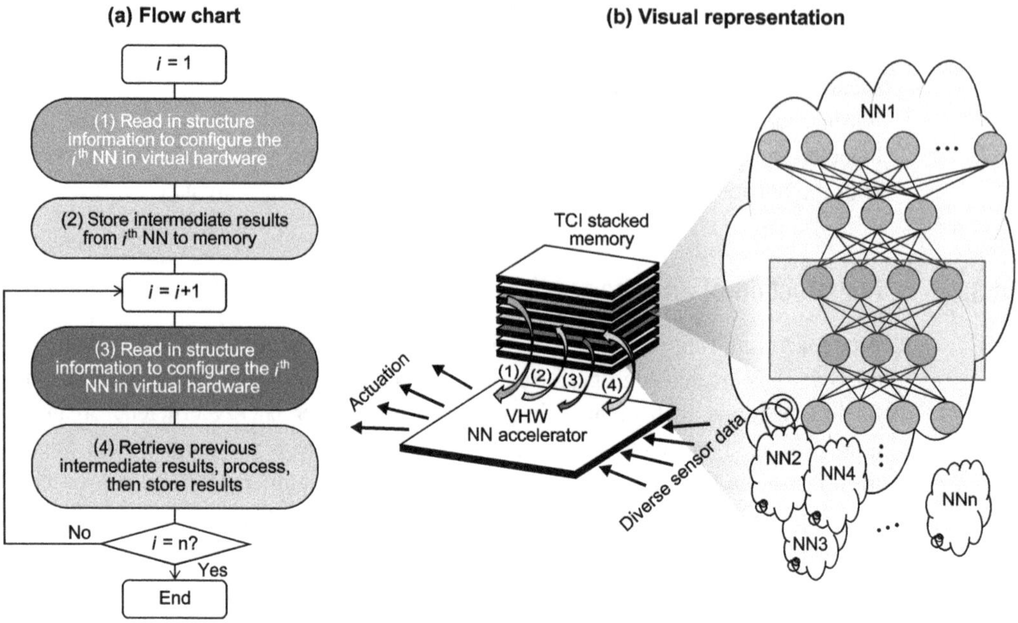

Figure 4.16 Operation of a VHW-NN accelerator.

accelerators (Table 4.1). Nevertheless, being a wired logic computer, the scale of problems it can solve is limited. By adopting a reconfigurable FPGA for the neural network accelerator as discussed in Section 4.2, a virtual piece of hardware (VHW) can be created by integrating multiple virtual processors in time, each realized with the same FPGA, to implement a large procedural structure for solving problems that are many times bigger than the size of the hardware. The adoption of TCI for the stacked SRAM ensures high energy efficiency and small form factor while enabling bottleneck-free virtual integration through low-latency and high-bandwidth memory access to retrieve configuration information and exchange data (Figure 4.15b).

Figure 4.16 illustrates the operation of a VHW neural network accelerator that consists of n virtual processors, each implementing a different neural network. As described in the flow chart, the TCI stacked memory is used to store both configuration/structure information and for passing intermediate processing results between virtual processors. It is therefore necessary for the memory access to be low latency and high bandwidth, as well as highly energy efficient to ensure smooth virtualization in time. There is no theoretical limit on the number n of virtual processors that can be implemented. The practical limit is determined by the capacity and thickness of the memory chips. The chip thickness also determines the size of TCI coils. As memory manufacturing technology scales and chip thinning capability improves, more and more capacity will become available to support more and more virtual processors, creating a bigger and more capable right brain.

4.3.2 A Palmtop eBrain

The TCI solution from [13] offers improvement between 9% and 3.4×, depending on the performance metric, over alternate state-of-the-art solutions of the time (Table 2.12). It was demonstrated in 2014 using a 180 nm CMOS process. If the solution is implemented using the most advanced process and the latest thinning technology, it can result in a further quantum leap in performance. This is illustrated in Table 4.7 comparing a 7 nm process implementation with that from [13]. Performance of the 7 nm implementation is estimated through simulation using Predictive Technology Model from [14]. The implementation is projected to deliver 28–65× improvement in energy and area efficiency respectively. More aggressive chip thinning will further improve performance. Since coil diameter D is proportional to communication distance Z, which in turn is proportional to chip thickness, reduction in chip thickness results in a quadratic improvement in area and hence area efficiency. It is noted in Section 2.5.3 that a chip thickness of a few microns is feasible, and the prototype in Section 4.1.2 is built using a 10 μm thickness. A state-of-the-art implementation of the interface in [13] can therefore be more than an order of magnitude faster and smaller while consuming more than an order of magnitude less energy. This can open up a whole new world of possibilities, including the possibility of creating a palmtop eBrain.

Figure 4.17 illustrates how a palmtop eBrain may be constructed using TCI and TLC. Each function – left brain, right brain, storage, interface – is implemented as a standalone module using TCI to improve its energy efficiency and reduce its form factor to that of a smartphone, which is equipped with a standard TLC connector interface. An eBrain can then be constructed by stacking a certain number of each module type and connecting them together through TLC.

Table 4.8 lists the target performance of the palmtop left and right brain respectively, based on projection from the latest data. It represents quantum leaps in computing performance compared to the current state. The left brain delivers 1/5,000× the processing performance of Fujitsu's K supercomputer, which was ranked by TOP500 as the fastest supercomputer in 2011 and was among the top 20 fastest supercomputers as recent as 2019. Its processing efficiency is about 6 times that of the top Green500 computer in 2019. Meanwhile, the right brain has 1/1,000× the neuron count of the

Table 4.7. TCI performance headroom.

	Past [13]	**Near future**	**Headroom**
Process	180 nm CMOS	7 nm CMOS	–
Chip thickness	40 μm	25 μm	–
Data rate per coil	8 Gb/s	64 Gb/s	8×
Area efficiency	0.17 Tb/s/mm^2	11 Tb/s/mm^2	65×
Energy efficiency	710 fJ/bit	25 fJ/bit	28×
Aggregate BW (in 8 mm^2)	1.4 Tb/s	91 Tb/s	65×

Table 4.8. Target performance of palmtop eBrain.

	Left Brain	Right brain
Architecture	Von Neumann	Wired logic
Type of Processing	Logical/Arithmetic	Intuitive/recognition/ learning
Hardware Feature	256 Cores, 512 GB/s DRAM	10 million neurons (1/1,000× of human brain)
Processing performance	1/5,000× processing performance of Fujitsu's K supercomputer	–
Processing efficiency	100 GLOPS/W (6× of top Green500)	10× of GPU
Form factor	Similar to a smartphone (66 cm^3)	

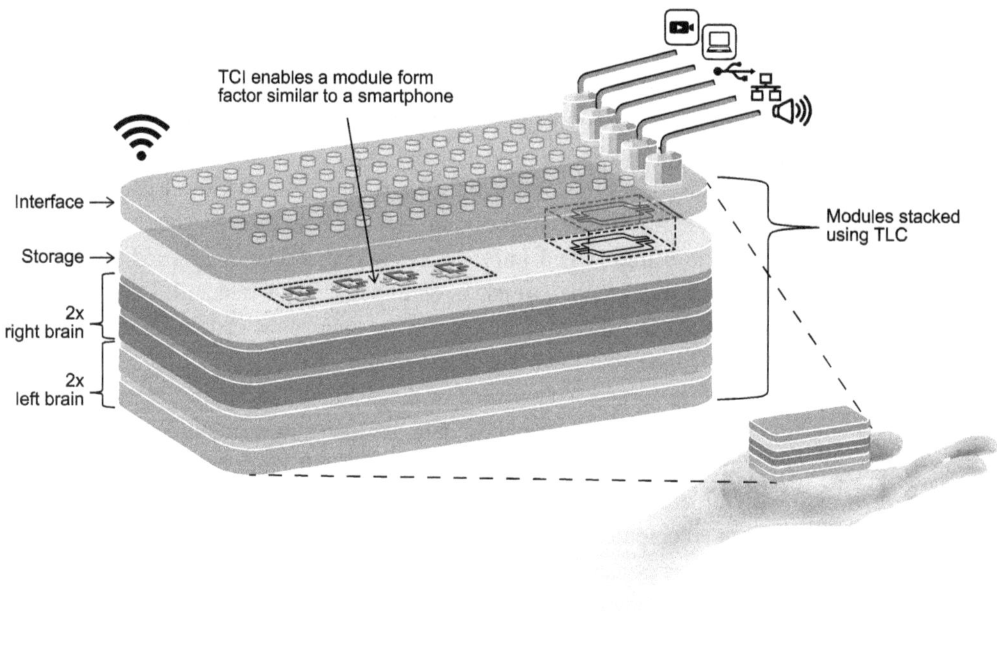

Figure 4.17 Construction of a palmtop eBrain.

human brain with 10 times the processing efficiency of a GPU. These are performances delivered by a device that fits on your palm.

4.4 Closing Thoughts: Predicting the Future of Computing

As American computer scientist Alan Kay likes to say, "The best way to predict the future is to invent it." Therefore, to predict the future of computing, let's contemplate the future we want to invent. Let's first take a look at recent trends in computing, and then examine how to innovate computing to address these trends.

4.4.1 Recent Trends in Computing

In 2016, the Japanese government coined the term Society 5.0 to describe its concept of the future of society. It captures the societal megatrend of the transition from a capital-driven to a data-driven and knowledge-based society ([15], Figure 4.18a). The transition is started with the explosive growth in the amount of data generated in the current information society (Society 4.0), and will be completed in Society 5.0 with the seamless fusion of physical and cyberspace (Figure 4.18b), where information is collected from the physical space, then converted into digital raw data, which are finally processed to extract knowledge. The knowledge will then be applied to solve big problems society faces, such as an aging society, accelerating regional depopulation, and social inequality. The proliferation of IoT devices and smartphones equipped with a multitude of sensors has made information-collecting devices abundant and ubiquitous, gathering voluminous amount of raw data from across vast physical space and diverse environments. Meanwhile, the advance in AI technology has enabled automatic processing of large amount of data to extract valuable knowledge. Together, they have enabled us to take the first step toward a knowledge-based society.

As a result, data are the "capital" of Society 5.0. From data, knowledge is created, based on which services can be individually tailored and delivered to provide maximum value-added to the consumer. Products – hardware and software – are still important for fusing physical with cyberspace to create services. But the value added is directly delivered by the services instead of products. Consequently, to realize Society 5.0, computing must enable efficient extraction of knowledge for the generation of data-driven, value-added services, by delivering solutions for the collection of information from the physical world, conversion to digital data, AI processing, and communication. The larger the amount and the more diverse the nature of the collected information, and the better the AI processing is, the more useful and relevant the extracted knowledge becomes – hence, the more value the knowledge-based services can deliver.

To increase the amount and diversity of information collected, collection needs to take place in a ubiquitous fashion. The ultimate goal is information collection by everything, everywhere. On the other hand, to improve AI processing and service delivery efficiency, both computational performance and energy efficiency must be improved. An implicit requirement is for this to be accomplished at low cost both in monetary and power consumption terms, which includes the cost of data communication. As a result, while centralized, batch data processing in the cloud is more cost effective, when data communication cost is factored in, the optimal solution calls for a balanced amount of local, distributed processing at the edge as well.

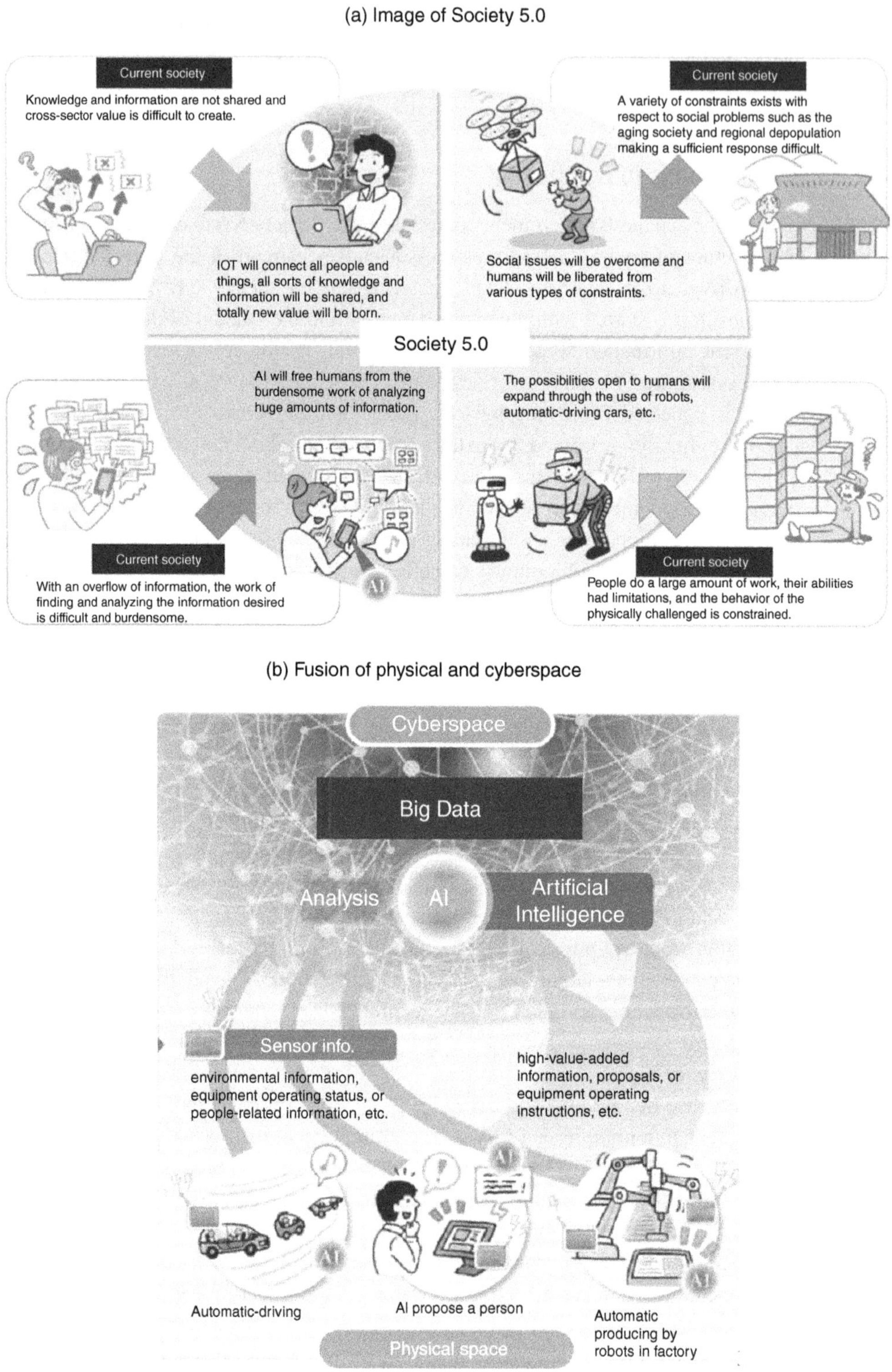

Figure 4.18 Concept of Society 5.0. Source: The Cabinet Office home page (www8.cao.go.jp/cstp/english/society5_0/index.html).

4.4.2 Recent Trends in the IC Industry

What are the implications for the IC industry? We need ICs for diverse operating environments, from inside the moist and dirt-filled soil on a farm to the thermally stressful vacuum in space. From a specifications standpoint, we need ICs that fit into a wide spectrum of form factor and price point, and operate on a power source of diverse nature, from energy harvesting to a power plant, and under wide range of environmental parametric values, including temperature, humidity, and mechanical stability. It is not difficult to come to the conclusion that we need specialized semiconductors in order to be able to efficiently address such diverse requirements.

Figure 4.19 depicts the shift of the IC industry between low-mix, high- volume and high-mix, low-volume production in the last 50 years of its history. It has a cyclical nature with roughly a 10-year half-cycle. Low-mix, high-volume production is propelled by device innovations such as the emergence of CPU, DRAM, GPU, and flash memory. Its growth fuels large capital investment and fierce competition, eventually resulting in industry realignment and consolidation. Manufacturers compete on cost and derive profit from production efficiency. Meanwhile, while its initial wave was triggered by new end products in the form of digital calculator, watch, and TV, high-mix, low-volume production is generally driven by innovation in design methodology such as auto layout, logic synthesis, and high-level synthesis to lower development costs in order to enable a large number of specialized chip designs. Manufacturers compete on unique value added and derive profit from product differentiation.

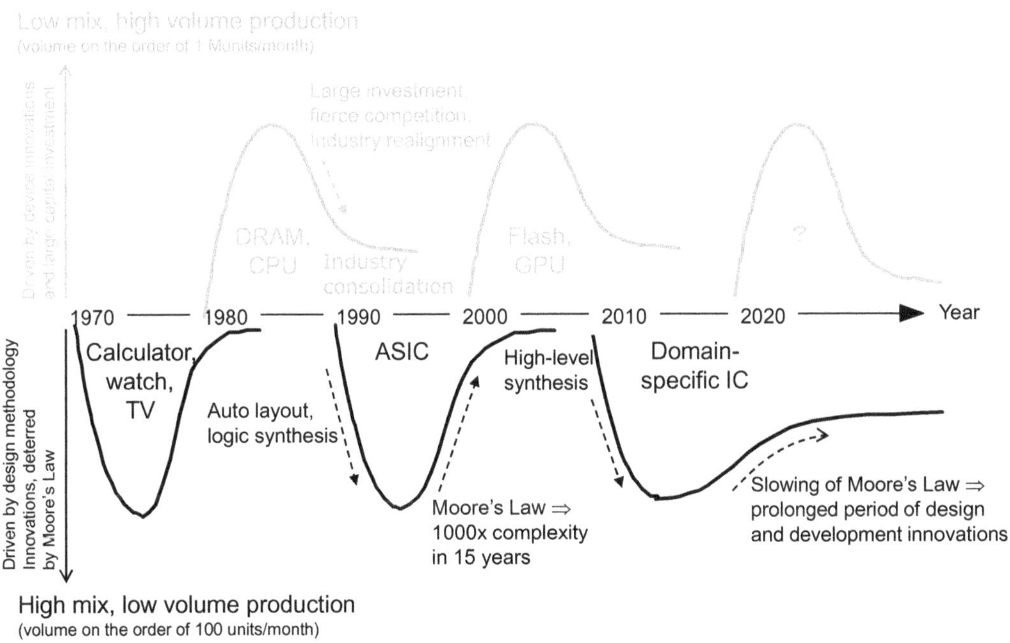

Figure 4.19 Cyclical history of the IC industry in production mix and volume [16].

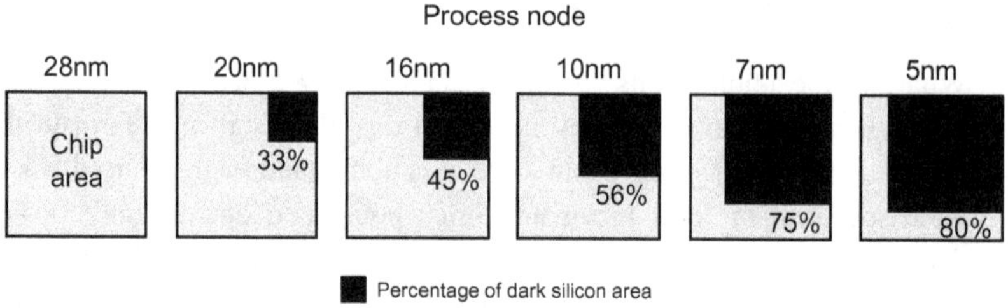

Figure 4.20 Percentage of dark silicon area vs. process node.

Moore's law, which has been able to increase chip complexity 1,000-fold in 15 years and incentivize manufacturers to shift production toward high-volume designs, has been a deterring force for chips with low volume. Therefore, conversely, the recent slowing of Moore's law has contributed to the resurgence of high-mix, low-volume production. However, unlike previous cycles, without a successor to Moore's law in sight, the current trend toward high-mix, low-volume production may enjoy prolonged growth, especially with the advent of a knowledge-based, service-oriented economy that increases demand for specialized chips.

In other words, specialized chips, which by nature are low volume, are emerging to address the transition to Society 5.0, while benefiting from the slowing of Moore's law. Furthermore, they are motivated by the increasingly large percentage of dark silicon area as estimated and depicted in Figure 4.20. Dark silicon refers to the part of an operating IC that is powered off to keep the chip within the thermal envelope of the application. Without breakthroughs in thermal solutions to drastically increase system power budget, continuous process scaling will only increase the percentage of dark silicon. Therefore, to take full advantage of increased device integration from process scaling, it is necessary to create innovative use of the dark silicon area while staying within the power budget. Heterogeneous computing is one such innovation where a conventional general-purpose processor is augmented with special-purpose cores such as GPU or DSP, which have better energy efficiencies for their respective specialized functions. A particular example is presented in [17]. Specialized chips are at the end of the spectrum where the entire chip is designed for a specific application with limited functionality compared to general-purpose chips. As a result, they can be optimized to maximize chip-level energy efficiency with one to two orders of magnitude improvement.

However, to enable the creation of a large number of specialized chips to address the diverse requirements of Society 5.0, the industry must once again innovate design methodology to reduce development time and costs. The reduction of development costs is especially important for specialized chips since the costs are amortized over a smaller volume. One emerging strategy is the use of domain-specific chips, as introduced in Section 4.2.2. In addition, improvement in design automation is imperative. Unfortunately, automation is generally achieved at the expense of computational performance and energy efficiency. As a result, computational performance and energy efficiency must be improved not only for their own sake, but also to enable design automation in order to address the trend of increasing specialized chip development. The IC industry must therefore transition from being capital intensive to drive down manufacturing costs, to being knowledge intensive to improve design methodology, while simultaneously boosting chip computational performance and energy efficiency.

4.4.3 Role of TCI and TLC Wireless Interfaces

The transition to Society 5.0 and the shift of the IC industry to specialized chips can benefit from the use of TCI and TLC wireless interfaces in multiple ways. While TCI enables wireless chip integration in 3D, 2.9D, 2.5D, and potentially 2D, TLC realizes wireless communication between modules. Together they make it possible to implement contactless interconnection of the building blocks of a computing system. The key merits of such a computing system are recapitulated in Table 4.9, which include low energy, small form factor, robustness, and implementation flexibility. Given these merits, the technologies will benefit a large variety of computing systems in diverse applications at a wide range of price points. The application of TCI to wideband DRAM, stacked flash memory, 3D network-on-chip, and DNN has been discussed and demonstrated. The lineup in Table 4.10 of TLC derivatives that have been prototyped illustrates possible physical constructions and applications of the

Table 4.9. Key benefits of TCI and TLC wireless interface technologies.

Merit	Contributing factors
Low energy	• Simple digital circuits • Low loading capacitance • Minimal data movement • Scaling with process
Small form factor	• Elimination of physical connections enables short communication distance • Better scaling of connection terminal size • TCI: Elimination of thermomechanical stress enables thin die • TCI: 2.5/2.9D integration eliminates interposer from stack height • TLC: Elimination of connector mechanical housing

Robustness	• Elimination of physical contact contributes to
	  ○ Robustness against shock and vibration
	  ○ Robustness in harsh physical environment through sealing of terminals
	• Elimination of thermomechanical stress contributes to robustness in harsh thermal environment

Implementation flexibility	• AC-coupled simplifies connection of devices with different supply voltages • Supports wide range of manufacturing processes – achieves high performance even when implemented in mature IC processes • TCI: 2, 2.5, 2.9, 3D chip integration • TLC: flexible physical shape and implementation medium – rigid PCB and flexible printed circuit (see Table 4.10)

Table 4.10. Current lineup of TLC derivatives.

Derivative	Characteristic	Image	Application
R-TLC	Rotatable		FA, robot, security camera
T-TLC	Twofold for Small Footprint	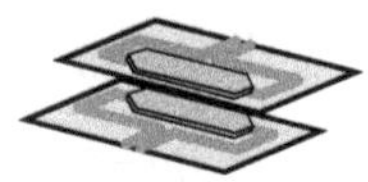	Smartphone
BD-TLC	Bidirectional for multimaster, multidrop bus	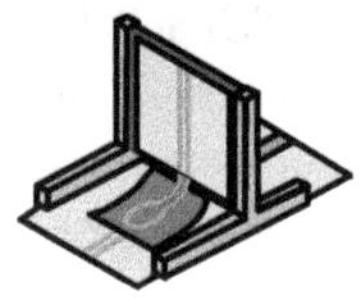	Satellite
SDC-TLC	Single-ended to differential conversion	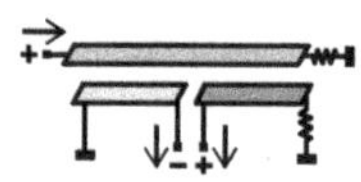	Memory production test
C-TLC / EM-Clip	Clip on twisted pair		Automotive
VD-TLC	Simultaneous dual channel transmission	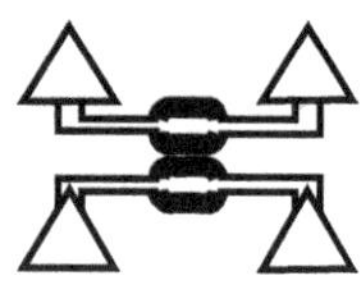	Display
EE-TLC	Energy equipartition for master–slave, multidrop bus	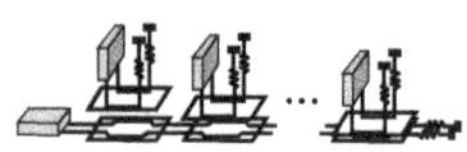	Memory system

technology, including factory automation (FA), robotics, and security camera, in addition to smartphone, satellite, automotive, display, and memory applications (DRAM system and testing and SSD) introduced in Chapter 3. But the potential application areas are far more numerous and diverse.

Low energy and small form factor means that we can create a tiny standalone computing device that requires a small power supply, making it applicable to edge computing in an environment deprived of adequate power supply infrastructure. This will enable, for instance, mass deployment of low-cost IoT devices over a large physical area, such as the installation of environmental condition monitoring devices on the farm. As an illustration, if we take the palmtop eBrain described in Section 4.3.2 and shrink it by a factor of 64, reducing its volume to about 1 cm^3, the left brain still contains four processing cores providing tens of GFLOPS of computing power comparable to an iPhone 7, and the right brain still consists of 160,000 neurons, all literally on your fingertip. This minimum computing unit can then be scaled up to tailor toward applications of different scales.

On the other hand, with robustness, the low energy and small form factor of a TCI-TLC–enabled computing system can be introduced into different operating environments, performing computation even in light of user mishandling, under thermal stress, subject to shock and vibration, or in the presence of contaminants.

Last but not least, the implementation flexibility offered by TCI and TLC facilitates design modularization, making it easier to create a large variety of building blocks that can be mixed and matched and interconnected to build computing systems to different specifications. In addition to being able to address different applications, this will also contribute to design reuse and lower development costs.

While Moore's law has slowed down, there is still room for further IC scaling in all three physical dimensions in the foreseeable future. Since TCI and TLC can both benefit from process scaling, the performance they deliver will also continue to scale, driving continuous improvement in energy efficiency and form factor.

4.4.4 A Future of Digital Inclusion

It is conceivable that scaling will eventually result in a low-power, miniature eBrain that can be implanted into the human body. We would then have the capability to send the likes of an artificial neuroscientist or medical doctor into our body to better study – monitor, track, analyze, diagnose, and understand – our brain, our organs, and even our cells. Local processing is important to minimize communication with the outside world, which can consume a dominating amount of energy. This will not only revolutionize personalized healthcare, but also make it more affordable and hence available to the masses.

The possibilities are unbounded; the sky is truly the limit. Figure 4.21 depicts just a small sample of some other possible applications of eBrain, including biological science, visual perception, mobility, security, medical diagnostic, and factory automation. For instance, an eBrain-equipped microscope can use its right brain to identify a certain microbe, which the left brain can closely track and monitor to automatically generate a massive amount of behavioral data, which the right brain can then analyze to extract behavioral patterns.

Society 5.0 will continue to digitize everything, everywhere, promoting digital inclusion where everyone can benefit from the advance of digital technology. The smartphone has already moved us a step closer to digital inclusion, where we can see grandparents who may not be tech savvy swiping their smartphones to access the Internet in order to communicate with their grandchildren. Imagine a future where grandparents can have a digital grandchild by their side at all times to look after them. Grandma may just be mumbling to herself, "Wait, where did I put it?" But her intelligent hearing aid, which has been by her side to understand the context, interprets it as grandma looking for her wallet. In response, a familiar image of her grandchild, who is always looking after her, appears in her smart glasses to tell her that she has left it on the desk where she was before. That will be another big step forward toward digital inclusion.

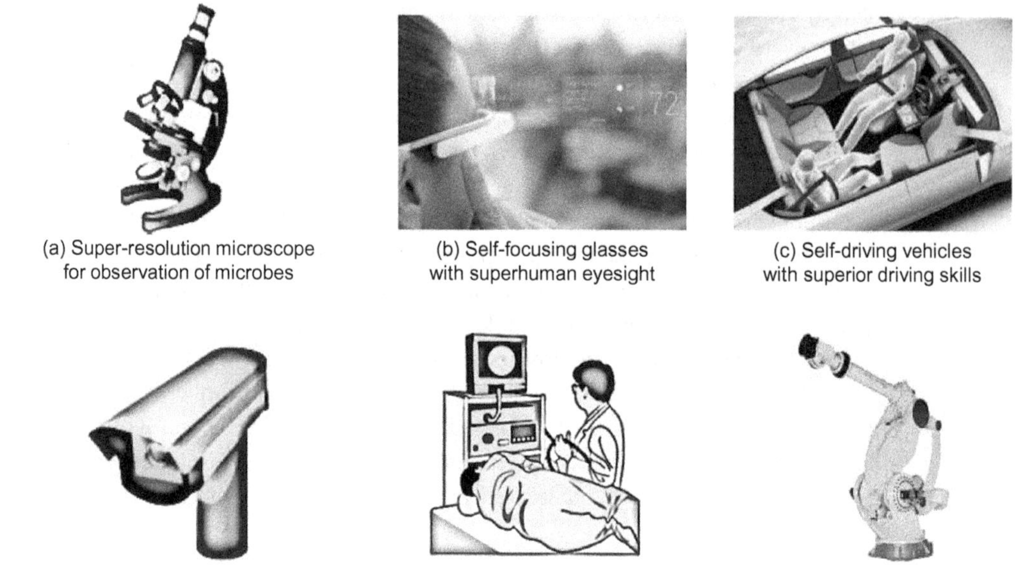

(a) Super-resolution microscope for observation of microbes
(b) Self-focusing glasses with superhuman eyesight
(c) Self-driving vehicles with superior driving skills
(d) Surveillance camera with expanded range for security and disaster prevention
(e) Medical diagnostic equipment with insightful diagnosis results
(f) Intelligent robots with human-like maneuvering

Figure 4.21 Possible eBrain applications.

Richard Feynman's famous declaration in 1959 that "There's plenty of room at the bottom" referred to the tremendous opportunity created by the ability to manipulate individual atoms, which subsequently drove explosive advance in the electronics industry. With TCI and TLC as More than Moore enablers to scale computing performance for diverse applications beyond what is possible from scaling of physical dimensions alone, we believe that "There is plenty of room at the top" to explore application of the technologies to various specialized functionalities to deliver unprecedented capability, gain new insight, and create a richer, safer, healthier, and more efficient life in a world of digital inclusion.

References

[1] K. Ueyoshi, K. Ando, K. Hirose, S. Takamaeda-Yamazaki, M. Hamada, T. Kuroda, and M. Motomura. (2019, Jan.). "QUEST: Multi-purpose log-quantized DNN inference engine stacked on 96-MB 3-D SRAM using inductive coupling technology in 40-nm CMOS." *IEEE Journal of Solid-State Circuits. 54*(1), pp. 186–196.

[2] D. Kim, J. Kung, S. Chai, S. Yalamanchili, and S. Mukhopadhyay. "Neurocube: A programmable digital neuromorphic architecture with high-density 3D memory." *2016 ACM/IEEE International Symposium on Computer Architecture*, pp. 380–392, Jun. 2016.

[3] M. Gao, J. Pu, X. Yang, M. Horowitz, and C. Kozyrakis. "Tetris: Scalable and efficient neural network acceleration with 3D memory." *2017 ACM International Conference on Architectural Support for Programming Languages and Operating Systems*, pp. 751–764, Apr. 2017.

[4] K. Ueyoshi, K. Ando, K. Hirose, S. Takamaeda-Yamazaki, J. Kadomoto, T. Miyata, M. Hamada, T. Kuroda, and M. Motomura. "QUEST: A 7.49TOPS multi-purpose log-quantized DNN inference engine stacked on 96MB 3D SRAM using inductive-coupling technology in 40nm CMOS." *2018 IEEE International Solid-State Circuits Conference*, pp. 216–218, Feb. 2018.

[5] O. Russakovsky, J. Deng, H. Su, J. Krause, S. Satheesh, S. Ma, Z. Huang, A. Karpathy, A. Khosla, M. Bernstein, A. C. Berg, and F. F. Li. (2015, Dec.). "ImageNet large scale visual recognition challenge." *International Journal of Computer Vision. 115*(3), pp. 211–252.

[6] D. Shin, J. Lee, J. Lee, and H. J. Yoo. "DNPU: An 8.1 TOPS/W reconfigurable CNN-RNN processor for general-purpose deep neural networks." *2017 IEEE International Solid-State Circuits Conference*, pp. 240–241, Feb. 2017.

[7] B. Moons, R. Uytterhoeven, W. Dehaene, and M. Verhelst, "Envision: A 0.26-to-10 TOPS/W subword-parallel dynamic-voltage-accuracy-frequency-scalable convolutional neural network processor in 28nm FDSOI." *2017 IEEE International Solid-State Circuits Conference*, pp. 246–247, Feb. 2017.

[8] S. Yin, P. Ouyang, S. Tang, F. Tu, X. Li, L. Liu, and S. Wei. "A 1.06-to-5.09 TOPS/W reconfigurable hybrid-neural-network processor for deep learning applications." *2017 Symposium on VLSI Circuits*, pp. C26–C27, Jun. 2017.

[9] CACTI version 6.5. [Online]. Accessed on Sep. 27, 2018. Available: www.hpl.hp.com/research/cacti.

[10] M. Motomura. "AIコンピューティングがアーキテクチャにもたらすもの [Changes in computer architecture driven by AI computing]," in Japanese, Hokkaido University, Sapporo, Hokkaido Prefecture, Japan, Oct. 2018.

[11] J. L. Hennessy and D. A. Patterson. (2017). *Computer Architecture: A Quantitative Approach*, 6th ed.. Cambridge: Morgan Kaufmann.

[12] S. Trimberger, D. Carberry, A. Johnson, and J. Wong. "A time-multiplexed FPGA." *The 5th IEEE Symposium on Field-Programmable Custom Computing Machines*, pp. 153–160, Apr. 1997.

[13] A. R. Junaidi, Y. Take, and T. Kuroda. "A 352Gb/s inductive-coupling DRAM/SoC interface using overlapping coils with phase division multiplexing and ultra-thin fan-out wafer level package." 2014 Symposium on VLSI Circuits, pp. 29–30, Jun. 2014.

[14] Arizona State University. Predictive Technology Model [Online]. Retrieve Dec. 8, 2015. Available: http://ptm.asu.edu/.

[15] Cabinet Office of Japan. Society 5.0. Retrieved on Apr. 10, 2020. Available: www8.cao.go.jp/cstp/english/society5_0/index.html.

[16] T. Kuroda. "Semiconductor industry in 2025," Panel Discussion (presented but not published), *2010 IEEE International Solid-State Circuits Conference*, Feb. 2010.

[17] N. Goulding, J. Sampson, G. Venkatesh, S. Garcia, J. Auricchio, J. Babb, M. B. Taylor, and S. Swanson. (2010, Aug.) "GreenDroid: a mobile application processor for a future of dark silicon." *Hot Chips 22*.

Index

For EU product safety concerns, contact us at Calle de José Abascal, 56–1°, 28003 Madrid, Spain or eugpsr@cambridge.org.

www.ingramcontent.com/pod-product-compliance
Lightning Source LLC
LaVergne TN
LVHW062207200926
848484LV00006B/1338
* 9 7 8 1 1 0 8 8 4 1 2 1 4 *